Acknowledgements

MWPS-30, Sprinkler Irrigation Systems

MWPS Committee
The MidWest Plan Service prepares publications under the direction of agricultural engineers and consulting specialists. It is an official activity of the following universities and the U.S. Department of Agriculture. The MidWest Plan Service Committee is responsible for the selection and supervision of all projects:

T.L. Funk; Y. Zhang; J. Siemens
University of Illinois, Urbana IL 61801

D.D. Jones; D.E. Maier; S. Parsons
Purdue University, West Lafayette IN 47907

J.D. Harmon; S.J. Hoff; S.W. Melvin
Iowa State University, Ames IA 50011

J.P. Murphy; R.G. Maghirang; J.P. Harner
Kansas State University, Manhattan KS 66506

W.G. Bickert; H.L. Person
Michigan State University, East Lansing MI 48824

W.F. Wilcke; K.A. Janni; J.M. Shutske
University of Minnesota, St. Paul MN 55108

J.M. Zulovich; N.F. Meador
University of Missouri, Columbia MO 65211

G.R. Bodman; D.P. Shelton; R. Koelsch
University of Nebraska, Lincoln NE 68583

K.J. Hellevang; T.F. Scherer; J. Lindley
North Dakota State University, Fargo ND 58105

R.C. Reeder; R. Stowell
The Ohio State University, Columbus OH 43210

S.H. Pohl; V. Kelley
South Dakota State University, Brookings SD 57007

D.W. Kammel; R.T. Schuler
University of Wisconsin, Madison WI 53706

L.E. Stetson
USDA-ARS, University of Nebraska, Lincoln NE 68583

B.K. Rein
USDA, Washington DC 20250
CSREES

MWPS Headquarters
F.W. Koenig, Engineer
J. Moore, Manager
B. Beach, Graphic Designer
L. Wetterhauer, Graphic Designer
L. Miller, Editor

ISBN 0-89373-077-7

For additional copies of this book or other publications referred to, write to: Extension Agricultural Engineer at any of the listed institutions.

MWPS-30
1st Edition, 1st Printing, 5M, 1999

Sprinkler Irrigation Systems

Table of Contents

Table of Contents

Chapter 3 continued

Table of Contents

Table of Contents

Table of Contents

Chapter 8 continued

Table of Contents

List of Figures

Chapter 2

Chapter 3

List of Figures continued

Chapter 4

Chapter 5

List of Figures continued

Chapter 5 continued

Chapter 6

Chapter 7

List of Figures continued

Chapter 7 continued

Chapter 8

Chapter 9

List of Tables

Chapter 2

Chapter 3

Chapter 4

Chapter 5

List of Tables continued

Chapter 6

Chapter 7

Chapter 9

List of Tables continued

Chapter 9 continued

List of Equations

List of Equations continued

Chapter 8

Chapter 9

List of Examples

Chapter 2

Chapter 7

Chapter 8

Chapter 9

Chapter 10

Chapter 1

Planning a System

Chapter 1 Contents

This guide to sprinkler irrigation systems provides you--a producer, student, agricultural consultant, engineer, equipment dealer, or government agency employee--a planning, reference, and design manual for the application of sprinkler irrigation systems and methods to manage an irrigation system efficiently. The information allows you to evaluate water needs and determine a minimum recommended system capacity. Also included are methods to evaluate the water supply and type of sprinkler system that best fit your application needs. Sprinkler performance characteristics are given as well as the pump and water delivery system requirements. The selection of sprinklers for applying liquid waste also is discussed for use in special situations. Finally, design examples take you step-by-step through the planning and designing processes for various sprinkler irrigation systems.

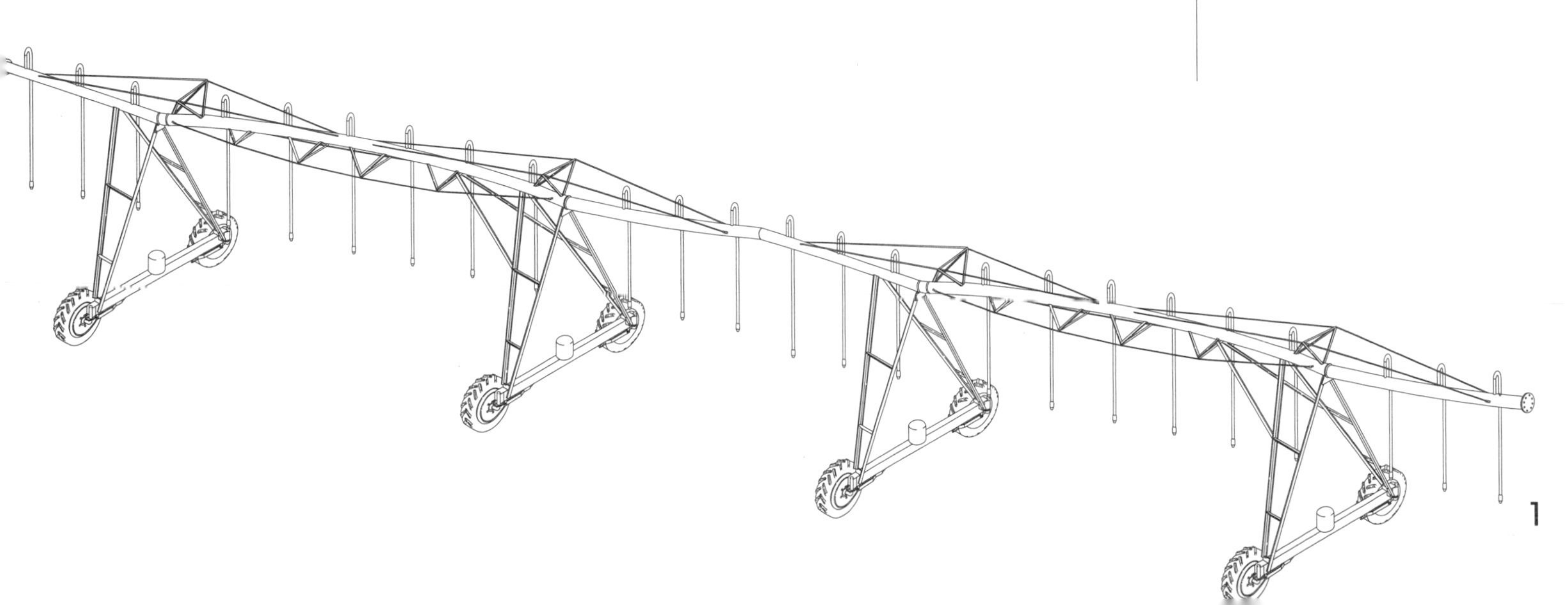

Why Do We Irrigate?

Irrigation is the process of applying water to the soil to meet crop water demands. The role of irrigation is to improve production and input efficiency in areas where the climate limits production potential. In some areas, irrigation plays a vital role by helping to leach salts that accumulate in the crop root zone. Another application is irrigation of selected high-value crops for frost protection. The result of irrigation growth has been the conversion of many rain-fed areas in the United States and throughout the world into important irrigated crop production areas.

Irrigation will play a continued and increasing role in global economics to meet the demand for food as the population continues to grow. Irrigation makes it possible for the production of a more stable food supply.

Irrigation methods have changed dramatically during the past 30 years. The number of surface-irrigated acres has declined by more than 5 million, while the number of sprinkler-irrigated acres has increased from less than 7 million to approximately 25 million acres. A portion of the increase in sprinkler irrigation is due to converting surface systems to sprinkler systems. This trend will continue as a result of the improvements in water application efficiency and labor reduction associated with sprinkler systems.

Recently, sprinkler irrigation has increased as a way of using effluent from municipalities and from animal and food production systems. The process of applying effluent to the land can improve crop production, supply nutrients and water, and use liquid effluent effectively. Sprinkler conversions and application of liquid waste can be accomplished using environmentally safe methods.

To meet the challenges of the future, sprinkler irrigation systems must be managed to optimize water supplies to meet crop water demands. Irrigation system improvements and management must be based on techniques that sustain soil and water resources by efficient irrigation. These improvements must protect the environment and the world's natural resources needed to produce raw food products.

Irrigation System Planning Steps

Below is an irrigation planning checklist that can be used when installing a new irrigation system or changing an existing system. Following this checklist can help match the needs of the crop and the characteristics of the soil with the design of the irrigation system to provide an efficient and environmentally-friendly sprinkler irrigation system:

- Select a site.
- Evaluate soil characteristics.
- Evaluate crop characteristics.
- Determine water supply availability.
- Evaluate water supply quality.
- Calculate the system's capacity.
- Select the sprinkler irrigation system.
- Select pumps, power units, and piping.
- Address environmental concerns.
- Plan management and maintenance.
- Assess the economics of the system.
- Address safety concerns.

Selecting a Site

Developing an understanding of the site always should be the first step in the design process. Obtain a soil survey map to learn the different soil types, drainage needs, and topography of the land. This information will be used to design the sprinkler irrigation system. The soil survey map also can be used to make a scaled drawing of the proposed system. If information is difficult to find or not available, contact the local Cooperative Extension or Natural Resources Conservation Service (NRCS) for assistance.

Evaluating Soil Characteristics

Evaluate the land based on a number of characteristics, including soil texture and field slope, to determine the suitability of the land for irrigation. Use soil survey information to determine the water-holding capacity of the soil. Base the water-holding capacity on the depth of soil and active crop rooting zone at the proposed irrigation site. Obtain information to estimate the soil infiltration rate for different soil textures in the field so the application rate of the irrigation system can be matched. A soil test, in addition to determining nutrient levels, can be used to evaluate the salinity level in the soil. This information and results of a water quality test of the water supply can, in some cases, help avoid increasing the salinity level of the soil and define any needed subsurface drainage or salt-leaching requirements.

Field slope will influence selection of the irrigation system and the choice of sprinkler devices to avoid potential runoff. Knowing field slope also can help find field surface drainage problems and can be used to evaluate the potential for soil erosion. Additional information on soil and soil-water-plant relationships can be found in Chapter 2, *System Design*.

Evaluating Crop Characteristics

Define the planned crop and crop rotation for the field that will be irrigated. The crops grown will determine the total crop water use during the growing season as well as peak water requirements for sprinkler design. Knowing the planned cropping patterns is necessary to match system capacity to the water needs of the crop, thus reducing the risk of water stress. The climatic conditions that affect crop production also are important factors to evaluate. This includes the length of the growing season to determine the number of frost-free days. Determining crop water use to design the sprinkler irrigation system is discussed in Chapter 2.

Determining Water Supply Availability

After locating a water source, determine the supply's pumping capacity, and develop an understanding of the legal requirements that must be met to use the water for irrigation purposes. If groundwater will be used, drill and pump test holes to confirm the quantity, quality, and depth to water. If surface water will be used, determine the average quantity of water that will be available for the duration of the growing season. For a system that depends on surface water, a pumping pit may be needed for temporary storage.

Development of the irrigation water supply is discussed in more detail in Chapter 3, *Water Sources*.

Evaluating Water Supply Quality

To determine whether water quality is an issue, test a water sample for the amount of dissolved and suspended minerals in it. For groundwater, a good time to collect a sample is when test holes are being drilled. The test can indicate both the toxic levels that might be reached over time, as well as beneficial nutrients the crop can use. For more information on water quality, see Chapter 3.

Calculating the System's Capacity

After identifying the site to be irrigated, the water source, and the crop, begin designing the irrigation system. For a description of different types of sprinkler irrigation systems, see Chapter 4, *Sprinkler Systems*.

After selecting a system type to fit the design criteria, determine system capacity. At this point, obtain as much information as possible about the crop's water requirements and the water-holding capacity of the soil. Also, knowing precipitation levels and the probability of precipitation occurrence will assist in designing storage structures, irrigation design capacities, and runoff capabilities. More information on irrigation system capacity is in Chapter 2.

Selecting a Sprinkler Irrigation System

The selection of sprinkler irrigation systems is based partially on the performance and selection of individual sprinkler devices. New sprinkler devices that operate at low pressure require different management techniques than do high pressure systems. For a description of different sprinkler irrigation systems, see Chapter 4. Chapter 5, *Sprinkler Characteristics*, explains sprinkler performance characteristics.

Selecting Pumps, Power Units, and Piping

The cost of owning and operating a sprinkler irrigation system efficiently depends on selecting the correct pump and power unit.

The pipe must be sized correctly to reduce pressure loss when conveying water to the field. Cost, maintenance, and convenience of the pumping unit should be considered in combination with the selection of an energy source when choosing the preferred method of pumping water. Chapter 7, *Pumps, Piping and Power Units,* details selecting and installing pumping plants and piping.

Addressing Environmental Concerns

Be sure to address the system's impact on the environment during the planning phase. Agricultural chemicals used for crop production can move with the soil or may be suspended and may move with irrigation water. The potential for runoff due to high-water application rates poses a concern for both water and soil that leaves the field. In addition, excessive rainfall or water application depths can cause deep percolation of water below the root zone.

Not only is water moved into the groundwater, but some chemicals in solution also may move through the soil profile and beyond the root zone. If over-irrigation continues, the process of deep percolation may move selected chemicals down in the soil profile and into the groundwater aquifer. Chapter 8, *Chemigation*, discusses techniques for applying chemical and fertilizer through a sprinkler system.

If effluent is to be applied through a sprinkler system, check any legal constraints that may affect your plans. To prevent litigation at some future date, be sure to check with state and federal offices to confirm that critical habitat is not being altered.

Collect information from around the site in relation to the sensitivity of the proposed irrigation design to the environment. For a new system, this might include the soil's vulnerability to wind or water-induced soil erosion and leaching. Determine where runoff will go if it leaves the field site. Soil erosion or movement of chemicals in the wastewater off a field might cause environmental problems or human interference that need to be addressed. Chapter 9, *Sprinkler Application of Manure and Effluent*, discusses techniques for applying effluent through a sprinkler system.

Planning Management and Maintenance

Sprinkler irrigation systems must be managed according to the type of system installed. Some systems, such as a center pivot or lateral move, are automated. However, a hand-move, solid-set system, side roll, or large-volume sprinkler requires more monitoring and manual changing. Any irrigation system requires periodic maintenance to extend the life of the equipment.

Assessing the Economics of the System

After the design is complete, use an economic analysis to determine the feasibility of the overall irrigation development project. Evaluate total labor input, capital inputs, and annual irrigation costs, and compare them to the expected return. Make sure the design uses accurate and adequate information collected during the planning process.

Addressing Safety Concerns

As the irrigation system is being designed and later operated, keep safety in mind. Make sure precautions have been taken to reduce the possibility of injury because of electrical shock, drowning, or unshielded PTO assemblies.

Systems used to distribute animal manures, and municipal or food processing effluents should be designed and operated to prevent application to surface water. In addition, systems used for chemigation should consider the positive and negative effects on wildlife and beneficial insects.

Chapter 2
System Design

Chapter 2 Contents

All irrigation system design should begin with an evaluation of the soil. The soil acts as a reservoir, holding water and nutrients for plant use. Plants take water from the soil through the root system and release most of the water to the atmosphere through the leaves. The amount of evapotranspiration is greatly affected by the amount of sunlight (solar radiation), humidity of the air, air temperature, and wind speed. The sprinkler irrigation system capacity should be designed to meet peak water demands, or designed to an acceptable level below peak demand with an understanding of the risks incurred and a willingness to implement appropriate water management strategies.

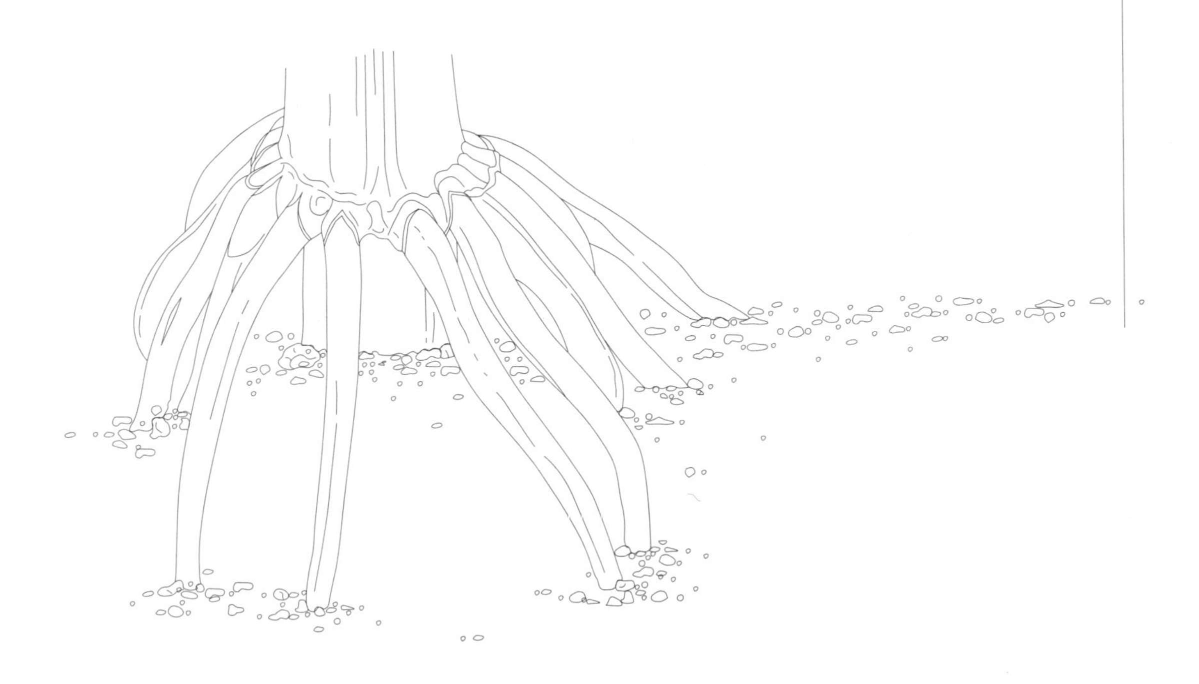

Soils

Soil contains mineral particles from weathered rock, organic matter from decayed plants and animals, water, nutrients, and various gases. All of these soil components are important to plant growth. A typical volume of soil will be about 50% solids and 50% pore space, depending on the degree of compaction, Figure 2-1. The pore space is filled with either air or water, which varies as the soil goes through wetting and drying cycles. Many of the nutrients needed for good crop growth are bound to the soil solid-water interface. Therefore, a proper balance between soil air, soil water, and the soil solids is necessary.

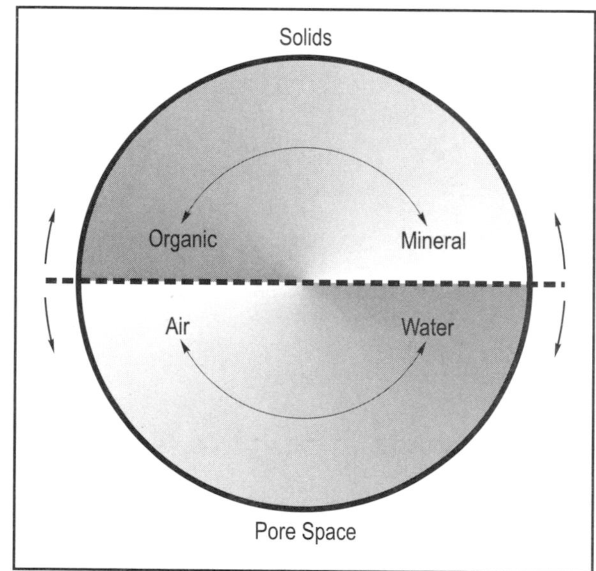

Figure 2-1. Soil composition.

Soil Texture

Most soils contain a mixture of clay, silt, and sand particles. To classify soils according to their *texture*, soil solids are separated into three major groups based on particle size.

The smallest group is clay, classified as any solid particle that will pass through a sieve whose openings are 0.000079 inches (0.002 mm) across. Silt is any solid particle that will pass through a sieve whose openings are 0.002 inches (0.05 mm) across but will not pass through the clay sieve. Sand is any solid particle that will pass through a sieve whose openings are 0.079 inches (2.0 mm) but will not pass through the silt sieve. Anything larger than 0.079 inches is considered to be gravel.

When sand particles dominate, the soil is classed as coarse-textured (sandy loam, loamy sand). Soils high in clay content are fine-textured. Soils with a higher silt content are medium-textured. The texture of the soil has a large influence on soil water and air movement and on chemical transformations.

Soil Structure

Soil structure is the arrangement of soil particles into groups or aggregates. Unlike soil texture, structure is easily influenced by nature and man's activity. Different structures result from combinations of the following:

- Root penetration.
- Wetting and drying.
- Freezing and thawing.
- Organic matter.
- Clay content.
- Tillage and other field operations.
- Irrigation droplet impact.
- Rainfall.
- Compaction.
- Chemical additions to the soil.

Soils with relatively uniform-sized particles have larger spaces between particles. Favorable structure enhances the movement of water and air and the development of crop roots.

Clay is plate-like in shape and tends to bind soil particles into groups. Clay holds more nutrients and water than silt or sand particles. Many clay minerals expand and contract as water is added or removed, thus affecting a soil's physical properties. Adding organic matter to these soils reduces the effect of the clay as a binding agent. The pores in fine-textured soils are very small and flat. This type of pore tends to hold water tightly and in larger quantities, much of which is unavailable for plant use.

Soil structure is most difficult to develop in fine-textured soils (high clay content). Good management techniques are necessary on fine-textured soils to develop and maintain soil structure. Many separate but closely-bound soil particles make up each aggregate in a fine-structured soil.

Organic matter helps stabilize the aggregates. A structured soil may be loose, open, and friable (easily pulverized) if aggregates of the proper size and nature are developed and maintained. Favorable structure in a fine-textured soil is essential for maintaining the satisfactory movement of water and air. However, improper handling of fine-textured soil (for example, traffic or tillage when wet) often causes aggregates to break down into a tight, consolidated condition that impedes satisfactory movement of water and air.

Organic matter is present as solid particles and as a coating on soil solids. It is more chemically active than clay minerals, holding more water and supplying more plant nutrients than an equivalent amount of clay. Organic matter, a source of nitrogen, also helps cement soil particles together and causes the darker color of topsoil.

In a coarse-textured soil (high sand content), most soil particles function separately (single-grained structure) but may be bound into aggregates (granular structure). Coarse-textured soil is loose and open with large pores that allow easy movement of air and water.

Medium-textured soils (high silt content), such as a loam, act like a combination of single-grained and aggregated states. Medium-textured soils may have some sand particles that act independently along with sand, silt, and clay granules that are aggregated. The resulting large pores (macropores) allow easy movement of water and air; the many smaller pores (micropores) effectively retain water.

Soil Water

Soil water is classified as gravitational, capillary, or hygroscopic, Figure 2-2. Gravitational water, sometimes called free water, moves by gravity in a soil with unobstructed drainage. Pores temporarily hold more water than their capillary capacity, and the excess water drains downward. Plants take up only part of the water because roots have access to it for only a short time.

Capillary water, soil water in small pores that is available to the plant, is the most important for crop production. When in the root zone, capillary water is available for uptake by plants.

Hygroscopic water, on the other hand, is held tightly to soil particles and is not available for plant use. It is held in soil as a fine coating on individual soil particles and in pore spaces.

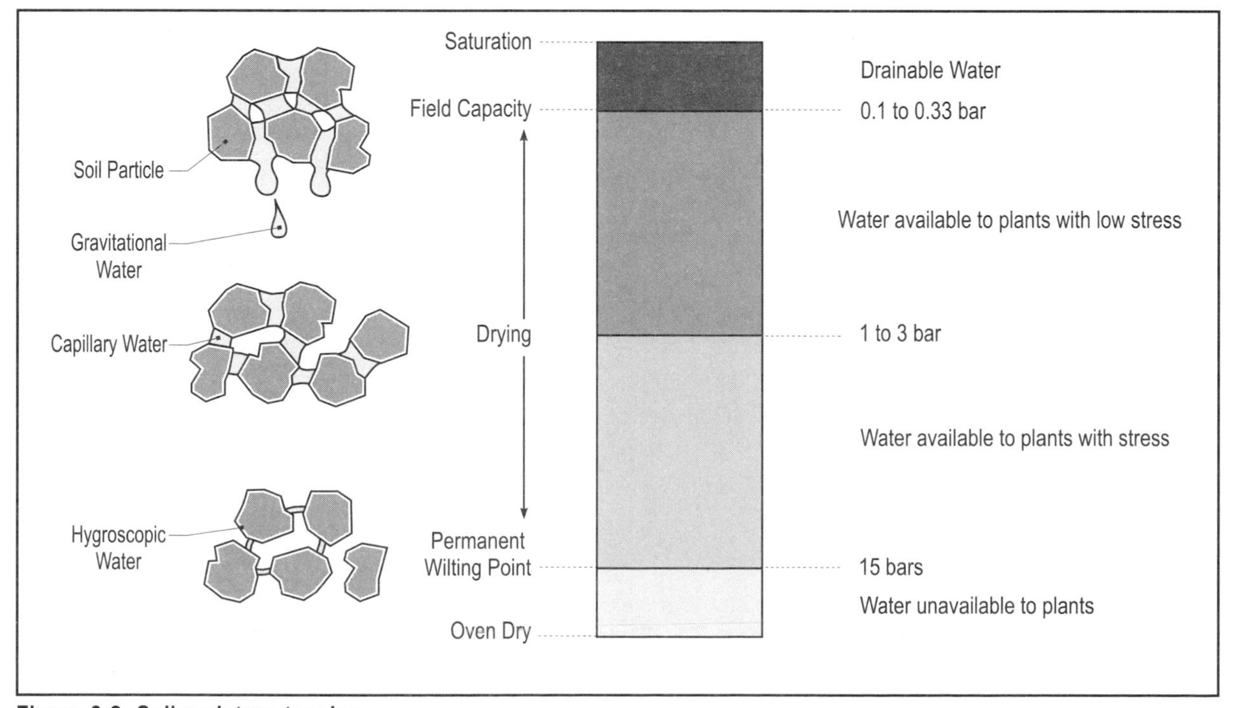

Figure 2-2. Soil moisture tension.

When rain stops or irrigation is finished, water continues to move through the soil due to gravitational and/or capillary forces. Differences in soil water tension produce capillary forces that cause water movement around soil particles and aggregates into and through tiny pores. Capillary forces move soil water upward, downward, or laterally.

Approximately 24 to 48 hours after a rainfall or irrigation that fills the soil with water (time is dependent on soil type), a well-drained soil will be at the upper limit of its plant-available soil water range. This upper limit is called field capacity.

At field capacity, most capillary pores are full of water, although some pores contain trapped air. The lower limit of the available soil water range is referred to as the permanent wilting point. At the permanent wilting point, the capillary water that was available to the plants has been exhausted. Soil water that is available for plant growth is in the range between field capacity and the permanent wilting point, that is, plant-available water (%) = field capacity (%) minus permanent wilting point (%), Figure 2-2.

Soil Water Storage

The amount of water a soil can store for plant use (its available water capacity) depends on the size and arrangement of the soil particles, the amount and type of organic matter, and the depth to which roots can penetrate, Table 2-1. The soil water stored within the plant root zone (the volume of soil containing plant roots) is available for plant use.

Table 2-1. Approximate available soil water.

Soil texture	Available soil water	
	in/in depth[a]	in/ft depth[b]
Coarse sand & gravel	0.03 to 0.06	0.4 to 0.7
Sand	0.06 to 0.08	0.7 to 0.9
Loamy sand	0.08 to 0.11	0.9 to 1.3
Sandy loam	0.10 to 0.16	1.2 to 1.9
Fine sandy loam	0.14 to 0.18	1.7 to 2.2
Loam	0.15 to 0.22	1.8 to 2.6
Silt loam	0.17 to 0.25	2.0 to 3.0
Clay loam & Silty clay loam	0.17 to 0.22	2.0 to 2.6
Silty clay & clay	0.15 to 0.20	1.8 to 2.4

[a]Inches of available soil water per inch of soil depth

[b]Inches of available soil water per foot of soil depth

During a plant's vegetative growth stage, excess irrigation or rainfall can discourage deep root development by having a constant supply of water at a shallow depth. A shallow root zone will limit the amount of total available soil water later in the plant's growing season.

At the other end of the spectrum, allowing water stress during vegetative growth will not encourage deep root development unless adequate soil water is available for growth. The amount of available water that a soil holds is important to the irrigator for the following reasons.

- It governs the amount of water that can be applied effectively at each irrigation.
- It influences the time interval between irrigations.

Soil Water Retention

The force or tension that holds water in the soil determines how easily plants can extract (or use) the water they need, Figure 2-2. The soil water tension (or suction force) is a measure of the energy required to remove the water from the soil. The most common units of measure for soil water tension are either bars or atmospheres. One atmosphere is the standardized barometric air pressure at mean sea level.

1 Atmosphere = 14.70 psi
= 1.013 bars
= 32.8 feet vertical column of water

The higher the soil water tension, the more difficulty plants have extracting water from the soil. Most plants can extract water from the soil at tensions from zero to 15 bars, and most irrigated crops reach the permanent wilting point (when the plant is dead) when the soil water tension approaches 15 bars. For most crops and soils, maintaining soil water tension below one to two bars reduces the effect of water stress on yields. Soil water tensions greater than this level will affect yield or biomass.

Drought tolerant plants will survive at soil water tensions greater than 15 bars; how- ever, these tensions will reduce plant vigor.

Soil water tension can be measured with a tensiometer, Figure 2-3, a device that has a hollow, cylindrical, porous, ceramic cup attached to a water column. The ceramic cup is in contact with the soil. As soil water tension increases in the soil, water is drawn from the column through the porous cup, forming a vacuum at the top of the column. This vacuum (negative pressure) indicates the energy or force a plant must exert to remove water from a soil.

A fine-textured soil holds more water at a given soil water tension than does a coarse-textured soil. Figure 2-4 shows the relationship between soil water content and soil water tension. More soil water is available for plants to use at low soil water tensions than at higher levels. In fact, plants obtain most of their water at soil water tensions of less than one bar.

If the soil profile has a water table, the surface of the water table is at zero tension. Below the water table, soil water is at a positive pressure. Above the water table, soil water has a negative pressure or tension. A high water table in the root zone can affect crops by transferring water and salts upward into the root zone by capillary action.

Figure 2-3. Tensiometer in use.

Capillary action moves water and salts upward a distance dependent on soil texture and soil structure. If the plant roots can access this water, salts are left to accumulate in the soil. Salts in soil increase soil water tension, thus making the plants work harder to obtain water. By working harder to obtain water, the plants are put under stress that affects their growth rate.

Plant and Water Relationships

The weight of actively growing plants can be more than 90% water. Without enough water, plants gradually stop growing, wilt, and die. Moisture stress during any portion of the growth cycle can reduce crop yield.

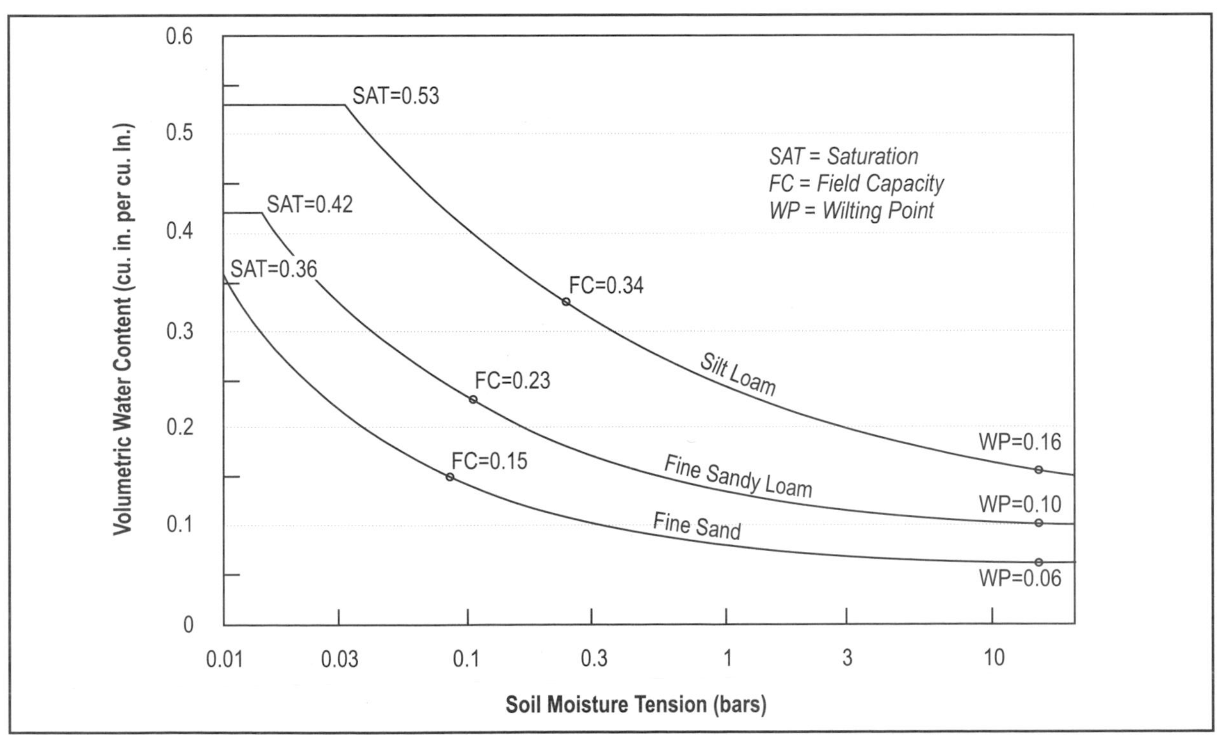

Figure 2-4. Soil water content-moisture tension relationship.
Source: Irrigation Systems Management.

Plants are most vulnerable to water deficiency at the reproductive stage: blossoming, heading, or pollinating.

Water transports plant nutrients from the soil into the plant through roots and throughout the plant. Water also maintains plant turgidity for the proper form and position of stems, leaves, and shoots to capture sunlight. Wilting, or the loss of turgidity, is the first visible effect of insufficient water. Several hours of wilting reduces the rate of cell division, which affects growth.

Plants use energy to get water from the soil. When soil water is at field capacity, water is easily extracted from the soil. As soil water content is reduced, the soil yields water to the plants more reluctantly until no water is available for plants, and the wilting point is reached. The harder a plant works to get water from the soil, the more slowly it grows.

Most of the water that enters the roots does not stay in the plant. Less than 1% is used in photosynthesis and to maintain turgidity. The rest moves to the leaves where it transpires into the atmosphere. The rate at which a plant uses water is controlled partly by genetic characteristics, but primarily by atmospheric and soil environmental conditions. High transpiration rates are not damaging if high absorption by the roots is maintained.

Plants extract only soil water that is in contact with roots. As soil water is removed, capillary action replaces that water with soil water farther away from the roots. If water is equally available throughout the root zone, plants will take more water from the upper part of their root zone than from the lower part largely because of root mass. Figure 2-5 shows typical plant-soil-water extraction. In uniform homogenous soil, more than 70% of the soil water is taken up in the top half of the plant root zone. In shallow soils and soils with impenetrable layers or compacted layers, roots will be concentrated near the soil surface and water management decisions are more critical. Under-irrigating a crop with a shallow root system can result in immediate water stress.

Plants absorb water mainly from root tips, Figure 2-6. Root tips are small, hairlike

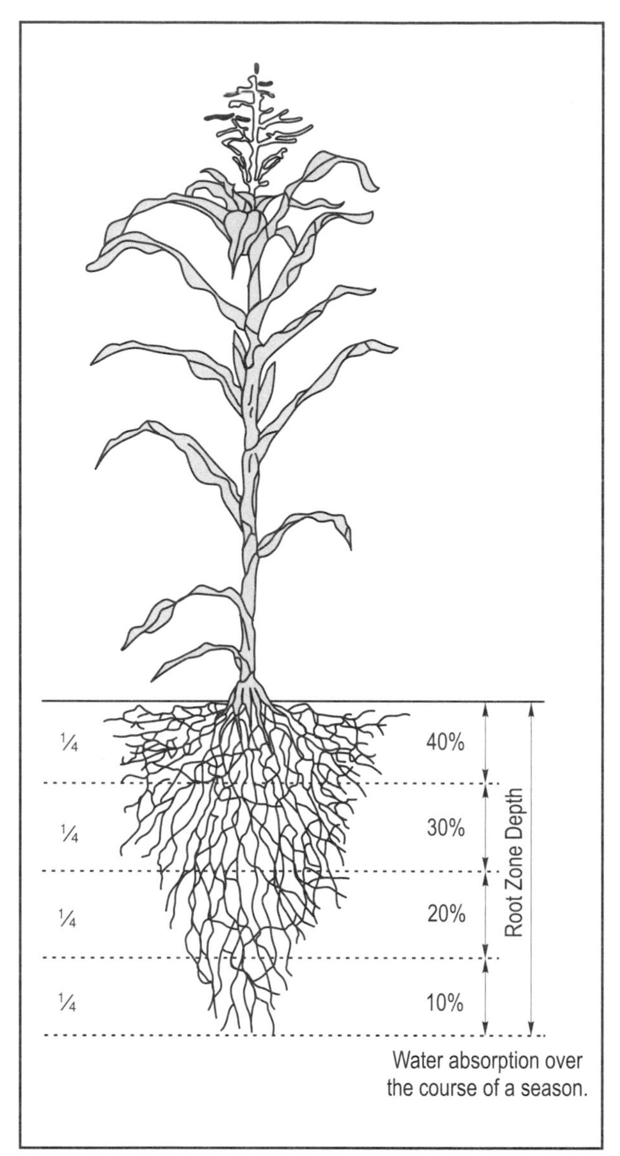

Figure 2-5. Plant-soil-water extraction.

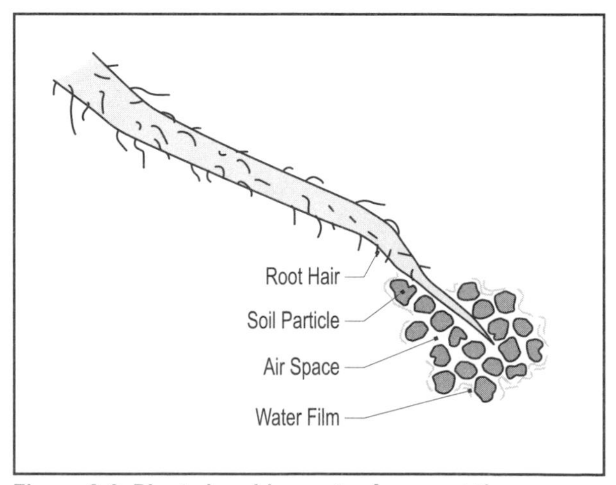

Figure 2-6. Plant absorbing water from root tip.

structures that extend through soil pores to contact and absorb water. Plants with a dense root system, such as grasses, may absorb all the available water in the root zone. Plants with low-density roots, such as vegetables, rely more on capillary action and remove water from a limited soil volume.

Thus, plants with low-density roots can be more susceptible to water stress, and irrigation water management is more critical.

Table 2-2 gives typical rooting depths of mature crops in deep uniform soils. High water tables, restrictive layers, excessive soil water, and compacted soils restrict root growth. Because plants take up water faster from the upper roots, irrigation water management typically targets the upper 60 to 80% of actual rooting depths.

Table 2-2. Rooting depths for crops grown in deep uniform soils.

Crop	Root zone depth[a] fully developed (in)	Irrigation mgmt. depth (in)
Broccoli & cauliflower	18 to 24	12 to 18
Blueberry & strawberry	18 to 24	12 to 18
Potato	24	18 to 24
Tomato & cantaloupe	12 to 36	12 to 24
Dry bean	30	18 to 24
Soybean	48 to 60	30 to 36
Small grain	42	30 to 36
Sweet corn & asparagus	36 to 48	24 to 36
Field corn, Sugar beet & sunflower	48	30 to 36
Established alfalfa	60	36 to 48

[a]Restrictive soil layer, high water table, excessive soil water, and compacted soil can restrict root growth.
Source: AE-792, North Dakota State University; G75-204 (revised 1984), University of Nebraska and others.

Crop Water Requirements

Crop water use is the amount of water used by a plant through transpiration of water through plant leaves and evaporation of water from the soil surface. This process is called evapotranspiration or ET. Actual values of daily ET vary by the crop type, stage of growth, crop canopy, soil moisture content, and climatic conditions. Primary climatic conditions affecting ET include solar radiation, temperature, humidity, and wind. Figure 2-7 shows an example of the seasonal ET for corn grown in central Nebraska and the variation that results from day-to-day changes in water use during the growing season. The curve in Figure 2-7 shows the average daily water use values for corn from planting to maturity.

This curve is developed on daily long-term average ET values. As a result, daily and annual variation are masked, and a smooth line is shown.

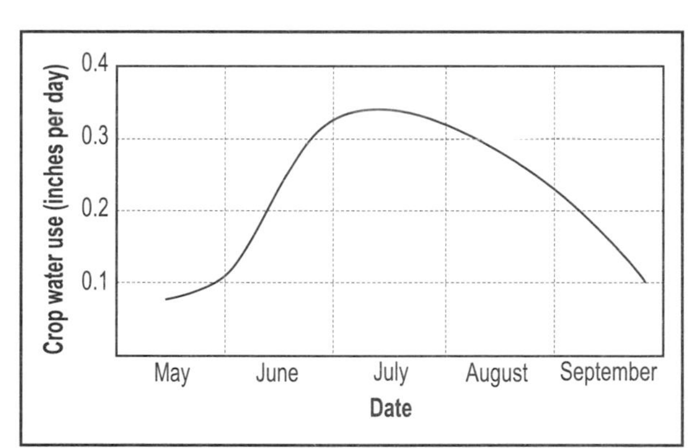

Figure 2-7. Average daily water use values for corn from planting to maturity.

Table 2-3. Seasonal evapotranspiration rate (ET) for selected states and crops within the Midwest.

Crop	Total ET (in)				
	North Dakota	Nebraska	Wisconsin	Minnesota	Kansas
Alfalfa	24 to 26	32 to 35	20 to 22	20 to 25	32 to 40
Corn	19 to 21	24 to 27	14 to 16	15 to 20	24 to 30
Soybean	17 to 19	21 to 23	13 to 15	14 to 18	18 to 24
Wheat	15 to 17	16 to 18	-	13 to 15	16 to 22
Dry Bean	15 to 17	15 to 17	-	12 to 16	16 to 22
Sugarbeet	22 to 24	23 to 25	-	19 to 22	-
Sunflower	19 to 21	20 to 22	-	15 to 18	16 to 22
Pasture	30 to 32	30 to 32	-	20 to 28	-
Potatoes	17 to 19	22 to 24	-	15 to 20	-
Sweet Corn	-	-	14 to 16	12 to 15	-
Strawberries	-	-	-	14 to 17	-

Each crop has its own water use curve that must be adjusted for planting, emergence date, expected cover, and time to reach crop maturity. Seasonal crop water use can range from approximately 10 to 35 inches depending on crop type and climate. Table 2-3, gives seasonal ET for selected crops and states and compares ET between similar crops grown in different regions of the Midwest. For ET data of crops grown in your area, contact your local Cooperative Extension.

Daily ET is difficult and expensive to measure for a specific field or area. A number of states have developed automated weather networks that provide estimated daily ET rates. The weather stations, Figure 2-8, measure air temperature, humidity, incoming solar radiation, and wind speed. From this information, reference ET is calculated, which represents water use from a reference crop of grass or alfalfa with a full cover. Individual crop ET can be estimated using the reference ET value and a crop coefficient. Crop ET often is less than reference ET due to the crop's stage of growth.

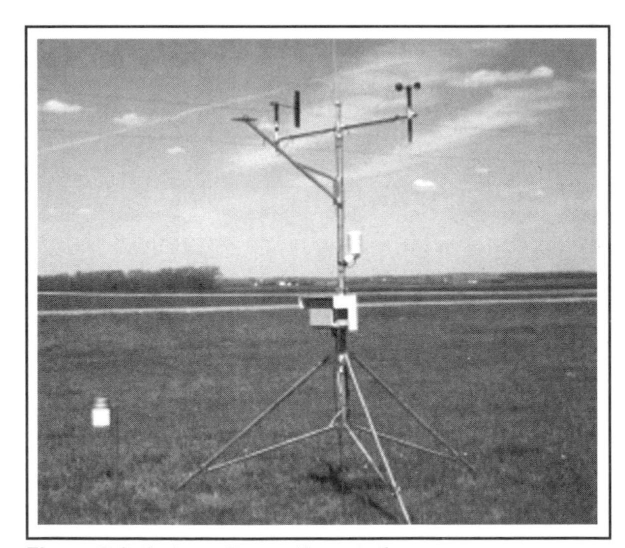

Figure 2-8. Automatic weather station.

Universities often provide estimates of ET for different crops through the news media, hotlines, or the Internet. The ET estimates represent crop water use without water stress. As water stress increases, crop ET will decline. Contact your local Cooperative Extension to find out what information is available for your area.

The automatic weather stations generally represent a geographic area that experiences similar climatic conditions. Although the actual ET rate on a given day in a specific field might vary from the regional ET estimates, over the full growing season regional estimates are excellent values for irrigation system design and management. Regardless of how ET values are obtained, periodically monitor and update soil water conditions in the field after the system is in operation. Soil moisture will vary as a result of differences in effective rainfall and irrigation efficiency. In addition, ET values are estimates of a very complex system involving the soil, water, climate, and plant. Thus, variations from field to field and within a field are possible.

Examples of typical ranges of net seasonal irrigation water application are in Table 2-4 for selected locations. Seasonal water requirements vary between years depending on ET rates and rainfall at a given location.

For example, in west central Minnesota, irrigation amounts varied over a 10-year period from less than 1 inch to more than 20 inches of water applied. To determine seasonal water requirements, make estimates of water needs using the information given.

Table 2-4. Typical range of seasonal net irrigation application (inches).

Crop	North Dakota	Nebraska	Illinois	Wisconsin	Minnesota	Kansas
Alfalfa	8 to 14	14 to 22	17 to 19	10 to 14	11 to 13	6 to 25
Corn	11 to 13	14 to 18	7 to 9	10 to 14	9 to 11	5 to 15
Soybean	6 to 12	4 to 12	7 to 9	9 to 12	5 to 8	0 to 13
Wheat	5 to 7	2 to 10	4 to 6	-	4 to 6	0 to 11
Dry bean	6 to 10	10 to 12	-	-	5 to 7	8 to 11
Sugarbeet	6 to 12	18 to 20	-	-	-	-
Sunflower	6 to 12	-	-	-	-	8 to 10
Pasture	6 to 14	10 to 20	-	-	-	0 to 24
Potatoes	8 to 12	16 to 18	-	-	8 to 12	-
Strawberries	-	-	-	-	6 to 10	-

Review other available information from area irrigators, irrigation dealers, local Cooperative Extension, and the Natural Resources Conservation Service (NRCS). An irrigation system should be able to meet the crop's total seasonal irrigation needs every year. Some irrigation systems may be economical only when designed to meet a crop's water needs eight or nine years out of 10 unless a high-producing well is available. Use this information to determine a design evapotranspiration rate (ET_d) rate.

Irrigation System Capacity

The irrigation system capacity, or flow rate, is the rate at which the system should apply water. Capacity usually is expressed in gallons per minute (gpm) or gallons per minute per acre (gpm/ac). The design flow rate can be calculated using Equation 2-1:

Equation 2-1. Discharge Capacity.

$$Q = \frac{450 \cdot A \cdot ET_d}{T_p \cdot E_A}$$

Where:
Q = discharge capacity (gpm)
A = area irrigated (acres)
ET_d = design ET rate (in/day)
T_p = allowable pumping time (hr/day)
E_A = water application efficiency (expressed as a decimal)

Water application efficiency (E_A) is the ratio of the average depth of irrigation water that infiltrates into the soil during irrigation to the average depth of irrigation water pumped. Factors that affect this ratio are evaporation from sprinkler droplets, wind drift, leaf canopy evaporation, runoff, sprinkler uniformity, and type of irrigation system. (See Chapter 5 for more details.) Depending on the situation, the application efficiencies can vary from less than 50% to more than 90% for a given irrigation event. Table 2-5 gives average sprinkler system efficiencies that can be used for design purposes. Check with your local Cooperative Extension or NRCS to determine whether more accurate information is available.

Table 2-5. Suggested irrigation sprinkler system design efficiency.

Sprinkler Type	Suggested design efficiency
Solid Set	75%
Hand Move Set	65%
Side Roll or Wheel Roll	75%
Traveler	75%
Center Pivot	85%
Linear Move	85%
LEPA System	95%

Allowable pumping time (T_p) reflects the average number of hours the irrigation system can be expected to apply water each day. Some systems (for example, a center pivot) can operate nearly continuously. To design a system, consider all potential down time such as sprinkler move time, equipment failure, power failure, available labor schedule, system maintenance, and electric power load management programs.

For a hand-move system, 8 to 16 hours per day are typically available for irrigating, while up to 23 hours per day are possible

Example 2-1. Determining the amount of water needed to irrigate a given area.

What volume of irrigation water is necessary for a 120-acre center pivot system to produce corn in central Minnesota?

Solution:

From Table 2-4, the average seasonal irrigation application amount is estimated to be 10 inches. The total volume of water needed for the season is found by multiplying the total acres by the total application depth:

120 ac x 10 in = 1200 ac-in or 100 ac-ft.

This is the total volume of water that might be pumped during the growing season to meet water needs of the corn.

with a center pivot that allows one hour of down time each day. If the system is gov-ern-ed by a load management program in which the power company can dictate available time, operating time can be reduced further.

Design evapotranspiration rate (ET_d) can be determined with different methods. Depending on crop and location, ET_d is typically 0.20 to 0.35 inches per day in the Midwest. For purposes of irrigation system design, average daily ET for a month can be used to estimate total crop water requirements.

Check with Cooperative Extension or NRCS for average daily peak design ET. If this information is unavailable, then use Figure 2-9 to estimate the average daily peak design ET. Keep in mind that long-term estimates like these may not account for short periods of higher than normal ET. This can result in under design of system capacity during these time periods.

Consider soil type, effective root zone depth, application efficiency, probability of rainfall, and allowable soil moisture depletion to avoid underestimating ET_d. In some states, more detailed maps or estimates of ET_d rates are available from Cooperative Extension or NRCS. Other factors such as total available soil water in the crop root zone, crop yield response, and irrigation schedule, influence an irrigation system's design discharge capacity. Incorporating these factors could help optimize the design discharge capacity by increasing or decreasing the specified ET_d.

Before final selection of ET_d, contact a local or state irrigation specialist, or check with Cooperative Extension or NRCS for guidelines based on experience in the area.

Table 2-6 gives the required system capacity (gallons per minute per acre) to meet 100% of crop water needs based on application efficiency, design ET_d (inches per day) and average pumping time (hours per day).

The rate of water use varies from year to year; therefore, the average water use curve is not always adequate. A frequency distribution of the water use rate for well-watered alfalfa is shown in Figure 2-10. The 50% line represents the average water use rate. The 90% line represents a rate that will be exceeded in only 1 of 10 years.

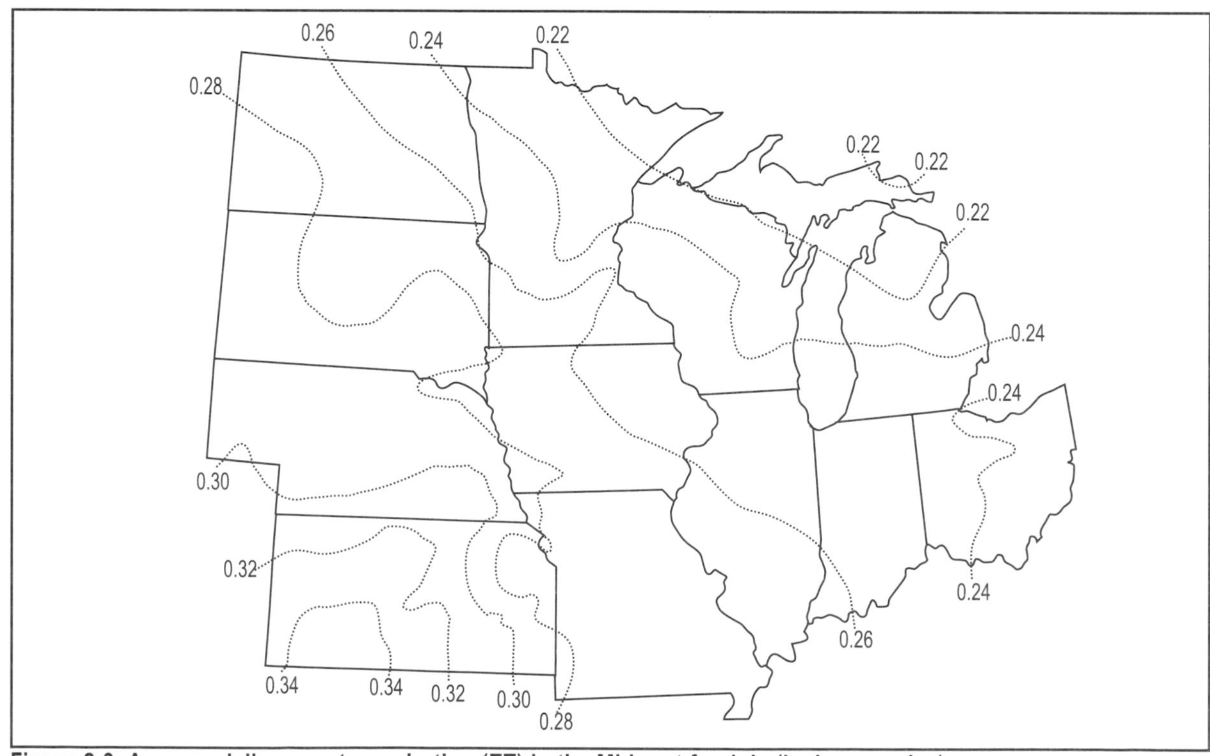

Figure 2-9. Average daily evapotranspiration (ET) in the Midwest for July (inches per day).

Table 2-6. Design system discharge capacity (gallons per minute per acre).

Design ET (in/day)	Average pumping time (hr/day)			
	12	16	20	24
Application efficiency=0.85				
0.40	17.8	13.3	10.7	8.9
0.35	15.5	11.7	9.3	7.8
0.30	13.3	10.0	8.0	6.6
0.25	11.1	8.3	6.7	5.6
0.20	8.9	6.7	5.3	4.5
Application efficiency=0.75				
0.40	20.1	15.1	12.1	10.0
0.35	17.6	13.2	10.6	8.8
0.30	15.1	11.3	9.1	7.6
0.25	12.6	9.4	7.6	6.3
0.20	10.1	7.6	6.0	5.0
Application efficiency=0.65				
0.40	23.2	17.4	13.9	11.6
0.35	20.3	15.3	12.2	10.2
0.30	17.4	13.1	10.5	8.7
0.25	14.5	10.9	8.7	7.3
0.20	11.6	8.7	7.0	5.8

Curves, such as the one in Figure 2-10, are useful in deciding the amount of risk that a producer is willing to accept. Producers can save water and money by reducing the amount of water applied, that is, applying based on an average water use rate. However, the manager would be more confident that the crop would not be stressed if a higher probability were being used.

Detailed system capacity information is available from state irrigation specialists.

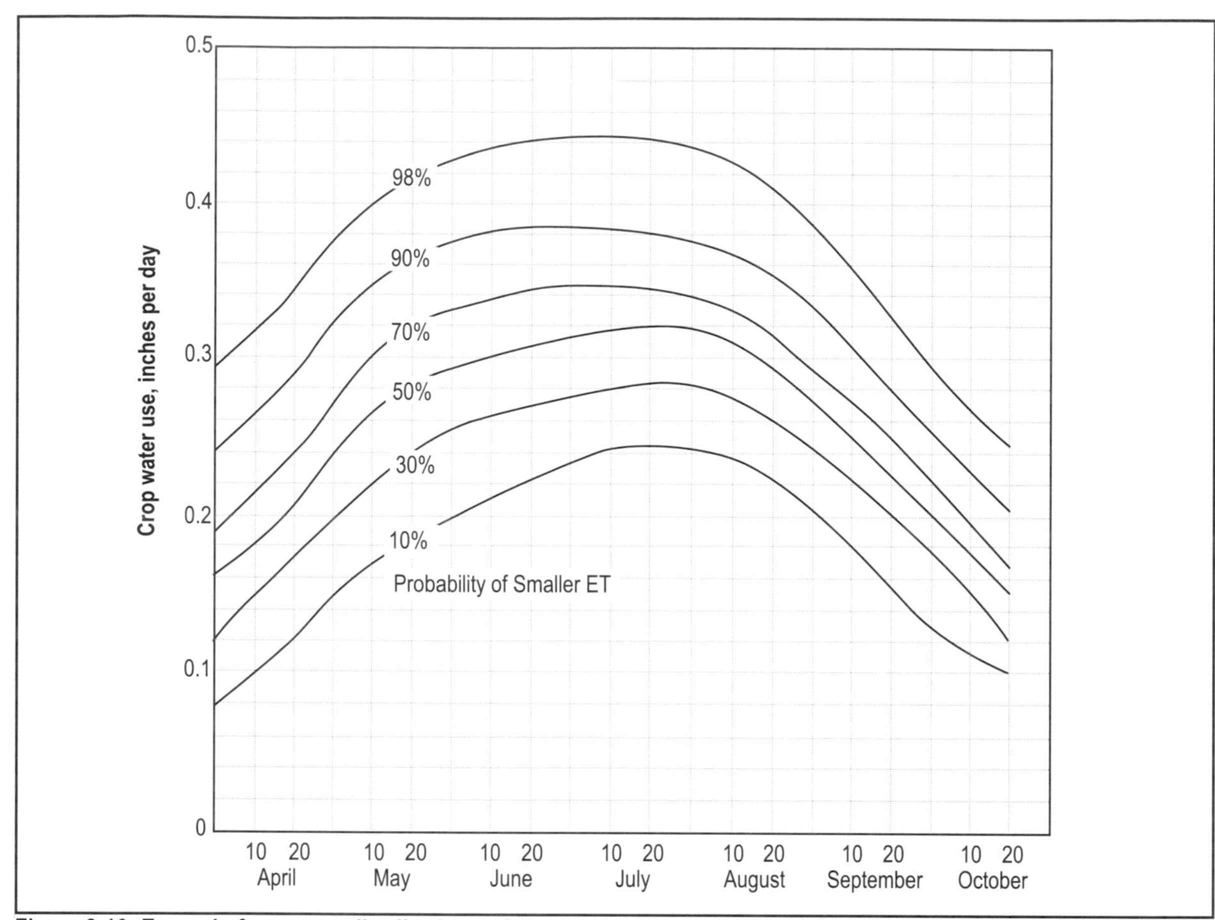

Figure 2-10. Example frequency distributions of daily water use for well watered alfalfa with full cover.

Example 2-2. Determining flow rate.

What is the flow rate requirement for a 125-ac center pivot in central Kansas given the following information?

Application efficiency: Assume 85% for a center pivot (Table 2-5).

Design ET_d = 0.32 in/day. Based on Figure 2-9.

The system is not under load control, therefore estimated average pumping time is 24 hours per day.

Solution:

Interpolating from Table 2-6, The design system discharge capacity = 7.1 gpm/ac

Flow requirement = 7.1 gpm/ac x 125 ac = 888 gpm

Chapter 3

Water Sources

Chapter 3 Contents

The water supply is the most important part of any irrigation system. Without a reliable source that provides good quality and a sufficient quantity of water, irrigation cannot be economically practiced or sustained. The amount of water needed to supply an irrigation system is presented in Chapter 2, System Design.

Legal Requirements for Irrigation

All states have laws that govern the diversion of water for irrigation. Irrigation water commonly is allocated using a permit system. An irrigation permit can be issued to an individual irrigator or to a group of irrigators such as the members of an irrigation district. A written application usually is submitted for an irrigation water permit. A filing fee may be required as part of the application.

Information about laws or regulations governing the allocation of surface water and groundwater for irrigation purposes can be obtained from local water authorities and/or from the state agency that handles irrigation water permits. Some states require a water permit to be issued before irrigation development can begin. In this situation, all necessary permits must be *in hand* before installing any irrigation equipment, including wells.

In addition to the water permit, other legal requirements may apply to irrigation equipment, well construction, or irrigation management practices. For example, most states have laws that require backflow prevention when chemigating and do not allow spraying water onto maintained roads. States also may require a minimum separation distance between wells or the use of flow meters. Certain areas within states also may regulate irrigation management by requiring use of Best Management Practices (BMPs). Many states have wellhead protection and recharge area or zone rules that protect wells supplying potable water for public water systems.

Irrigation Water Quality

The quality of some water sources is not suitable for irrigating crops. Irrigation water must be compatible with both the crop and soil to which it will be applied. The salt content, both the total concentration and types of salt, must be considered. An analysis of water for irrigation will measure the amount of principal salts: calcium, magnesium, sodium, iron, boron, bicarbonate, carbonates, sulfate, chloride and others specific to certain locations. In addition to these salts, the amount of nitrates also should be measured.

The two most important factors to look for in an irrigation water quality analysis are the **Total Dissolved Solids (TDS)** and the **Sodium Absorption Ratio (SAR)**. The TDS of a water sample is a measure of the concentration of soluble salts and is commonly referred to as the *salinity* of the water. The SAR of a water sample is a measure of its *sodicity,* the proportion of sodium to calcium and magnesium. Since it is a ratio, the SAR has no units.

Some irrigation water analysis reports also show the *adjusted SAR*. (See Figure 3-1) This value adjusts the SAR based on the effects of bicarbonates in the water. Irrigation water with a high TDS may affect plant growth, and a high SAR may affect soil structure.

Salt concentration (TDS) in the root zone influences plant germination, growth, yield potential, and soil physical conditions. Poor subsurface drainage or lack of precipitation can cause salt to accumulate in the root zone as a result of applying high-salt irrigation water. Soil salts increase as soil water is removed by plants and by soil surface evaporation. High SAR water frees up sodium ions that seal the soil surface by dispersing clay particles, thus affecting soil infiltration characteristics.

The salinity, or TDS, of a water sample is expressed in terms of its electrical conductivity (EC_w). Units used to categorize irrigation water salinity can be one of the following:

- millimhos per centimeter (mmhos/cm)
- deci-Siemens per meter (dS/m)
- micromhos per centimeter (μmhos/cm)

These units are related in that:

1 mmho/cm = 1 dS/m = 1000 μmhos/cm

The effect of irrigation water salinity on potential yield for various crops is shown in Table 3-1 (page 20).

Laboratories that perform irrigation water analysis may provide an irrigation suitability classification based on a water analysis system that was developed at the United States Salinity Laboratory in California. This classification system combines salinity and

Irrigation Water Analysis Report —— Lab Number 00

Date Received: 8-27-97 **Date Reported:** 8-28-97

Submitted by: Joe Irrigator
Box 56
Anywhere, ND

Identification: 1 **Water Source:** Test Well

ANALYSIS

pH	7.8		**SAR (sodium absorb ratio)**	1.87
Electrical Conductivity	854	**umho/cm**	**SARadj (for bicarbonates)**	2.35
Total Dissolved Solids	687	**mg/L**	**RSC (residual Na-carbonate)**	-0.13

Cations	meg/L	mg/L	Anions	meg/L	mg/L
Calcium (Ca)	4.12	82	Carbonates (CO_3)	0.00	0
Magnesium (Mg)	1.33	16	Bicarbonates (HCO_3)	5.32	325
Sodium (Na)	3.09	71	Chloride (Cl)	0.19	7
Potassium (K)	0.24	9	Sulfate (SO_4)	3.27	157

RECOMMENDATIONS

Soils Information Source: County Soil Survey

Soil Information:

Soil Series	%	ECmax	SARmax	Misc.
Valentine	30	3000	12	
Hecla	60	3000	12	
Hamar	10	3000	12	

The water is ☒ **Good** ☐ **Satisfactory** ☐ **Poor** ☐ **Unsuitable**
for irrigation of ☒ **Field Crops** ☐ **Crops with good salt tolerance**
on the soils that dominate the area to be irrigated as indicated above.

Factors that determined the recommendation:

	Acceptable	Moderate	High	Very High
Salinity	☒	☐	☐	☐
Sodium Hazard (SAR, SARadj, RSC)	☒	☐	☐	☐

PER John Smith

Figure 3-1. Sample irrigation water analysis report.

Table 3-1. Salt tolerance levels for various crops (micromhos per centimeter).

Crop	Rating[a]	Yield potential			
		100%	90%	75%	50%
Field					
Barley	T	5,300	6,700	8,700	12,000
Field beans	S	700	1,000	1,500	2,400
Corn	MS	1,100	1,700	2,500	3,900
Cotton	T	5,100	6,400	8,700	11,300
Cowpeas	MT	3,300	3,800	4,500	5,900
Sorghum	MT	2,700	3,400	4,800	7,300
Soybean	MT	3,300	3,700	4,100	5,000
Sugarbeet	T	4,700	5,800	7,300	10,000
Wheat	MT	4,000	4,900	6,300	8,700
Forage					
Alfalfa	MS	1,300	2,300	3,600	5,900
Red clover	MS	1,000	1,500	2,400	3,800
Orchard grass	MS	1,000	2,100	3,700	6,400
Perennial rye	MT	3,700	4,600	5,900	8,100
Sudan grass	MT	1,900	3,400	5,700	9,600
Tall fesque	MT	2,600	3,900	5,700	8,900
Tall wheat grass	T	5,000	6,600	8,900	12,900
Birdsfoot trefoil	MT	3,300	4,000	5,000	6,700
Crested wheat grass	T	5,000	6,000	7,300	10,000
Horticulture					
Apple and pear tree	S	1,100	1,500	2,200	3,200
Asparagus	T	2,700	6,100	11,100	19,400
Beans	S	700	1,000	1,500	2,400
Beets	MT	2,700	3,400	4,500	6,400
Broccoli	MS	1,900	2,600	3,700	5,500
Cabbage	MS	1,200	1,900	2,900	4,700
Cantaloupe	MS	1,500	2,400	3,800	6,100
Carrot	S	700	1,100	1,900	3,100
Cucumber	MS	1,700	2,200	2,900	4,200
Grape	MS	1,000	1,700	2,700	4,500
Lettuce	MS	900	1,400	2,100	3,500
Onion	S	800	1,200	1,900	2,900
Pepper	MS	1,000	1,500	2,200	3,400
Potato	MS	1,100	1,700	2,500	3,900
Radish	MS	800	1,300	2,100	3,300
Spinach	MS	1,300	2,200	3,500	5,700
Squash	MT	2,100	2,500	3,200	4,200
Strawberry	S	700	900	1,200	1,700
Sweet corn	MS	1,100	1,700	2,500	3,900
Sweet potato	MS	1,000	1,600	2,500	4,000
Tomato	MS	1,700	2,300	3,300	5,100

[a] S = sensitive, MS = moderately sensitive, MT = moderately tolerant, T = tolerant

Source: Adapted from Ayers and Westcot, 1976.

sodicity. For example, a water sample with an EC_w of 1,000 mmhos/cm and an SAR of 8 would be classified as C3-S2 (See Figure 3-2). It would have a high salinity rating and a medium sodium rating. The scale for sodicity is not constant because it depends on the level of salinity. For example, an SAR of 8 is in the S1 category if the salinity is from 100 to 300 µmhos/cm; S2 if the salinity is from 300 to 3,000 µmhos/cm, and S3 if the salinity is greater than 3,000 µmhos per cm.

Water with an EC greater than 2,000 µmhos per cm and an SAR value greater than 6 is not recommended for continuous irrigation. In cases where sporadic irrigation is practiced (for example, a particular piece of land is irrigated only one year out of three or more years and has good natural or artificial drainage), lower quality water may be used if rainfall is sufficient to leach salts from the root zone. However, the lower quality water should not have an EC that exceeds 3,000 µmhos per cm or an SAR greater than 9.

States have more specific guidelines, so check with the local Cooperative Extension or NRCS office.

Compounds containing calcium can be added to irrigation water to lower the SAR and reduce the harmful effects of sodium. However, this practice will increase soil salinity, thus requiring soil sampling every 3 years. The effectiveness of added calcium depends on its solubility in irrigation water and whether it will plug sprinkler nozzles. Calcium solubility is controlled by both the source of the calcium (e.g., calcium carbonate, gypsum, calcium chloride) and also the concentration of other ions in the irrigation water. Compared to calcium carbonate and gypsum, calcium chloride additions will result in higher concentrations of soluble calcium and be the most effective at lowering irrigation water SAR. However, calcium chloride is more expensive than calcium carbonate and calcium sulfate (gypsum).

Carbonates

Carbonate and bicarbonate ions in the water can combine with calcium and magnesium in the soil to form compounds that precipitate out of the soil water solution. The

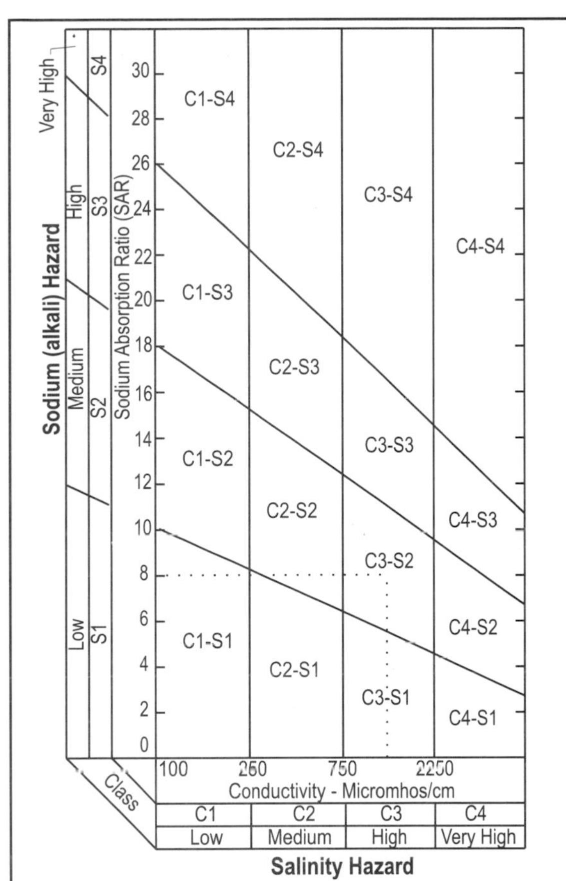

Salinity

- **C-1, Low Salinity Water** can be used for irrigation with most crops on most soils with little likelihood that soil salinity will develop. Some leaching is required, but this occurs under normal irrigation practices, except in soils of slow and very slow permeability.
- **C-2, Medium Salinity Water** can be used if a moderate amount of leaching occurs. Plants with moderate salt tolerance can be grown in most cases without special practices for salinity control.
- **C-3, High Salinity Water** cannot be used on soils with moderately slow to very slow permeability. Even with adequate permeability, special management for salinity control may be required and plants with good salt tolerance should be selected.
- **C-4, Very High Salinity Water** is not suitable for irrigation under ordinary conditions, but may be used occasionally under very special circumstances. The soils must have rapid permeability, drainage must be adequate, irrigation water must be applied in excess to provide considerable leaching, and very salt-tolerant crops should be selected.

Sodicity

- **S-1, Low Sodium Water** can be used for irrigation on almost all soils with little danger of the development of harmful levels of exchangeable sodium.
- **S-2, Medium Sodium Water** will present an appreciable sodium hazard in fine-textured soils, especially under low leaching conditions. This water may be used on coarse textured soils with moderately rapid to very rapid permeability.
- **S-3, High Sodium Water** will produce harmful levels of exchangeable sodium in most soils and requires special soil management, good drainage, high leaching, and high organic matter additions.
- **S-4, Very High Sodium Water** is generally unsatisfactory for irrigation purposes except at low and perhaps medium salinity.

Figure 3-2. Classification of irrigation water.
Source: USDA Agricultural Handbook: No. 60, 1954.

removal of calcium and magnesium increases the sodium hazard to the soil from the irrigation water. The increased sodium hazard often is expressed as *adjusted SAR*. The increase of adjusted SAR over the SAR is a relative indication of the increase in sodium hazard due to the presence of these ions.

Accumulation of carbonate minerals that precipitate out of the water can plug small-diameter sprinkler nozzles in some situations.

Boron

Boron is essential for the normal growth of all plants, but the quantity required is very small. Plants such as dry beans that are sensitive to boron require much smaller amounts than the plants such as corn, potatoes, and alfalfa that are tolerant to boron. In fact, the concentration of boron that will injure the sensitive plants often is close to that required for normal growth of tolerant plants.

Boron concentration greater than 2 parts per million (ppm) may be a problem for certain sensitive crops, especially in years that require large quantities of irrigation water.

Groundwater Aquifers

Groundwater exists nearly everywhere below the earth's surface. An aquifer is an underground water-bearing formation that provides water sufficient to operate a pump in a well. A saturated clay layer, even though its pores are filled with water, cannot yield groundwater rapidly enough and, thus, is not an aquifer. Different types of aquifer formations and associated water levels are illustrated in Figure 3-3.

Formations

Sand and gravel aquifers are unconsolidated formations that can be confined or unconfined. Shallow aquifers are typically unconfined, composed of sand and gravel, and often called "water table or surficial" aquifers, Figure 3-3. They often discharge to streams (springs on banks), or recharge deeper aquifers. Buried sand and gravel layers form confined (artesian) aquifers and occur where confining layers (clay or shale) exist both above and below water-producing layers.

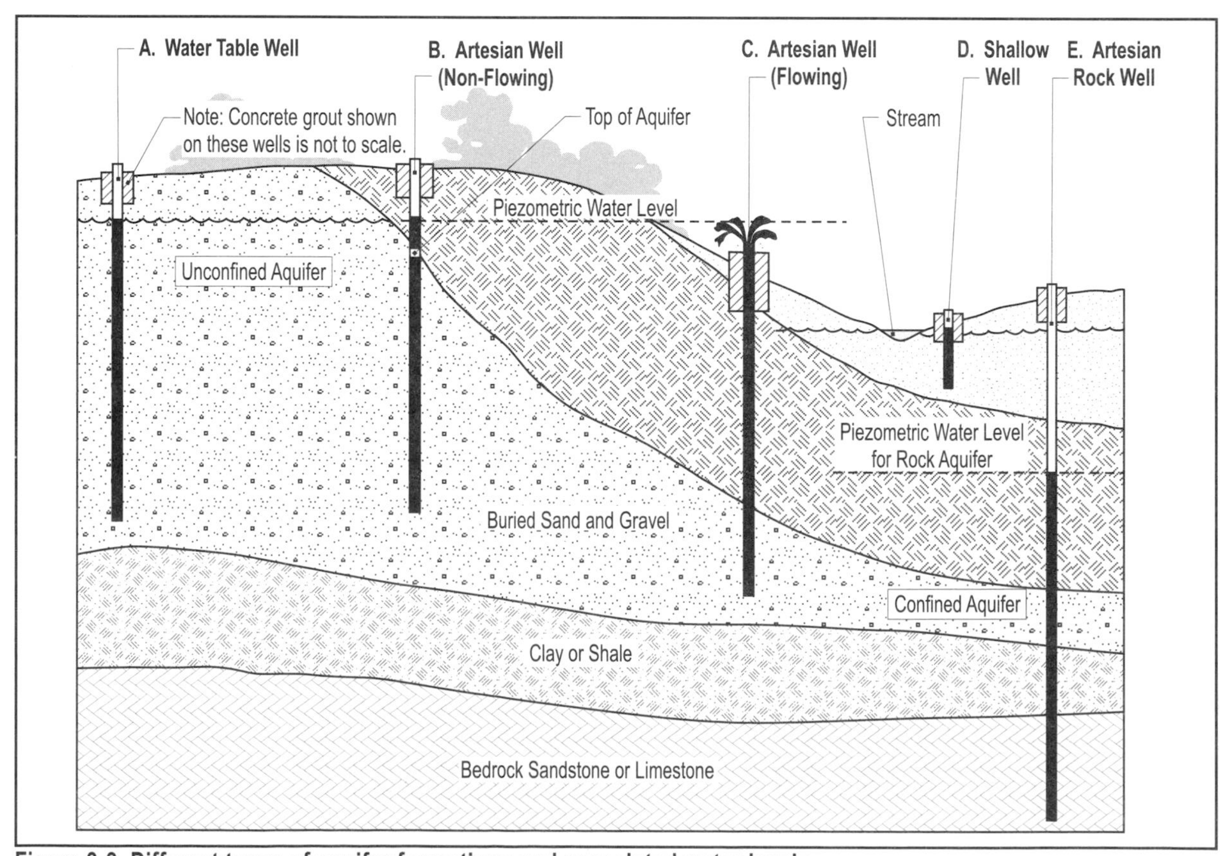

Figure 3-3. Different types of aquifer formations and associated water levels.

Bedrock aquifers are porous (fractured) rock formations generally composed of sandstone or limestone where water is readily available because it can move through fractures to reach a well.

Water Levels

The static water level in a wall is the piezometric surface of the aquifer in which the well has an opening. For irrigation wells, this would be the aquifer where the screened portion of the well is located. A piezometer measures the pressure of a fluid, thus the piezometric surface indicates the water level from the pressure in the aquifer. The piezometric water level commonly is called the *static* water level in a well.

For water table wells, the static water level is at the same elevation as the water in the surrounding formation. The upper water level is at atmospheric pressure, so minor changes in piezometric pressure head occur due to changes in atmospheric pressure. For artesian aquifers (confined aquifers), the static water level is typically above the top of the aquifer.

Frequently, irrigation wells are constructed in areas with more than one aquifer. It is not unusual to have a water table aquifer underlain by one or more confined aquifers. The static water level for the lower confined aquifers may be below, equal to, or above the water level of the upper aquifer. A piezometric water level above the ground surface results in a flowing artesian well.

The static water level will fluctuate over the course of a year due to pumping, weather, precipitation amounts, and other factors. Generally, the static level in an irrigation well will be greatest in spring and the lowest in July and August. Large changes in static level will affect the production potential of some wells.

In Figure 3-3, wells A and D are water table wells. Wells B, C, and E are artesian even though only C is a flowing well. Note that the static water levels in separate aquifers are often different, for example, wells B and E. To obtain the water level shown in well E, it must be cased to the bedrock and not have a screen in the sand and gravel layer.

Planning a Well

Irrigation wells produce from 50 gallons per minute (gpm) to more than 2,000 gpm. Wells can be constructed in unconsolidated (loose) sand and gravel, Figure 3-4, or in consolidated (rock) formations such as limestone or sandstone, Figure 3-5. Much of the following discussion is for sand and gravel aquifers; however, many of the principles discussed, such as well development, apply to all kinds of wells. Good design and construction procedures are important to achieve an efficient well.

Most states require irrigation well drillers to be licensed and to follow state well drilling codes. Good communication between the driller and irrigation equipment supplier allows matching the flow rate from the well with irrigation needs. An important step toward constructing an efficient irrigation well is selecting the well drilling contractor. The well driller generally designs, constructs, and tests the well. Because there are many

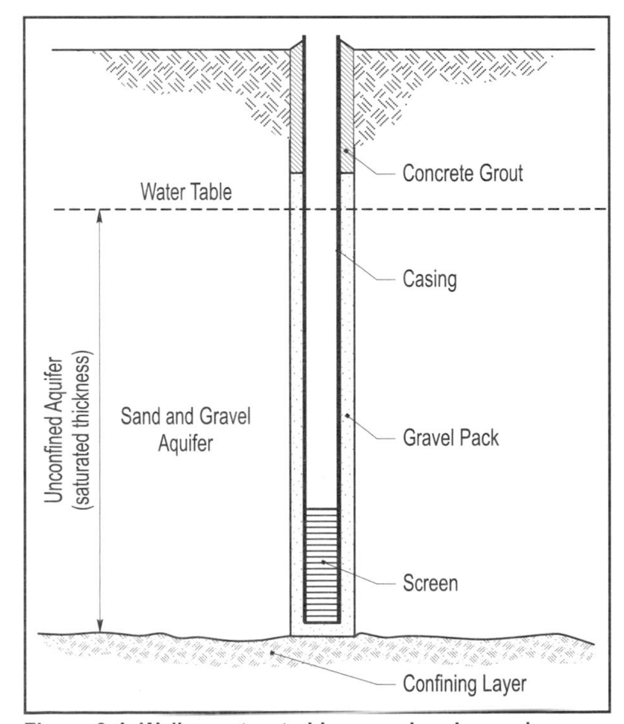

Figure 3-4. Well constructed in a sand and gravel formation. A casing and screen are always used. A gravel pack is optional.

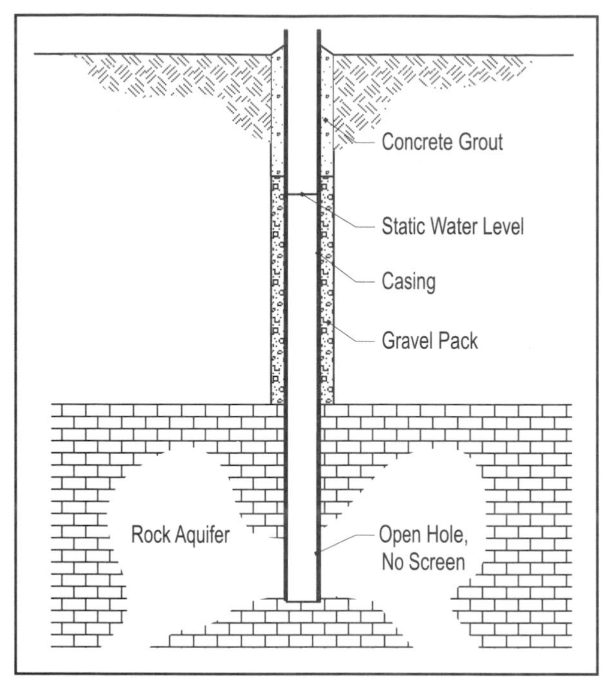

Figure 3-5. Well constructed in a rock formation. A partial casing is used, and the well is generally not screened. No gravel pack is used unless it is screened.

ways to cut corners and reduce the cost of well construction, do not select a driller on price alone. Constructing a high-producing, efficient irrigation well is more than just putting a hole in the ground. Lack of experience, expertise, or care can result in a well that pumps sand, uses more pumping energy than necessary, or doesn't yield the required flow rate. Additional initial investment can result in a better well and can pay off in yield, efficiency, and well life.

Follow these steps for an efficient irrigation well:

- Preplan and drill a test well.
- Design the well before drilling.
- Drill and construct to design specifications.
- Use a good well development method.
- Test pump the well and treat it with chlorine.
- Maintain records on use and performance.

A good reference source for construction of irrigation wells is Engineering Practice, EP400.2T, *Designing and Constructing Irrigation Wells,* from ASAE.

Preplanning

The success of a well depends largely on what happens before the drilling rig arrives. No two wells are the same, and there are several ways to construct an irrigation well suitable for a particular aquifer formation and water quality. Collect information for each proposed well location. Some issues to consider during well planning are these:

- Legal water rights and separation code. Contact the state water use agency to determine requirements.
- Estimated irrigation system water requirements, Chapter 2, *System Design*.
- Aquifer type and characteristics. Look for published groundwater studies, or consult the state water use agency. This information can answer the following questions: What type of aquifers are under the field? What is the thickness of the water table aquifer and the number and thicknesses of artesian aquifers? Are they sand and gravel, or bedrock? Will water quality influence design choices?
- Anticipated well depth and diameter. The required pumping flow rate affects casing size and, thus, diameter.

To learn as much as possible about the aquifer system, consult with neighboring landowners, well drillers, Cooperative Extension, and government agencies, such as the state and local health departments, state water authority, and the United States Geological Survey. Groundwater surveys are especially helpful if they are available. If little is known about the aquifer, use test wells to determine if an adequate water supply is available.

Location

Locate the well so there is no chance of it being inundated by surface waters from a flooding stream or surface runoff channel.

Ideally, the well should be located where the least amount of piping is required to get the water to the sprinkler irrigation system. This not only reduces the capital investment, but also reduces the annual energy costs of pumping the water. For example, the well for a center pivot sprinkler irrigation system should be located near the pivot point.

Test Drilling

Test wells are temporary, small-diameter holes, usually 4 to 6 inches, drilled into the aquifer for exploratory purposes. Test wells

- Locate the depth and thickness of the water bearing formation.
- Provide the geologic and aquifer formation attributes needed for the well design.
- Provide a water sample to determine the quality of the water for irrigation.
- Can be used to estimate the irrigation well capacity (potential sustained flow rate).

Test wells should be drilled where there is the best chance of getting a good water supply or where the irrigation well will be located, for example, in the middle of the field. Unless the area is of known uniform geology, test wells are necessary to determine if a high-capacity irrigation well can be successfully installed. Some test holes can be converted to water level observation wells, but the majority should be filled and sealed following state recommendations.

Well Characteristics

When water is pumped from a well, the water level in the well drops from the static water level to a pumping water level. The difference between the static and pumping water levels is the drawdown, Figure 3-6. Pumping water from a well always causes drawdown. The greatest portion of the drawdown occurs within the first hour after the pump is started. After several hours, the drawdown generally is stable at a certain pumping water level. It will continue to increase at a very slow rate.

The pumping (dynamic) water level depends on several factors including the rate of pumping, the amount of time since pumping began, the type of aquifer, and well construction. Pumping lowers the water table in the aquifer around a well, but the piezometric water level in the aquifer outside the well is always above the pumping level. This elevation difference causes water to move into the well and forms a "cone of depression" of the water level in the surrounding aquifer, Figure 3-7.

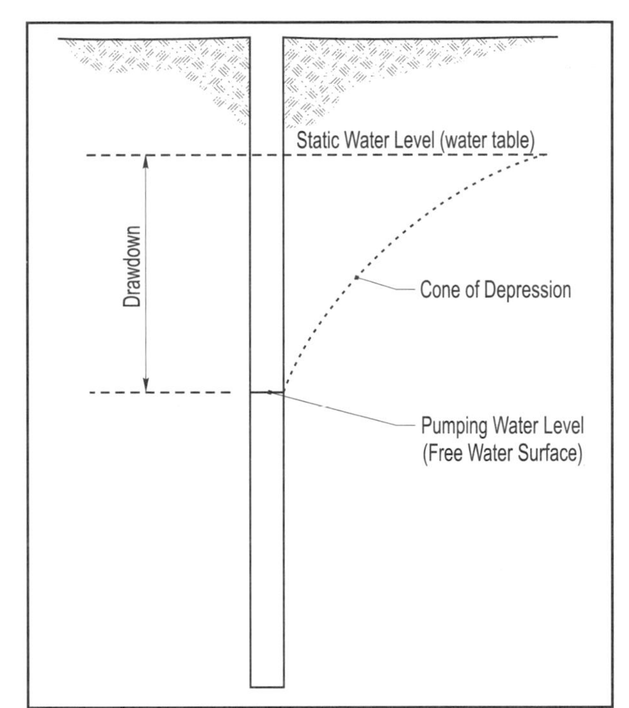

Figure 3-6. Well hydraulics.

Well Efficiency

An efficient well has an actual measured drawdown close to the theoretically predicted drawdown for that specific aquifer. Good design, construction, and development will produce an efficient well. An efficient well will have the least possible drawdown while pumping at a particular flow rate, resulting in the lowest energy consumption.

Well Design

Accurate sampling of the aquifer formation materials is essential for predicting yield and selecting the size of the well screen(s). Wells that obtain water from rock formations may not require a screen, but the formation material still should be sampled.

The driller should collect samples of the aquifer formation at a number of depths (usually 5-foot increments) during test drilling. In aquifer formations requiring a well screen, samples are sieved through various sized screens to determine the proportion of particle sizes of sand and gravel at each depth. Sample analysis determines

- Screen length, type, and slot opening. Screens may not be needed in rock aquifers.

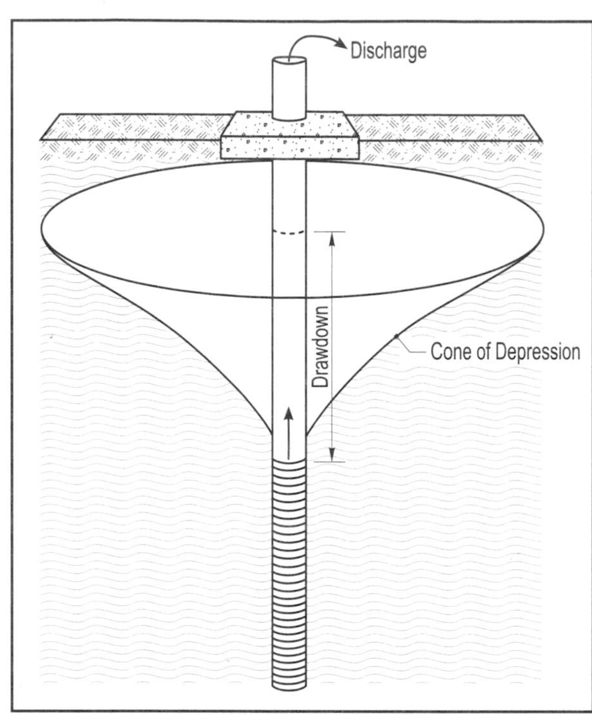

Figure 3-7. Cone of depression in the water table around a pumping well.

- Gravel pack size, when used.
- Required well development methods or techniques.
- Finishing steps needed, such as grouting, treatment, and test pumping.

The aquifer water quality can influence the selection of well construction materials. Corrosive water affects screen and casing materials. Water with high carbonate content can form deposits, called encrustation, in the aquifer near the well and in the screen openings, in part due to the water velocity entering the well.

Screens with a large open-area percentage will reduce the entrance velocity, which helps minimize encrustation. Acid treatment can sometimes remove encrusted material after it has formed, but an acid-resistant screen and casing must be used if acid treatment is expected.

Well Casing

The well casing diameter should be selected at least one size larger than the pump bowl size, however, two sizes larger is preferable. The desired pumping rate in gallons per minute (gpm) in Table 3-2 gives approximate well casing diameters for various well flow rates.

Table 3-2. Nominal casing sizes.

Pumping rate (gpm)	Optimum casing size (in)	Minimum casing size (in)
Less than 100	6	5
75 to 175	8	6
150 to 400	10	8
350 to 600	12	10
600 to 1,300	16	14
1,300 to 1,800	20	16
1,800 to 3,000	24	20

Casings should extend 1 to 2 feet above grade; however, state well construction standards provide more detail. The concrete seal, pad, and an access tube are shown in Figure 7-5 on page 107. As that figure shows, the well must have access to the inside of the casing from above the ground surface. The access performs two very important functions: it allows measurement of the static and pumping water levels, and it provides a way to introduce chlorine into the well. The access should be of sufficient diameter to introduce the quantities of chlorine needed for treatment and allow easy measurement of water levels.

Water level measurement can be made with a measuring line or tape, electric sounders, air line devices, or pressure-sensing equipment. To make these measurements easier, 3/4-inch diameter polyethylene (PE) tubing should be attached to the pump column when the pump is installed and the end slid through the access pipe. Water level measurements can then be made throughout the life of the well without encountering obstructions. The end of the tube in the water should terminate near the pump inlet. This end should be bent over or plugged and 1/4-inch holes drilled into the tube every 2 feet to above the static water level. The holes allow water to enter and leave as drawdown occurs. The plugged end prevents water level sensors from being drawn into the pump.

Casing Materials

The casing keeps the borehole open and provides support for the pump. Selection of casing material is based on water quality, well depth, borehole diameter, and state or local regulations. The most common casing material is steel, but fiberglass, thermoplastic (PVC), and concrete also are used. In some rare cases, stainless steel casing is used when the water is highly corrosive.

Deep wells (over 100 feet) may pose problems for PVC casing. Due to the pressure of the overburden and the possibility of casing collapse, some states require PVC casing to be a certain size. SDR is the standard dimension ratio of the pipe outside diameter divided by the wall thickness. A low SDR value (18 to 25) indicates a large wall thickness, and a high SDR value (30 to 45) indicates a small wall thickness.

Well Screen

The well screen is where water enters a well. Screens are made of steel, PVC, stainless steel, fiberglass, and other materials. The choice of screen material is based on three factors: water quality, strength requirements (overburden pressure), and potential presence of iron bacteria. A water quality analysis can determine if the groundwater is corrosive, encrusting, or both. Well depth and the nature of the formation materials will indicate forces the screen might experience.

Iron bacteria are a nuisance that cause plugging of the screen. Strong solutions of chlorine and/or acids are needed to control this organism. If a well is to be installed in a formation with iron bacteria or cascading water, the well screen material must be able to withstand repeated exposures to these chemicals.

Screen Slot Size

Well screen slot size depends on the size of the sand and gravel in the aquifer (for a natural gravel pack), or on the size of the gravel (when a gravel pack is added). Slot size openings must be selected large enough to let water flow freely into the well, but small enough to keep out sand after the well is developed.

Many types of screen openings are available, Figure 3 8. Cutting slots in the casing with a torch is not recommended. Choose a screen with a large total open area

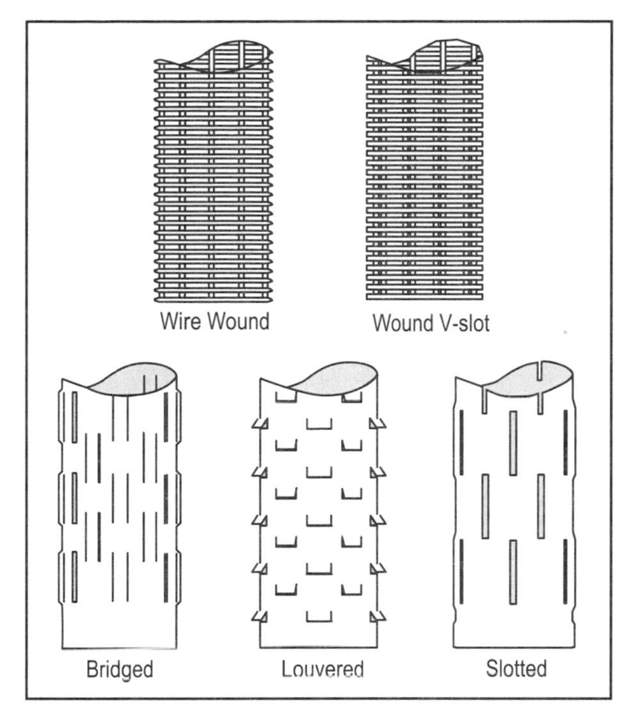

Figure 3-8. Well screen slots.

for a given slot size. Screen open areas vary from less than 1% to more than 50%. A general rule is that a screen open area of 15% or greater will not restrict the production of a well for a given diameter screen.

Screen Length

Longer screens have more total open area making it easier for water to enter, however, there are optimum lengths of screen for aquifers. Water table aquifers will produce the optimum amount of water when 30 to 40% of the saturated depth is screened. An artesian aquifer will produce the optimum amount of water when 80% of the aquifer thickness is screened, Figure 3-9. However, less can be screened when the pumping capacity is less than optimum. Usually, the local experience of the well driller will establish the screen length. For either type of aquifer, additional screen does not increase the flow rate from the well significantly, but it does increase the cost.

Some wells are constructed to obtain water from two or more artesian aquifers. The entire thickness of each aquifer often is screened. **Connecting two aquifers using a well may not be legal in some states.**

Do not over-pump a well; keep the pumping water level above the top of the screen. Over-pumping causes screen plugging or encrustation, and can shorten the expected life of a well. Over-pumping also can cause increased sand pumping due to the increase in the velocity of the water entering the well.

Gravel Pack

A filter pack may be needed if the formation has uniform fine sand, or alternate layers of coarse and fine sand. The filter pack commonly is called a gravel pack, composed of clean, chlorinated, quartz-based, well-rounded grains of uniform size. The gravel pack is placed between the borehole and the outside of the screen. The size of the gravel pack is chosen to retain most of the formation material. The size of the well screen openings are then selected to retain at least 90% of the gravel pack. The gravel pack prevents sand from being pumped through the screen and into the well. Any screened well can be drilled oversize and then gravel packed.

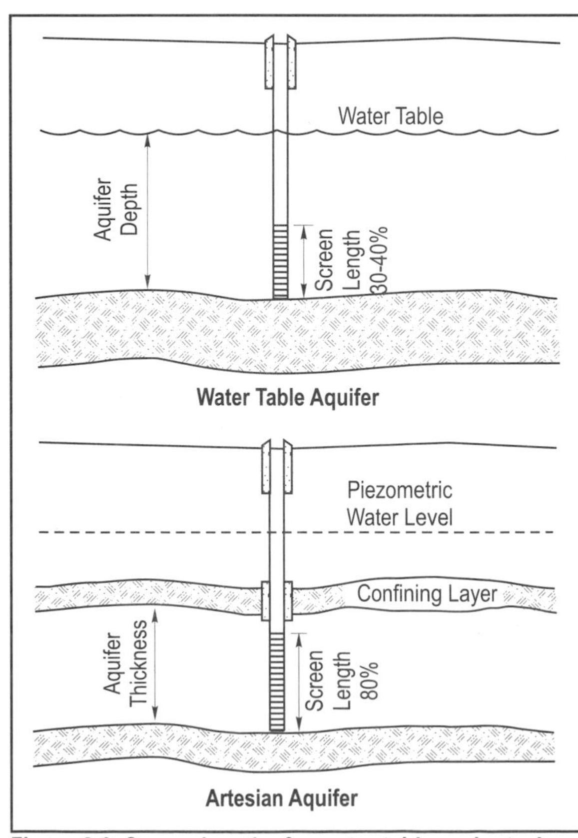

Figure 3-9. Screen lengths for water table and artesian aquifers.

A gravel pack also is used in formations that consist of alternating layers of fine, medium, and coarse sediment, and in weak sandstone where sand particles are prone to flake off and enter the well. Gravel pack wells require development to remove the fines from the aquifer formation near the gravel pack, improve well yield, and remove drilling mud (if used) from the well borehole.

In some wells, no filter material is added; rather, a natural gravel pack is developed around the well screen, Figure 3-10. The screen in a naturally developed well is nearly the same diameter as the drill hole and is next to the formation material. Careful development removes fine material near the screen, forming a size gradation from large to small away from the screen. Properly developed, natural gravel pack wells can be highly productive.

Drilling a Well

A durable, successful well yields the needed amount of water, requires a minimal amount of maintenance, and is dependable. The driller is responsible for the following:

- Properly placing of the gravel pack and screen.
- Producing a plumbed and aligned (straight) well that does not pump sand.
- Taking precautions to prevent polluting or contaminating the groundwater.
- Destroying bacteria in the well and equipment.
- Complying with local and state well drilling codes.
- Taking samples and recording geological information including formation makeup, depth to different formations, and depth to water-bearing formation.
- Developing the well.

Drilling the well includes boring the hole, setting the casing and screen (if needed), placing the gravel pack, filling the annulus between the borehole and casing to near the surface, placing a concrete plug at the soil surface, and completing the well

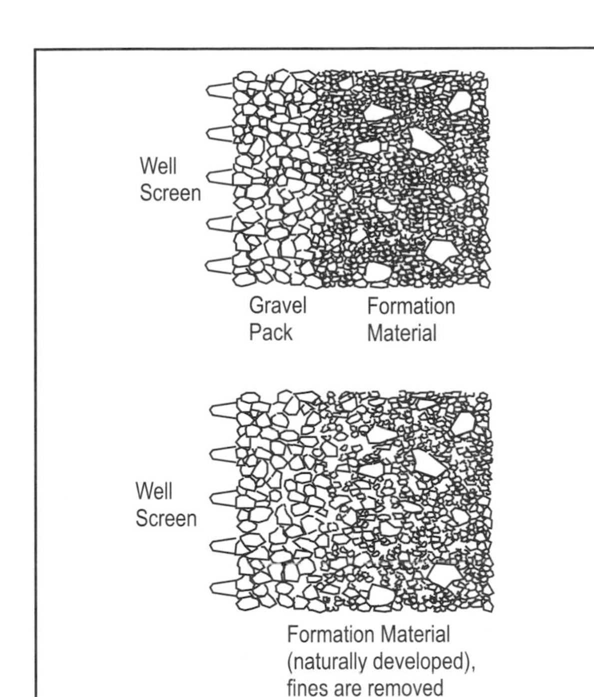

Figure 3-10. Gravel pack material vs. naturally developed well.

head for installation of the pump. Only items that can affect well performance are discussed here.

It is impossible to over-emphasize the importance of a plumbed and aligned (straight) well casing. These two important parameters are controlled by the drilling operation. Alignment is the most important. Pumps have operated in straight wells for many years that were not plumb, but wells with a dog-leg or S-curve can cause premature bearing and shaft failure in the pump. In some wells with severe misalignment, the pump could not be installed to its proper depth. The driller should continually check alignment and plumb during the drilling operation.

The driller must install the screen at the right depth and make sure the screen is centered in the borehole. The use of centering guides, especially in gravel-packed wells, is necessary to ensure the screen is centered.

The drilling method depends on several factors, including aquifer type, geographic area, and driller preference. The four most common methods are rotary, reverse rotary, cable tool, and auger bucket. These drilling methods are compared in Table 3-3. Other drilling methods include jetting, air rotary, and percussion rotary.

Some drilling methods, especially rotary drilling, use drilling mud, a thick, slurry-like mixture. The most common drilling mud is a mixture of water and bentonite powder. Bentonite is a naturally-occurring swelling clay. Drilling muds also can use man-made polymers or organic-based residues.

During drilling, the mud lubricates, helps remove cuttings from the well, seals the borehole wall against escaping fluids, and keeps the borehole from collapsing during open hole drilling. Drilling mud rarely is used with cable-tool or bucket auger drilling.

Because drilling mud can plug the water-bearing layers of the formation, its use must be followed with extensive well development.

Well Development

Good well development is a very important part of well construction. Well development is the cleaning of drilling mud, clay, and fine sand from the aquifer formation around the well screen after the well has been constructed. It should be the last step in the construction process and must be done before the new pump is installed. Development accomplishes the following:

- Repairs damage to the formation that occurs during normal well construction.
- Removes drilling mud or other materials used during construction.
- Increases the permeability of the formation around the well.
- Stabilizes the sand around the screen or gravel pack to produce sand-free water.
- Reduces sand pumping; increases pump and well life, and reduces maintenance.
- Improves the yield and performance up to 100%.

Well development is inexpensive relative to the higher efficiencies and lower operating

Table 3-3. Drilling method characteristics.

	Rotary	Reverse rotary	Cable tool	Auger bucket
Drilling	Fast drilling	Fast drilling	Moderately slow	Fast drilling
Well diameter	Small to medium diameter, 4 to 24 in.	Large diameter, 24 to 60 in.	Diameters less than 36 in.	Medium to large size over 18 in.
Applications	Wide range of drilling applications -- depth, size	Best suited for soft, unconsolidated formation and formation materials	Versatile - suited to all materials to a limited depth Easier to drill in a cold climate	Best suited to shallow water tables and gravel formations
Mud	Drilling mud generally used	Light mud used occasionally	Mud seldom used	Mud seldom used
Basic construction process	Casing, screen, and gravel pack set after hole is drilled; hole is kept open with drilling fluid	Casing, screen, and gravel pack set after hole is drilled; drilling fluid must be maintained at ground level to hold hole open	Casing set during drilling; blind casing and pull back procedure is needed to set screen and to place the gravel pack	Casing, screen, and gravel pack normally set after hole is drilled; water level must be kept at ground level to hold the hole open
Other considerations	Proper mud mixing and handling is critical to success	Need large quantities of drilling water	Little drilling water needed	May need large quantities of drilling water
Drilling operation	Drilling operation must proceed to completion after starting	Drilling operation must proceed to completion after starting	No time limit on drilling operation	Drilling must be completed or partial casing must be set
Common problems or requirements	Loss of circulation is possible Poorly suited for collecting formation or water quality samples	Loss of circulation is possible Poorly suited for collecting formation or water quality samples Large boulders or rocks are a problem High water table causes problems	Heavy steel casing is required Best suited for collecting water quality and formation samples and water level information	Loss of circulation is possible Cannot use in rock-type formations Fine sand or boulders are a problem High water table desirable to hold open
Development	Development is essential to clean mud from the formation	Development is highly desirable	Development is desirable	Development is desirable

costs obtained by the well. In practice, development should continue until no sand is left in the water while the well is being pumped. The well is not completely developed until the water is sand-free and the well yields the needed water with as little drawdown as possible. This prevents the new pump from being damaged by the sand removed by the development process.

Surging

Surging alternately forces water into and out of the formation through the well screen openings, Figure 3-11. A piston-like tool moves up and down in the well to create the surging action. The surging of the water through the well screen loosens the mud and fines in the well borehole and draws them into the well to be removed by pumping or bailing. Surging is especially suited to cable tool drilling. While common for bridge or louvered well screens, surging is not very effective with very deep wells or those with multiple screens.

Air Lifting

Air lift pumping forces compressed air through an air line to the bottom of the well, Figure 3-12. As air bubbles rise, they create a surging effect and carry water and fines out of the well. Air lift pumping is alternated with short periods of no pumping, which forces water out into the formation to help break up sand bridging around the screen. Well development by air lifting is only effective if the water is deep enough in the well to get the surging action. Air lifting does not work if lift to the surface is too great.

Double Packer Air Lifting

This development method concentrates air lift pumping on a small zone of the screen and formation that is isolated between the packers, Figure 3-13 (page 32). Raising and lowering the packer assembly develops the entire screened aquifer section selectively. Lower permeability zones can be developed more to improve production.

The double packer is more effective than regular air lift pumping, especially for longer screens, because less air is needed and equipment costs are lower. The method is well suited for certain screen designs, such as bridge or louvered screens.

Jetting

The best well development method is high pressure water jetting with simultaneous air lift pumping, Figure 3-14 (page 32). High velocity water jets through the screen and gravel pack into the formation to loosen and break down the fine materials. The jetting tool rotates slowly and is moved up and down inside the well screen. Air lift pumping removes the loosened sand and mud as they enter the well screen.

The jet stream can be directed at any part of the formation around the well for selective development. The screen must be open so that jets can act on the formation. Jetting often is the most costly development method.

Other development methods are available. Over-pumping (pumping the well at a higher rate then normal) has limited effectiveness because there is no surging action.

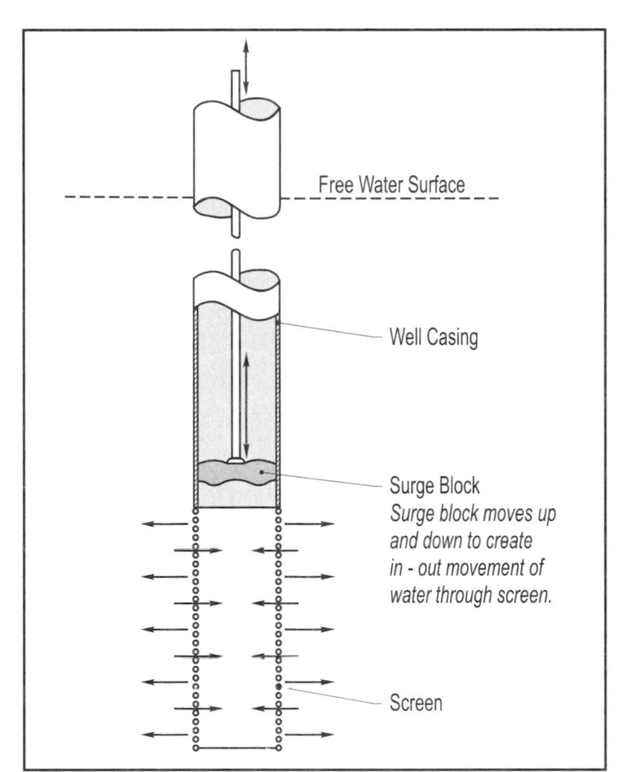

Figure 3-11. Surging.

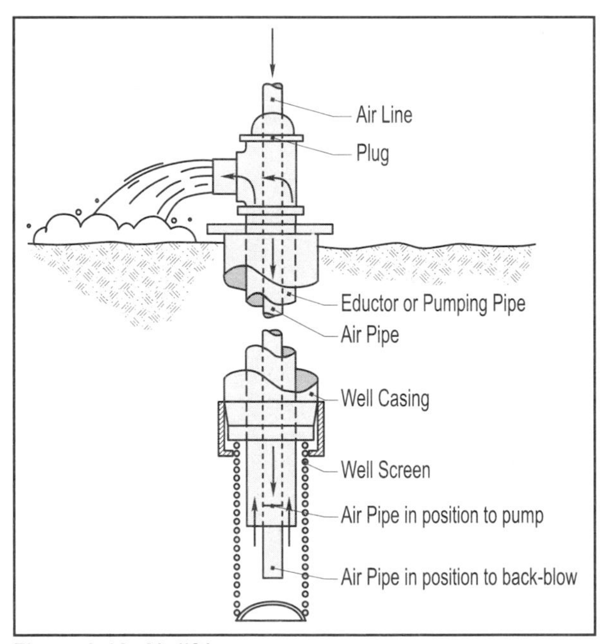

Figure 3-12. Air lifting.

Rawhiding is surging by starting and stopping the pump rapidly. This method is less effective than others and tends to decrease pump life. If over-pumping is used to develop the well, the driller's pump should be used, not the irrigation pump.

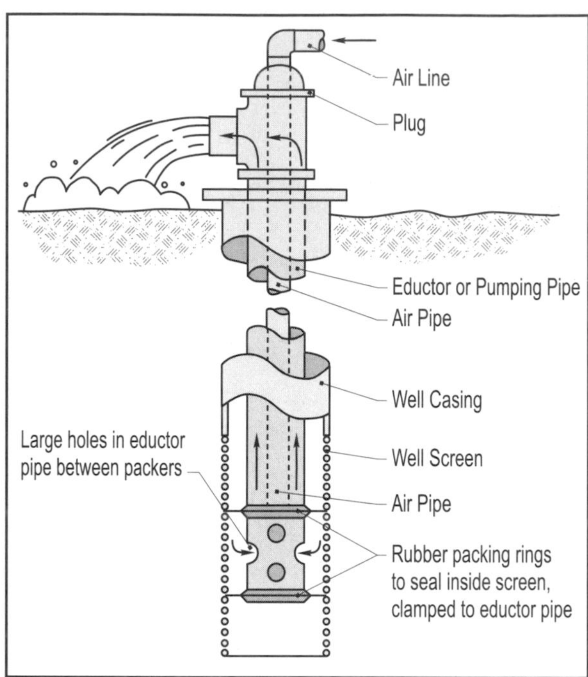

Figure 3-13. Double packer air lifting.

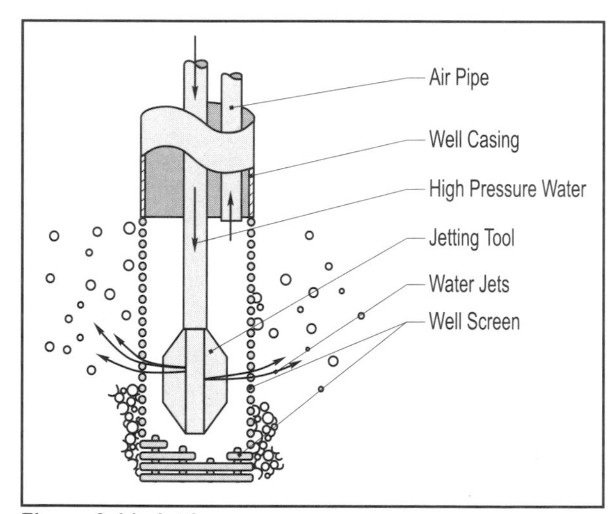

Figure 3-14. Jetting.

Treating the Well

A common problem in wells is the growth of iron bacteria. Iron bacteria do not cause disease but live on the iron in the water, pump, or pipe and form a red, slimy mass. Iron bacteria can plug the well screen and formation, and are very difficult to remove. Iron bacteria can be transmitted by well drilling tools and introduced into the well during drilling. Iron bacteria can be controlled during the drilling process by periodically disinfecting the equipment with a chlorine solution.

After development, the well should be disinfected by shock-chlorination. Each state may specify a minimum disinfection amount. However, the chlorine content of the water in the well should be raised to at least 200 parts per million and let stand for at least 24 hours. This will help prevent the growth of iron bacteria.

Test Pumping the Well

Test pumping the well should be included in the contract with the driller. This step is very important because it provides verification of well production (flow rate) at a certain drawdown. Knowing what the well can safely produce is the basis for selection of the pump or pumps, motors, sprinkler packages, pipe diameters, and many other parameters. Some states may require a pump test before granting approval of an irrigation permit. Request the following information from the well driller:

- Static water level before and after the test pumping.
- A range of pumping flow rates and associated drawdown with each.
- Complete drilling log and test-pumping data.
- The specific capacity of the well for each pumping flow rate.

The specific capacity of a well is the flow rate in gallons per minute (gpm) divided by the drawdown. This value provides an estimate of maximum production rate (at the time of drilling) of the well. The maximum drawdown in a well should be to the top of the screen. Knowing the maximum drawdown and the specific capacity provides the maximum flow rate of the well.

Test pumping information also helps determine the correct pump size, the number of stages or bowls, and the power unit required for a proposed irrigation system. Often an accurate well test will more than pay for itself in lower equipment and operating costs.

Measure flow rate during test pumping with a flow meter accurate to within 2%. Timing water flow into a barrel is not accurate enough for testing. Do not stop or interrupt pumping during the test period.

Pump until the pumping water level nearly stabilizes—from 4 to 72 hours. Test pumping for several hours or days produces a lot of water—plan to dispose of it outside the zone of influence of the well being tested.

Also get a complete set of specifications from the driller when the well is finished, as shown in the Well Specification Worksheet (following page). Well records are valuable for evaluating well problems and maintenance, and deciding whether the well can yield more water to meet future changes in the irrigation system.

Well Maintenance

Keep records of an annual measure of well water levels, flow rate, and specific capacity. Measure static water level at least before and after the irrigation season. Record the flow rate and pumping water level at the beginning, middle, and end of the irrigation season. A good well record helps spot potential problems early. For example, if the specific capacity drops below 80% of the new well, take action to restore performance.

Iron bacteria and mineral encrustation can seriously reduce well yield and life. Knowing well performance will help determine when these conditions are affecting the performance. Shock chlorination is an annual preventive maintenance procedure that can help control these problems and increase the life of a well. This procedure should be performed after the irrigation season and before winter. Many states have Extension bulletins that outline the chlorination procedure.

Mineral encrustation in the well screen can be removed by treating with a strong acid solution that chemically dissolves the encrusted minerals. Well drillers can rehabilitate many wells with acid treatment followed by redevelopment if trouble is noted early. Some drillers use a video camera to examine and record the screen condition before and after treatment.

Rehabilitation is performed by jetting the acid mixture into the well screen and pumping at the same time. This produces a fluctuating water flow through the well screen. The most common acids used to treat wells

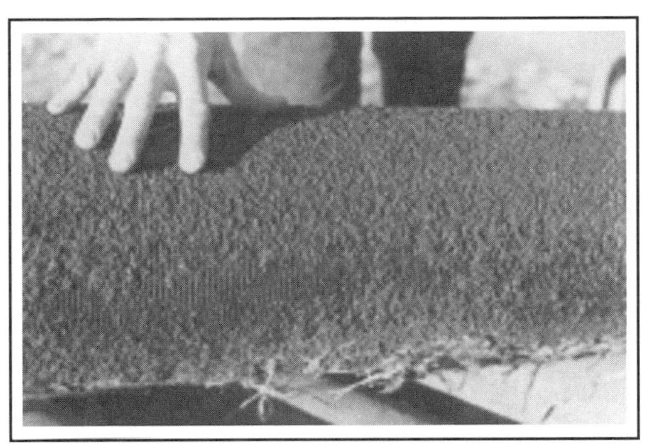
Figure 3-15. Encrusted well screen.

are sulfamic acid, hydrochloric acid (common name is Muriatic acid), and hydroxyacitic acid (also known as glycolic acid). Caution: only wells with fiberglass, PVC, or stainless steel screen should be repeatedly acid-treated.

Surface Water Sources

Surface water supplies for irrigation include small streams, rivers, drainage ditches, lagoons, dug-out ponds, natural ponds, and lakes. These supplies vary from quite limited to more than adequate in available volume and/or flow rate. For irrigation, many states may require minimum flows in streams or minimum lake levels.

A water appropriation method, based on an English law called the riparian doctrine, lets owners of property adjacent to surface water make reasonable use of the water on their own land. This doctrine says that enough water should be left for reasonable use by other users. Some states require a permit for use of the water. Also, some streams have flows protected to maintain certain downstream requirements such as aquatic life, municipal water supply, recreational flows, and industrial needs.

Surface water sources can contain floating debris such as wood, paper, logs, etc., as well as suspended sediments, algae, and other small particles. For these reasons, surface water sources can present filtration problems when pumped to sprinkler irrigation systems. Debris and sediment problems can be minimized by using one or more of the following practices:

Well Specification Worksheet

Well location ______________________________

Well depth (feet) ______________________________

Date completed ______________________________

Casing

Length (feet) ______________ Diameter (inches) ______________

Material ______________ Thickness (inches) ______________

Screen

Length (feet) ______________ Diameter (inches) ______________

Type ______________ Material ______________

Slot size ______________________________

Manufacturer ______________________________

Capacity (gpm at 0.1 fps velocity) ______________________________

Pump

Brand ______________________________

Model ______________________________

Number of bowls ______________ Column length (feet) ______________

Tail pipe length (feet) ______________ Depth to Pump Inlet (feet) ______________

Attach a pump performance curve

Pumping Test Data

Static Water Level (feet) ______________

Pumping Water Level (feet) ______________

Flow rate (gpm) ______________ after ________ hours

Drawdown (feet) ______________

Specific capacity (gpm per foot of drawdown) ______________

- Locating the pump inlet halfway between the water surface and the bottom of the water source.
- Screening the inlet.
- Filtering the water after it is in the pipeline.

Water-short Western states usually apply the doctrine of prior appropriation, which is first water users have first rights, and users must have a registered or determined water right.

Flowing Water Sources

When streams are at their lowest levels, water quality is usually poorest, and irrigation water demand is greatest. Knowing the minimum stream flow during drought conditions is important in determining water available for irrigation.

Government agencies such as the United States Geological Survey (USGS) and state water agencies have flow data covering several decades from many stream-gauging stations. Data are published by the USGS in their annual Water Resources Data book for each state and are available at most public libraries. Electronic access to these data also is available on CD-ROM and the Internet.

Local residents of the area along the stream are another important source of information. The volume or flow rate of available surface water has a direct impact on irrigation pumping capacity.

If stream flow data are not available, the flow in the stream must be measured. The stream flow should be measured in the summer during the greatest irrigation demand and the lowest natural flow rates. The stream flow rate can be measured directly by installing a flume or weir in the stream. Stream flow also can be estimated by measuring the cross-sectional area of the flowing stream and multiplying that by the average velocity in the stream.

One method to estimate the cross-sectional area of a stream is shown in Figure 3-16. The velocity in the stream can be measured using a current velocity meter or by the float method. The float method is probably the quickest and easiest way to obtain an estimate of the average velocity in a stream, but is less accurate.

To use the float method, select a straight stream section 100 feet long with uniform depth and width. Put a marker at the upstream and downstream ends of the 100-foot stretch. At the center of the selected stretch, measure the stream width perpendicular to the stream flow at the water surface from bank to bank. Use the method shown in Figure 3-16 to calculate the cross-sectional area of the stream.

To estimate the average velocity, place at least three small floats (usually small wood blocks) in the center of flow upstream from the measured section. Start a stopwatch when the first float passes the upstream marker and stop the watch when the first float reaches the 100-foot marker. It may not be the same float. Perform three or four trials and use the average travel time. Then use the following formulas to calculate an estimate of the stream flow rate:

Equation 3-1. Estimated Stream Surface Velocity.

$$\text{Surface Velocity} = \frac{(\text{Trial Section Length})}{(\text{Float Travel Time})}$$

Where:

Surface Velocity = Estimated average velocity on the surface, feet per second

Trial Section Length = Trial section length, feet, usually 100 feet

Float Travel Time = Average time for float to travel trial section, seconds

Equation 3-2. Estimated Stream Flow Rate.

$$\text{Stream Flow Rate} = K \cdot (\text{Surface Velocity}) \cdot (\text{Area})$$

Where:

K = A constant to convert surface velocity to average stream velocity. For most streams use 0.8, but for small streams that are marginal for irrigation, use 0.7

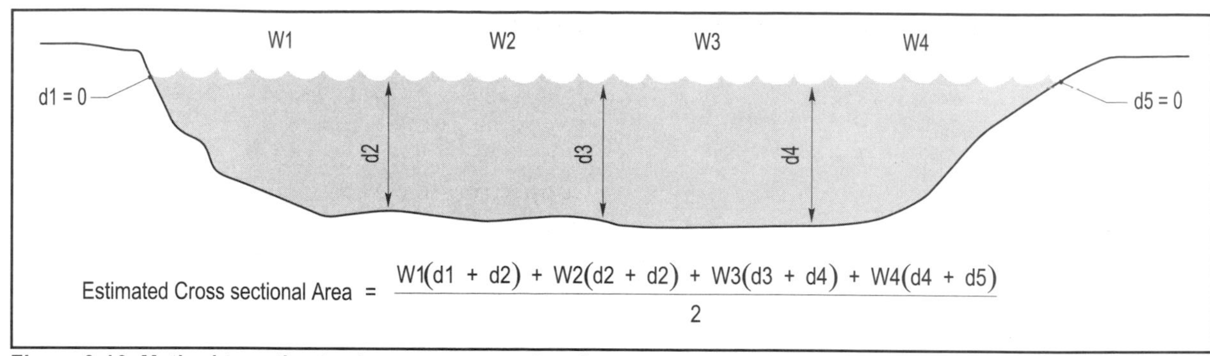

Figure 3-16. Method to estimate stream cross-sectional area.

Area = Average stream cross-section area in square feet (ft^2); use Figure 3-16.

Constructed Reservoirs for Storing Irrigation Water

If a lower cost water source (well, natural lake, flowing stream, or public water supply) is not available, irrigation water can be stored in a constructed reservoir. Types of irrigation storage reservoirs include runoff storage (dams), seepage pits, off-stream reservoirs, and wastewater lagoons.

The storage capacity of an irrigation reservoir must be sufficient to meet the irrigation water requirements, plus the volume lost to evaporation, seepage, and siltation. Water lost to evaporation may be considerable, especially from shallow reservoirs. Reservoirs with greater depths and smaller surface areas will have reduced evaporation losses. The required depth of water in a reservoir depends on the amount of rain received at that location, Table 3-4. The reservoir volume lost to evaporation can be estimated by multiplying the depth of the evaporation loss by the surface area of the reservoir. An estimate of evaporation depth loss can be calculated from local weather data or from local annual pan evaporation (available from the state agricultural climatologist).

Table 3-4. Recommended minimum depths of ponds and reservoirs.

Climate	Annual rainfall (in)	Minimum water depth in over 25% of pond surface area (ft)
Super-humid	Over 60	6
Humid	40 to 60	8
Sub-humid, moist	30 to 40	9
Sub-humid, dry	20 to 30	10
Semi-arid	10 to 20	12
Arid	under 10	14

Source: NRCS Engineering Field Manual

The volume lost to seepage from a constructed reservoir depends on the method of construction, types of soil used for the embankments, the geologic makeup of the substrata under the reservoir, and the hydrostatic head produced by the stored water. These data could be estimated before construction by means of deep soil sampling or by consulting the NRCS.

Sediment accumulation (siltation) is generally associated with dams or ponds used to store water from flowing sources. The accumulated amount depends on watershed characteristics such as soils, topography, land use, vegetation, slope, and the extent to which the reservoir is fed by surface water. In watersheds where land use is predominately agricultural, soil conservation practices such as terraces, contouring, and conservation tillage have a significant effect on reducing sediment in the flowing water. However, a sediment trap or grass filter strip will remove even more of the sediment.

A quick estimate of the water storage volume in lakes, ponds, and reservoirs can be obtained by using Equation 3.3:

Equation 3-3. Estimated Volume Water Storage in Lakes, Ponds and Reservoirs.

$$\text{Volume} = K1 \cdot (\text{Depth}) \cdot (\text{Area})$$

Where:

Volume = Estimated volume of water stored, cubic feet

K1	=	A constant to estimate average depth. Use 0.4 for most water bodies and use 0.5 for a dredged pond.
Depth	=	Greatest lake depth, feet
Area	=	Estimated surface area, square feet

Equations 3-4, 3-5, and 3-6 convert cubic feet of storage into other storage units.

Equation 3-4. Converting Cubic Feet to Gallons.

$$\text{Volume in gallons} = 7.48 \cdot (\text{Volume})$$

Equation 3-5. Converting Cubic Feet to Acre-Feet.

$$\text{Volume in acre-feet} = \frac{\text{Volume}}{43{,}560}$$

Equation 3-6. Converting Cubic Feet to Acre-Inch.

$$\text{Volume in acre-inch} = \frac{\text{Volume}}{3{,}630}$$

Dams

An earthen dam across a watercourse or small valley impounds surface runoff in rolling or hilly topography. Select a watershed above the structure large enough to supply the necessary volume needed for irrigation, Figure 3-17. For example, a reservoir with a volume of 450 acre-feet in central Missouri will require a watershed of 2,250 acres. For a more specific estimate, contact the local NRCS office. Many states and/or local water resource districts require a construction permit to build a dam (even small ones) across a watercourse.

If possible, design a storage structure to hold enough water to supply the irrigation water requirements for two or more consecutive dry years. Decide how much risk you can accept in budgeting the annual irrigation water required. For example, is it economical to be short of irrigation water one year in 10 or two years in 10? Storing the water in an impounding water reservoir

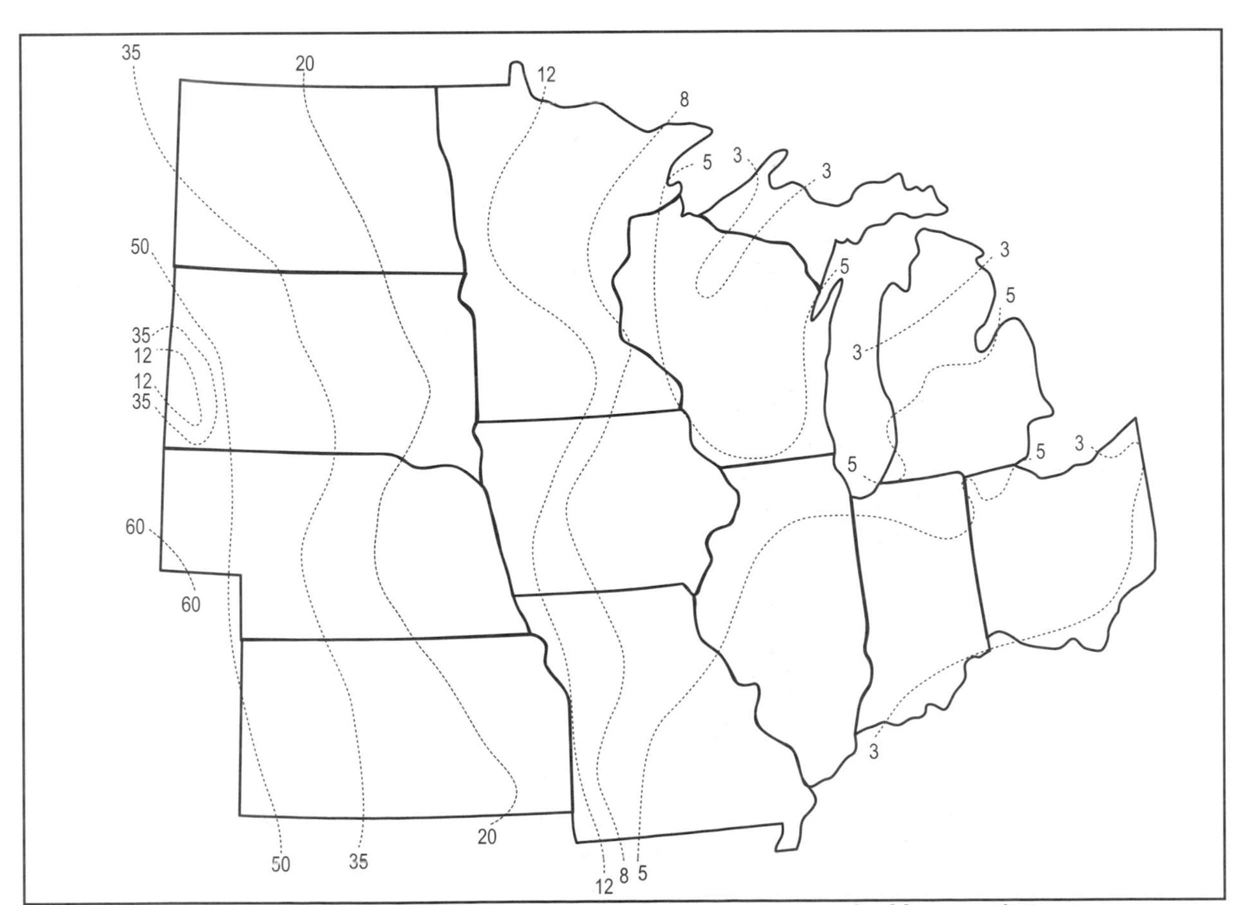

Figure 3-17. Guide for estimating approximate acres of drainage area per acre-ft of farm pond storage. These data may be conservative for large watersheds and reservoirs.

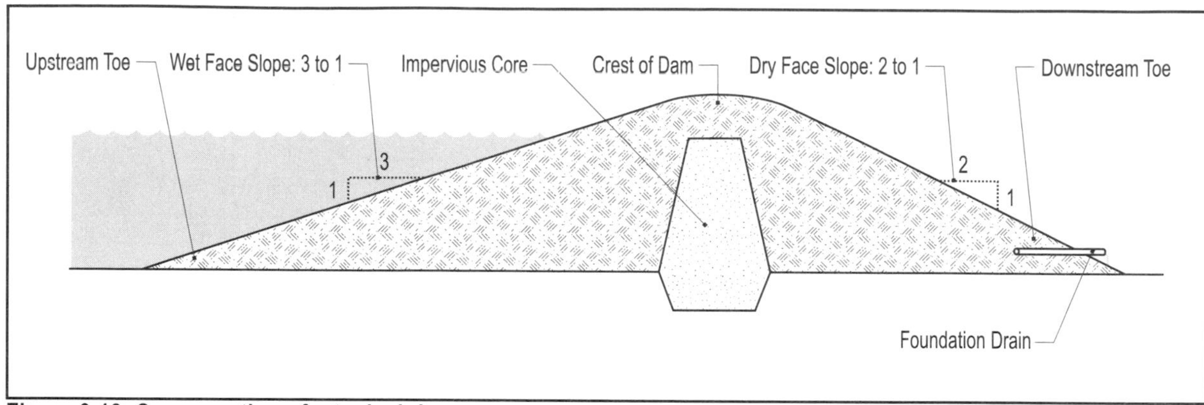

Figure 3-18. Cross-section of a typical dam.

is costly, particularly if the watershed is much larger than needed and the spillways must be built to bypass storm flows. Consider the cost of the land and the crop it could produce.

The earthen embankment's top width is usually at least 8 feet for dams up to 10 feet high; for higher dams, increase the top width 2 feet for each additional 5 feet of height. For safety reasons, the water side of the dam is usually built at a 3:1 (horizontal to vertical) or flatter slope and the backslope 2:1 or flatter. A flatter inside slope enables a person or animal that falls into the water to get out of the reservoir. Seepage through a dam can cause failure, which is why many are constructed with the drain tile placed in the toe of the dam's backslope, Figure 3-18. The water side of the dam should be protected from wave action by using rip-rap or other means.

The reservoir spillway outlet must be large enough to release excess water and prevent it from overflowing the embankment. Runoff reservoirs usually require principal (pipe) and emergency (grassed channel) spillways.

Place the pump intake pipe through the dam, if practical, so the water level in the reservoir is higher than the pump's level. The pipe should have a shutoff valve installed. It should be either locked or hidden to prevent vandalism. The withdrawal pipe for the pump and a bottom-withdrawal spillway pipe can be combined, Figure 3-19. Any piping through a dam (or any embankment) must have anti-seep collars installed at several locations along the length of the pipe to prevent water from traveling along the pipe and eroding a hole through the dam.

Seepage Storage Reservoirs

A seepage storage reservoir (Figure 3-20) is an excavated pit in a low-lying or high water table area. When pumped, it is replenished by underground water movement. A dependable excavated reservoir requires a high natural water table under adjacent land and a shallow, pervious sand and gravel layer permitting rapid lateral water movement that will be intercepted at practical excavated depths (usually 12 to 20 feet).

Some seepage reservoirs have enough daily inflow to provide continuous irrigation water. Others store annual inflow for the irrigation season. Designs for these systems will be based on trial and error. It is difficult to predict what a seepage reservoir will yield. Rely on local experience with similar sites. In some locations, it will be necessary to provide extra water from springs, tile lines, surface runoff, or a well.

Off-stream Storage Reservoirs

Off-stream storage reservoirs are used to store high stream flows due to spring snowmelt or large rainfall runoff events for later use. Flood water may be diverted by gravity but more commonly is pumped into the storage area.

In some locations where the watershed is too small to provide all the irrigation water, an earthen dam across a gully or small valley permits pumping a portion of the irrigation water into the reservoir. On level sites,

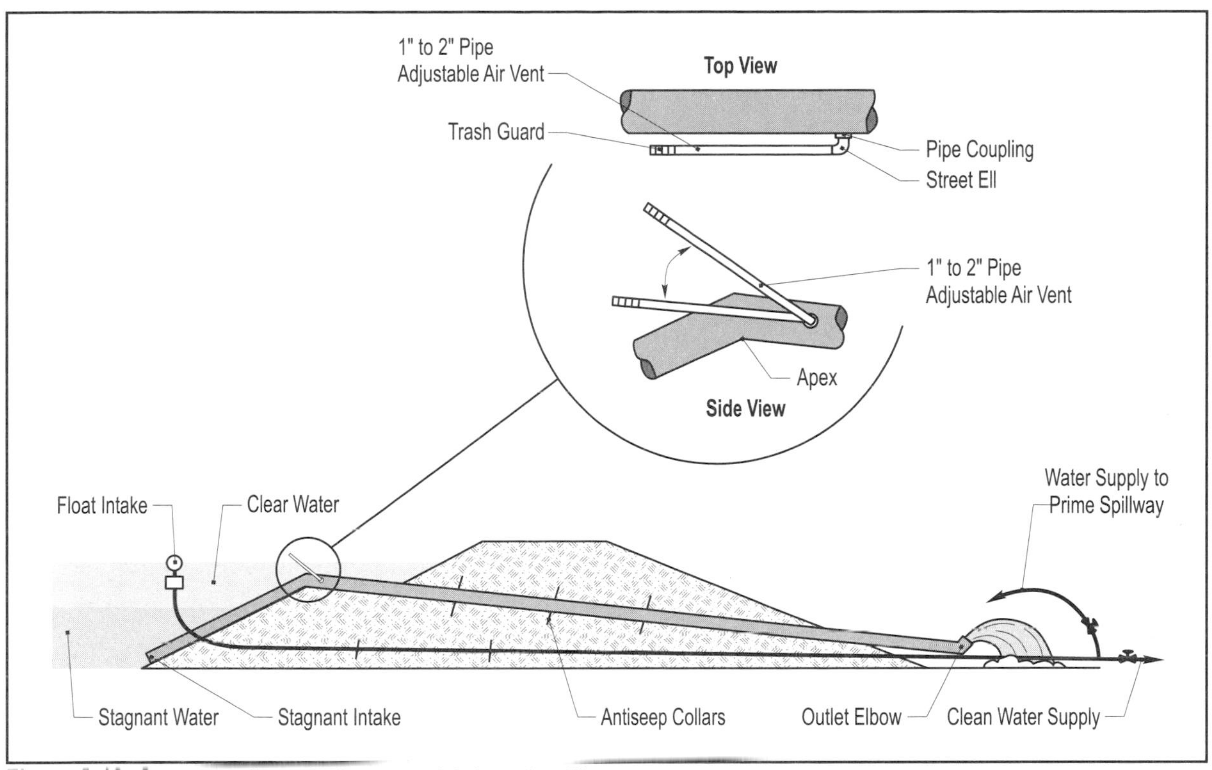

Figure 3-19. Cross-section of a bottom-withdrawal spillway.

Figure 3-20. Seepage reservoir.

excavate a pit, and build a levee around the pit with the soil. The height and top width of the levee will be determined by the size of the pond because the excavated soil is used to construct the levee. This requires a balance between the amount of soil removed and the amount required for the levee. A general soil cut/fill ratio is between 1:2 and 1:3 depending on soil type. The side slopes on both the water side and outside will be determined by the height of the levee and the soils used to build the levee. Wave action can create erosion problems in these types of reservoirs.

Off-stream reservoirs often are filled several months before the irrigation season. In areas where the water can be pumped for several weeks, consider using the irrigation pump and portable mainline pipe to fill the reservoir. It is convenient to have an automatic electric pump with float attachment that starts whenever water is available. In areas where the increased stream flow is available for only a short time, a low-head, high-capacity pump will be needed. These pumps often are powered by a tractor power takeoff. With these types of pumps, it is not unusual to fill a storage reservoir in less than 100 hours of pumping.

Regulating Storage Reservoirs

A regulating reservoir collects continuous small flows from wells, springs, or supply canals and stores it for a day to a year. It is usually an excavated pit and levee on a level site or an earthen embankment across a low area.

Effluent Water Sources

Some states encourage treated effluent application using irrigation with cost-share arrangements. Waste treatment costs are usually lower when land application is the final stage of the treatment process. Therefore, one possible source of irrigation water is from municipalities, large animal operations with lagoons, or food processing industries that need effluent application sites.

Determine the nutrient and salt loading rates on the soil, as well as appropriate crop mix under the irrigation system. A chemical analysis of the waters from these sources is important. Effluent water from some sources may require further treatment before being applied to the land. For more detailed information on these water sources, go to Chapter 9, *Sprinkler Application of Manure and Effluent.*

Maintenance of Constructed Storage Reservoirs

Prompt repairs can keep a minor problem from becoming a major reconstruction project or reservoir failure. This section applies to dams less than 35 feet high located where damage caused by dam failure is limited to farm buildings, agricultural land, and township or county roads. It also applies to any constructed storage reservoir.

Inspection

Inspect reservoirs and dams the first time they are filled. Dams also should be inspected after each large storm, especially one causing flow through the emergency spillway. Look for evidence of seepage flow on the downstream face of dams and erosion damage on both the wet and dry faces of earth embankments. On the wet faces of embankments, look for any evidence of wave erosion.

Erosion Control

Good vegetative cover on the embankments provides many benefits. It controls erosion on the embankments and emergency spillways of dams. Don't let rills and gullies form on embankments. Fill and reseed them as soon as they form. Fertilize, seed, and mulch the bare areas as soon as practical. Maintain a vigorous vegetative growth; fertilize annually if needed. Where fast restoration of vegetative cover is required, consider using sod on bare spots and stabilizing with staked synthetic mulch.

In the parts of the reservoir subject to fluctuating water levels, erosion from wave action must be controlled with rock rip-rap or other wave-reducing equipment such as floating logs.

Seepage

Unless special drainage features are part of the dam or reservoir, water should not pass through, under, or around it. If you notice seepage on the downstream side of an embankment, have the site inspected by someone who understands its design and construction, such as an NRCS area engineer.

Fences

Fence livestock away from dams, emergency spillways, constructed embankments, and the steeper land around the waterline of the pool. Livestock can do a large amount of damage such as creating unwanted drainage paths, destroying vegetation, and increasing embankment erosion, which increases siltation of the reservoir.

Chapter 4

Sprinkler Systems

Chapter 4 Contents

A variety of factors determine which type of sprinkler system is most appropriate for a given application. Included in these factors might be the number of acres to be irrigated, the shape and slope of the field, any physical barriers around the field such as windbreaks or streams, the crops that will be grown, and the available labor force. Considering the operation and maintenance characteristics of the different sprinkler systems available will help in selecting the best system for a particular situation.

Overview

A wide selection of sprinkler systems is available to meet various objectives. Each system design must be suited to the field layout, soils, topography, and crops to be irrigated, and should deliver water uniformly without generating runoff. The design should allow the operator to fulfill desired multiple management schemes such as frost protection, effluent distribution, or irrigation.

Sprinkler irrigation systems may consist of an individual sprinkler or a group of sprinklers installed along a lateral pipeline. For systems with a single sprinkler, travel lanes or sprinkler positions must provide sufficient overlap for uniform water distribution. Sprinklers operating as a group must provide overlap among individual distribution patterns to produce uniform water application along the lateral pipeline and between lateral positions.

Depending on the type of installation, sprinkler systems are operated in one of the following ways:

- Remain stationary (solid set).
- Periodically moved by hand or mechanically (side roll or hand move).
- Continuously moved around a pivot in a circle (center pivot).
- Continuously moved along a closed or open water supply (linear move).
- Continuously moved along a travel lane towed by a hose or cable (traveler).

Other sprinkler systems, such as booms and towlines, are available but currently are not manufactured in the United States.

Table 4-1 provides sprinkler pressures and labor requirements along with suitability of different sprinkler irrigation systems to different field shapes and suggested maximum slopes for operation. The maximum slope may not be the maximum possible for a system and certainly may not be the maximum slope that is feasible considering soils and runoff potential.

Stationary systems (solid set) may have delivery pipelines placed below ground permanently or on the soil surface to allow the system to be removed at the end of each cropping season. These systems are adaptable to any irrigation frequency. They can be designed and operated for frost and freeze protection and/or crop cooling in addition to meeting crop water needs. However, water distribution by stationary systems is affected by the prevailing wind conditions more than continuous-move systems. Consequently, spacing between sprinklers and/or laterals is based on wind speed.

Like side-roll and hand-move systems, periodic-move systems remain stationary while water is applied. Because they are stationary during water application, spacing between sprinklers and laterals is based on wind speed. When the desired depth has been applied, the system is moved by hand (hand move) or mechanically (side roll) to

Table 4-1. Operating characteristics of some sprinkler irrigation systems.

System type	Labor requirement hr/acre/irrigation	Sprinkler pressure[a] (psi)	Maximum slope[b] (%)	Field shape[c]
Stationary				
Hand move	0.6 to 1.0	40 to 60	10 to 15	Any shape
Side roll	0.2 to 0.6	40 to 75	5 to 7	Rectangle, square & trapezoidal
Solid set	0.1 to 0.5	40 to 60	15 to 20	Any shape
Moving				
Traveler	0.2 to 0.4	60 to 100	5 to 10	Any shape
Linear move	0.1 to 0.3	5 to 80	5 to 10	Rectangle, square
Center pivot	0.05 to 0.2	5 to 80	5 to 10	Circular, square, rectangle

[a]Pressure supplied to an individual sprinkler/nozzle. Pressure at the inlet to the system must be higher to compensate for friction losses and elevation differences within the field.

[b]Suggested maximum slope for operation of the system.

[c]A range of field shapes are possible, however, some shapes may limit the total acres that can be irrigated or the suitability of specific system types.

another portion of the field. Periodic-move systems are suited for areas requiring irrigation no more than every five to seven days. For frequent irrigations, solid-set or continuously moving systems are more adaptable.

Continuous-move systems can travel along a straight line (traveler or linear move) or in a circular pattern (center pivot). These systems are less adaptable to odd field shapes and may not be able to irrigate the entire field area. Water can be supplied through underground pipelines, by flexible or hard hoses, or in a surface channel or ditch. Wind direction and wind speed affect the overall field water distribution uniformity less in moving systems than in stationary systems.

Center Pivot

The center pivot is a self-propelled moving system that rotates around a center or pivot point, Figure 4-1. Center pivots have been adapted to operate on many different soils, to traverse variable terrain, and to provide water to meet a variety of management objectives. By varying the length of machine or adding a corner attachment, pivots can be used on almost any field size or shape. Existing systems are propelled using electricity or hydraulics (water or oil). Water drive systems are no longer being manufactured for use in the United States.

A center pivot making a complete circle irrigates an area equal to the area of a circle with a radius equal to the wetted length of the system. Figure 4-2 presents the area irrigated by each 10% of a system that is 1,320 feet long (1/4 mile). The first 132 feet (10%) of system length irrigates only a little over 1 acre, and the last 132 feet (10%) of system length irrigates 24 acres.

Figure 4-1.
Typical center-pivot sprinkler irrigation system.

Moving away from the pivot point, each tower must travel a greater distance to make a revolution. For example, a tower located 132 feet from the pivot travels 830 feet in a revolution while a tower 1,320 feet from the pivot travels 8,294 feet in a revolution. Thus, the time needed to apply a fixed amount of water decreases with distance from the pivot point. Combining the increasing number of irrigated acres with a decreasing amount of application time means that the outer portions of the system have the greatest water application rate.

The pivot or lateral pipelines are 3 to 10 inches in diameter made of steel or aluminum pipe. However, most systems have either 6- or 6 5/8-inch galvanized steel pipe. The pipe is supported by A-frame towers spaced 90 to 200 feet apart using trusses to support the pipe at a height of 8 to 16 feet above the soil surface. Rubber tires on each tower are used to transport the system along its circular path. On towable machines, wheels can be rotated 90° for moving the machine from one field to another.

Center pivots are available in sizes from one tower that can irrigate 3 to 13 acres, to multiple tower units that irrigate more than 500 acres. Machines with four towers or fewer often are towable for use on multiple field areas. The most common center pivot systems operate on quarter-section sized square fields irrigating 125 to 130 acres. The circular path leaves about 9 acres not irrigated in each corner of a 160-acre field if an end gun is not used. Because of farmsteads and other obstacles, some center pivots irrigate a portion of a circle. On rectangular fields, center pivots often are operated in a windshield wiper-fashion with the pivot point along one edge of the field.

Corner units are optional attachments to the end of the center pivot lateral to irrigate a larger portion of the corners. The corner tower section is attached to the outer end of the main system lateral. When the system

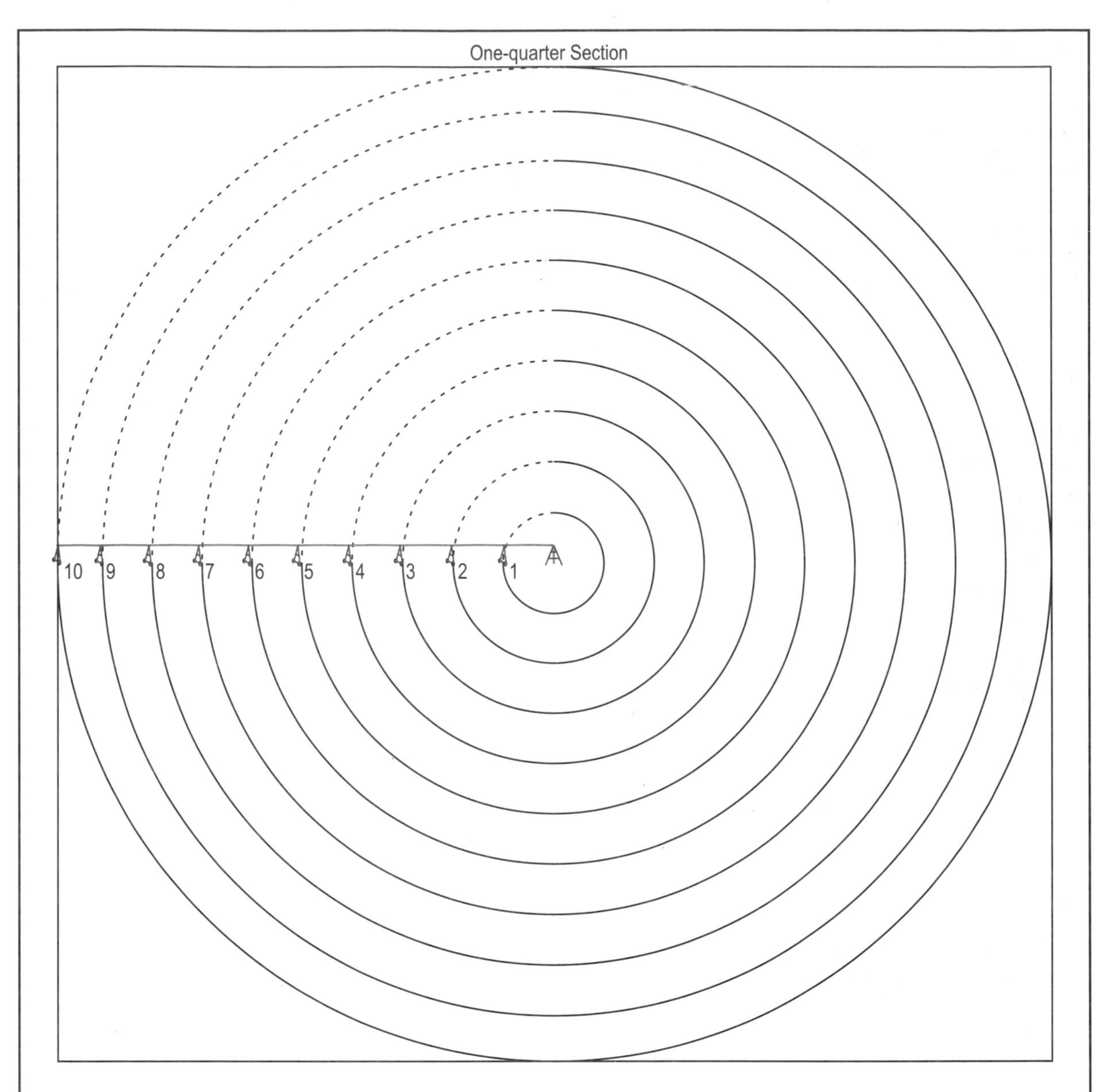

Tower	Tower travel distance (ft)	Section	Section irrigated (ac)	Total area irrigated (ac)
1	829	center to 1	1.26	1.26
2	1,659	1 to 2	3.77	5.03
3	2,488	2 to 3	6.28	11.3
4	3,318	3 to 4	8.80	20.1
5	4,147	4 to 5	11.3	31.4
6	4,976	5 to 6	13.8	45.2
7	5,806	6 to 7	16.3	61.6
8	6,635	7 to 8	18.8	80.4
9	7,464	8 to 9	21.4	102
10	8,294	9 to 10	23.9	126

Figure 4-2. Center-pivot irrigation system characteristics for each 10% of system length.

approaches a corner to be irrigated, the corner attachment swings out, and the main lateral slows down. As the system exits the corner, the corner attachment swings back into position behind the main system and the system returns to the previous travel speed. The most common guidance system for the corner unit is a buried electric cable. In the future, a global positioning system (GPS) may be used for guidance.

Corner attachments, like the one depicted in Figure 4-3, allow an additional 5 to 7 acres to be irrigated in each corner of a 160-acre field. But the cost per irrigated acre is more than double that of the remainder of the field. Thus, corner systems commonly are used to irrigate high-value crops such as vegetables, hybrid seeds, or other specialty crops. These systems also are used where irrigated land is in short supply.

Figure 4-3. Corner attachment on a center-pivot system.

Control Systems

The controls for center-pivot systems offer several options that improve and simplify system management. Some manufacturers have available new computerized control panels that allow system speed and direction and sprinkler operation to be varied several times during each revolution. Such changes in operation are based on position in a 360° circle determined at the pivot point. Control panels that enable changes in operation are particularly valuable in fields with drainage channels that should not be irrigated, obstacles that prevent a full system rotation, or different crops requiring different water management strategies.

Another option available for center pivots is remote monitoring of system operation. Using methods incorporating radio or telephone, the center pivot operation can be monitored from a vehicle or the farm office. Set points can be verified and changed using this equipment. This equipment is useful when the center pivot is located a long distance from the farm headquarters, when numerous systems must be monitored simultaneously, or when the system irrigates sensitive crops.

Each option adds to the total cost of the system, so the options should be economically justified and used often during the growing season. Less elaborate equipment may provide many of the needed controls. However, since some options require specific equipment, it may be less costly to purchase options when the system is first installed rather than retrofitting it at a later date.

Design

Center-pivot design begins with establishing a sufficient source of high quality water to meet crop water needs for the irrigated area. Water quality concerns are addressed in Chapter 3, and crop water demands are addressed in Chapter 2. Determination of system energy requirements is presented in Chapter 7.

A buried pipeline must be installed to transport water from the water supply to the pivot point if a suitable water supply cannot be located near the pivot point. Pipeline drains or pump-outs are required unless the pipeline is buried below the frost level. The design requires selecting the length, diameter, and material of this mainline pipe. In addition, all pipe fittings (elbows, tees, flow meters, and valves) installed in the delivery system must be accounted for to determine friction loss in the pipeline. Chapter 7 contains a detailed discussion about friction loss.

The friction loss in the mainline and fittings and in the pivot lateral must be considered in the design. These, together with elevation changes, contribute to the total head for the pumping plant (see Chapter 7). Since water is discharged along the center pivot lateral, friction loss is less than if all the water were discharged from the end. Most water in a center pivot is discharged at the outer sections because most of the land area is covered under them. Friction loss in the center lateral can be estimated by taking 54% of an equivalent length of pipeline with all the discharge at the end. A typical 1,300-foot center pivot will have a friction loss of 8 to 15 psi. Final design, however, should rely on the manufacturer's printout for a given gpm and length.

Center pivots are adapted to field terrains ranging from flat to rolling (Table 4-1). However, as slope increases, towers must be moved closer together so the lateral will more nearly follow the topography. This is particularly important for systems equipped with drop tubes extending below the trusses. Failure to consider this aspect of the design could cause the nozzles to drag on the soil surface, reducing their expected life and the uniformity of application.

Investigate the tradeoffs between initial installation cost and the long-term energy costs. Center pivots that irrigate large areas may require varying lateral sizes to have a design that is economical. For example, a center pivot that irrigates more than 300 acres may require some 10-inch, some 8-inch, and some 6 5/8-inch pipeline. This is necessary to optimize system pressure distribution and pumping energy requirements. It also minimizes the potential damage to system components from water hammer.

A three-phase power supply or a generator typically will be required for electric and oil drive units. Single phase power with a phase converter also may be used to power the electric motors if used solely for tower motors. A separate engine can be used to power the generator, or the generator can be driven from the engine powering the pump.

If the generator is powered by the pump power unit, increase the brake horsepower requirement to account for power use by the tower motors and the end gun booster pump. Also, equip the engine with a clutch between the engine and pump so the generator can be powered without operating the pump. This allows the system to be moved without applying water. Size the drive pulley for the generator and power unit so the generator speed is maintained at the manufacturer's specified revolutions per minute.

The outside tower controls the speed of the system. A system of alignment controls keeps the other towers in line between the end tower and the pivot point. The manufacturer should supply a table showing the depth of water applied (gross) over a range of travel speeds or timer settings. These figures are based upon zero tire slip. Hence, to be most accurate, the speed of travel at each 10% increment of timer setting should be measured in the field to establish actual application depths.

For a set depth of water application, the revolution time for a center-pivot system increases with system length and decreasing pump capacity. Thus, for the same application depth, a long system requires more time to make a revolution than does a short system. This could limit the system length for fields having low water-holding capacity soils or systems with limited water supplies.

With the center pivot, the rate of water application is the same for a given flow rate at the pump, regardless of rotation speed. Runoff or water movement under the system may require a change in operation or management. Increasing the rotation speed decreases the depth of water applied and reduces the potential for runoff. For some nozzle packages, providing added soil surface storage, such as furrow diking, may be the only option for reducing the runoff potential.

The system should be equipped with safety shutoffs to stop the pump and pivot operation due to low pressure conditions, and if the pivot lateral movement ceases for an extended time. Provisions for shutting off an end gun along field boundaries are required. If the system will be used for distribution of effluent, air-actuated valves, rather than water-actuated valves, should be installed with an end gun or any other sprinkler that will run intermittently.

Outlets for 110- and 220-volt power should be installed at the pivot point for use with chemical injection equipment. These outlets also may be helpful if power tools are needed for repairs near the pivot point.

Sprinkler Packages

Selecting a sprinkler package involves carefully matching sprinkler operating characteristics with the field conditions (Chapter 6). A vast array of different types and configurations of sprinkler packages is available to meet different goals. A detailed discussion of sprinkler characteristics is presented in Chapter 5.

Field-based information is required to accurately select a sprinkler package. Failure to collect and evaluate key field information could result in an inefficient irrigation system. Begin by obtaining soil survey information showing soil mapping units and an aerial photograph of the field from the Natural Resources Conservation Service.

Determine the preliminary system size and layout from an aerial photograph. Summarize soil infiltration rate, slope, water-holding capacity, and surface area for each soil mapping unit (Table 4-2). Develop a field map to establish elevation changes, the exact pivot point, and pipeline routing, and to ensure the machine will clear obstacles.

This can be accomplished with a GPS, a laser transit, or manually using a measuring tape and transit.

Spacing between sprinklers should be selected to ensure a sprinkler pattern overlap in excess of 100% between sprinklers. Sprinkler spacing depends on where the sprinkler is mounted on the system (on top of the lateral, or on drop tubes below the truss or near the soil surface), the type of sprinkler, and the operating pressure. Impact sprinklers often are spaced at equal distances with the sprinklers increasing in capacity with distance from the center point, Figure 4-4.

Another method uses the same capacity sprinkler along the entire length of the lateral line with a decreasing sprinkler spacing with distance from the pivot point. Various combinations of sprinkler sizes and spacing are common today. Regardless of the type of sprinkler package selected, your dealer should provide a computer printout showing the locations, sizes, and flow rates for all sprinklers installed on the system.

Table 4-2. Example summary of soil mapping unit information for a quarter-section of land, Pierce County, Nebraska.

Soil series mapping symbol	Field slope (%)	Intake family number	Water-holding capacity (in/in)	Field area (ac)
Co	0 to 1	0.3	0.20 to 0.23	42
He	0 to 1	1.0	0.21 to 0.23	24
CsC2	1 to 7	1.0	0.20 to 0.23	11
HhC	0 to 7	1.0	0.21 to 0.23	5
MoC	1 to 7	0.5	0.19 to 0.22	37
CsD2	7 to 11	1.0	0.20 to 0.23	28
NoD	7 to 11	1.0	0.20 to 0.23	2
CsE2	11 to 17	1.0	0.20 to 0.23	11

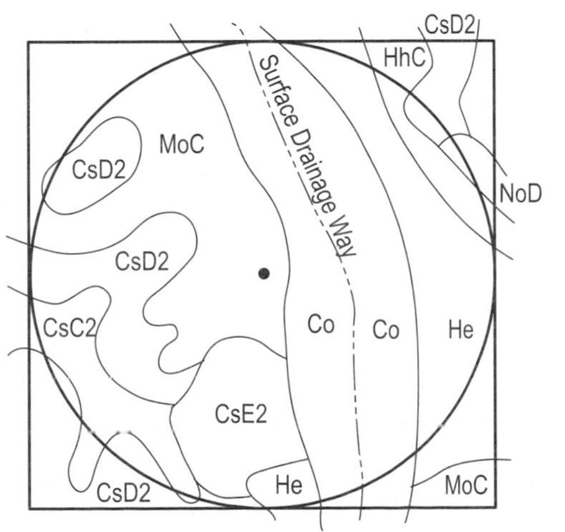

If pressure regulation is justified (see Chapter 5), select a device that will deliver water to each nozzle or sprinkler at the design pressure. Pressure regulators are available with ratings ranging from 5 to 50 psi, but typically are rated for a certain range of flow rates. Thus, the same rated pressure regulator might not be installed at each nozzle or sprinkler position along the pivot lateral.

Whether pressure regulators or flow control nozzles are selected, the system operating pressure must be increased to account for the friction loss of the device. For pressure regulators, the pressure loss ranges from 2 to 5 psi. Your dealer has this information, and the computer printout provided for the package will account for this additional pressure requirement.

Center pivot systems often use a large volume sprinkler (end gun) at the far end of the lateral line to help irrigate more area in the corners of a field. On low-pressure systems, an auxiliary pump is used to boost the water pressure to the end gun by 20 to 40 psi. This may require up to a 5 BHP (brake horsepower) booster pump. The uniformity of application under this unit usually will be much less than that under the main lateral sprinklers. Wind and the angle of the gun rotation can influence water application significantly. End guns should not be used during chemigation of pesticides.

Operation and Management

Center pivots are popular because of the ease of use and minimal labor required. Properly designed center pivots should have few maintenance problems. Full-circle pivots usually operate in one direction unless there is more than one crop type with different water needs in the field. They also can apply a portion of some crop's nitrogen requirements by chemigation (see Chapter 8).

Part-Circle Pivots

Where center pivots are installed on fields with obstructions that prevent full-circle operation, a portion of the circle will not be irrigated, and the system is operated

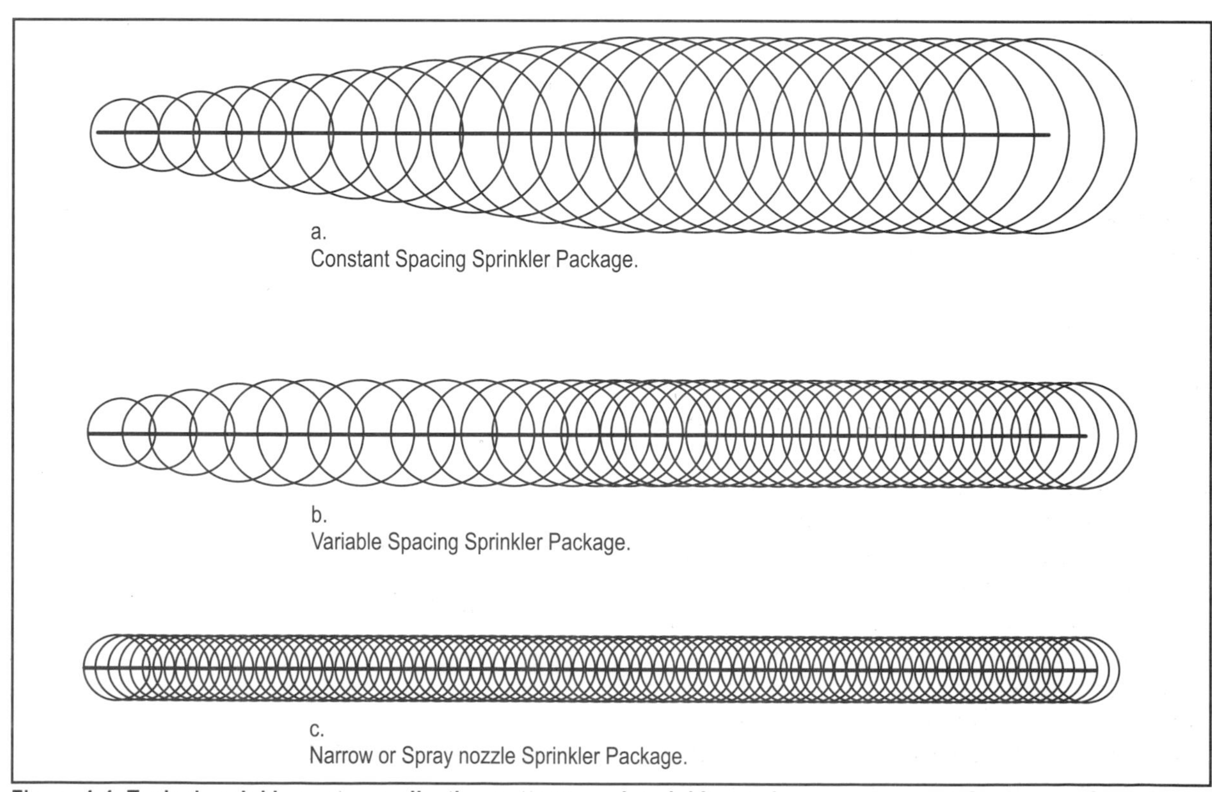

Figure 4-4. Typical sprinkler water application patterns and sprinkler package arrangement for center-pivot irrigation systems.

in a windshield-wiper fashion. In this situation, the pivot normally irrigates in one direction until it encounters an end-of-field stop placed in the path of the second-to-last tower. This is also true for a pivot set up to make a half circle.

Figure 4-5 shows options for managing a part-circle pivot system. Water can be applied in one direction and the system returned dry to the start to be ready for the next pass. Where an internal combustion engine powers the pivot tower motors with a generator, the engine may need to run up to 20 hours to return the system to the start position, unnecessarily using energy.

To prevent this waste of energy, most part-circle systems are managed by irrigating in both directions. First, the system makes a forward pass. Then the direction is reversed, and water is applied as the pivot returns to the starting position. This approach presents two problems:

1. The system will apply water to a portion of the field that was irrigated most recently.
2. Wheel track ruts may be developed near both start and stop positions.

Starting the reverse pass often is dictated by the rule that no part of the field should experience water stress. Since many systems may require three or more days to make a pass, nearly six days could elapse before the system is back in the start position if a full irrigation application were applied in both directions. This would not be suitable for soils with low water-holding capacity that may need a return pass to start immediately to prevent stress at the start position. Then irrigating the portion of the field that was most recently irrigated could cause over-watering and deep wheel tracks.

To prevent over-irrigation, the system can be managed in several ways. For example, an application of 0.75 to 1.25 inches could be applied in the forward direction. The system could then be reversed wet at the highest speed setting, where the total depth of water applied by forward and reverse passes is the same or more than the desired total application depth. This strategy returns the system to the start position without wasting energy while reducing both the potential for over-watering and the potential for plant stress that can occur by delaying water application.

Wheel track ruts occur when the combination of soil texture and soil water content reaches the soil's load-bearing strength. Also, wide spacing between towers increases the pressure applied to the soil due to the extra weight each wheel carries. To reduce these problems, travel speed should be increased as the pivot approaches the stop position.

When the system is reversed, the speed setting should start out slow and increase until the total water applied is equal for the whole field. For example, if the total water application depth desired is 1.25 inches, the system speed for most of the forward pass is set to apply 1.0 inches. During the last several hours of the circle, the speed could be set to apply 0.5 inches. In the reverse direction, the speed for the first several hours is set to apply 0.75 inches. Then the speed could be increased to apply 0.25 inches back to the start position. Delaying the start of the reverse pass could further reduce wheel track problems but could result in water stress for crops near the beginning position.

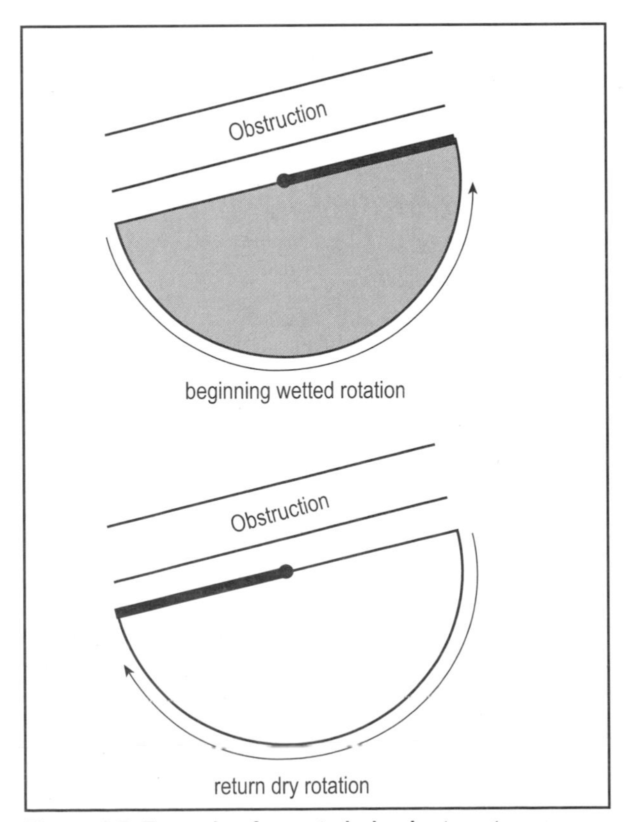

Figure 4-5. Example of a part-circle pivot system.

Startup Cycles

If an operator starts and stops the center pivot at the same time and location in the field and uses the same speed setting for every irrigation, each position of the field will be irrigated at essentially the same time of the day for each irrigation. Those portions of the field irrigated at night will typically receive more water, and the application uniformity will be higher than areas irrigated during the day. This is because wind drift and evaporation losses would be lower because air temperatures and wind speeds are lower and humidity is higher during the night. If the difference in efficiency between daytime and nighttime were 5 to 10% for each irrigation event, yields may be lower in parts of the field that are always irrigated during the day.

To reduce the impact of the time of day at which each portion of the field is irrigated, the system start time should be altered for each irrigation event. Altering the start time by several hours for each subsequent irrigation will ensure that different portions of the field are irrigated at different times of the day. When the system needs to be operated continuously to meet high crop demand, select speed settings that are not multiples of 24 hours.

Management of Systems Under Load Control

To keep electric power costs low, power companies typically offer power at reduced rates if the operator is willing to allow power to be shut off during peak usage periods. A variety of options for control are available depending on the electric supplier. Some offer one-day control per week, two-day control, or any day control. Other suppliers may control any time of day on any day, or they may limit irrigation to specified hours of given days. The systems may be controlled on a rotating basis to ensure that all systems are controlled about the same number of hours per season.

Data collected in Nebraska show that power was interrupted for the everyday control systems an average of about 100 hours per season, and control was implemented on 10 days during the cropping season. Other suppliers have similar control patterns. Thus, participating in a load control program should not adversely affect yields unless weather conditions are especially unfavorable.

The net effect of electric load management on an irrigation system is the reduction in effective system capacity. For example, a center pivot with a system capacity of 700 gpm can apply 1.0 inches of water (gross) to a 125-acre field in about 80 hours if operated continuously. Each 6 hours of down time due to electric load control will decrease the gross water application by about 0.08 inch per revolution because now water is applied for only 74 hours. Another way of looking at it is that the system flow rate would need to be increased by 60 gpm to apply 1.0 inch of water during an equivalent 80-hour (74 hours in operation) revolution.

Many center pivots have extra system capacity because the water supply was available when the system was installed. The system capacity that is required for load control options depends on the soil water-holding capacity, prevailing climatic conditions, the peak water use rate of the crop, and the type of load management option.

Sandy soils may have only 0.5 inches of water available per foot of rooting depth, while a silt loam may have 1.25 inches per foot. Peak water use rates for corn range from 0.28 to 0.30 inches in Iowa to 0.35 inches per day in western Nebraska. Parts of Iowa receive 36 inches of annual precipitation while the panhandle of Nebraska receives only 16 inches. Each of these factors dictates whether a system is likely to provide enough water to meet crop water needs 90 to 95% of the time. See Chapter 3 or check with an irrigation specialist for recommended system capacities for your location.

A system with reduced capacity due to load control must start each irrigation sooner than systems that can operate continuously. On sandy soils, the operator must plan further ahead for when irrigations need to occur. Irrigations may need to start before the soil is able to hold all the water that is being applied to prevent water stress in other portions of the field. In short, systems under electric load control must be managed more intensively than systems that run continuously. The system also must be more reliable because down time for repairs only increases the potential for under-irrigating some portions of the field.

System Maintenance

Most center pivots are designed to apply water with an application uniformity of 90% or greater. Sprinklers that malfunction and normal wear on the nozzles may cause the application uniformity of the system to decrease with time. New center pivot systems each can have more than 200 spray heads and 200 pressure-regulating valves. This creates more than 400 opportunities for a water distribution component to malfunction. For the system to deliver water uniformly, each nozzle must be installed in the correct position on the system. Check each nozzle against the computer printout showing the correct nozzle position.

Watch the system respond to water entering the pipeline. Water may be entering the system too rapidly if water reaches the end of the system, and the system flexes up and down for a short period. This symptom indicates water hammer in the pipeline. This happens most frequently with pumps using electric motors because pump output can be quite high after the system starts and before the pivot is pressurized. Systems with long delivery pipelines are most susceptible to water hammer problems. Installation of a throttling valve at the pump will help reduce the impact of water hammer problems. (See Chapter 8 for more information).

Be sure safety features such as pressure relief valves, automatic pressure shutoffs, engine monitoring switches, vacuum relief valves, and chemigation safety equipment are installed and working properly. Closely follow manufacturers' recommendations for maintenance, and make any needed repairs before the growing season.

The end of each cropping season brings the need to winterize the system by draining water from the pipelines, checking the cooling fluid for the engine, and making sure rodents cannot gain entry into electric control boxes or motors. The center pivot should be parked so it faces into or downwind of the prevailing wind direction. Generally, it should be parked facing northwest or southeast to minimize the opportunity for wind damage.

Linear Move

The linear-move system is mechanically similar to the center pivot, but instead of traveling in a circle, the system travels in a straight line across the field. Water can be fed into any point along the system but usually is at an end or the center. Some systems, such as the one shown in Figure 4-6, follow a ditch and take water directly from the ditch while others are supplied through a buried pipeline with risers. Linear-move systems generally have their own power source mounted at the main drive tower. If water is obtained from an open ditch, a motorized pump is mounted at the drive tower.

Linear-move systems can have as few as two towers, or they can have multiple towers (up to 2,600 feet) that can cover more than 300 acres. Two- to eight-tower systems often are moved between adjacent fields to cover larger areas.

Design

The linear system is best suited to rectangular or L-shaped fields with a length/width ratio of at least 2:1; that is, the lateral travel distance is more than twice the irrigation lateral length (Figure 4-7). The system is not well adapted to fields with rolling terrain because it can cause alignment problems.

Figure 4-6. Typical linear-move sprinkler irrigation system.

The drive mechanisms can be electric or hydraulic motors at each tower. Towers are generally powered by an engine mounted on the machine, but may get electric power by dragging a cord from the machine. A power unit for the water pump also may be needed.

Linear-moves have a guidance system to maintain the direction of travel. Guidance systems may use a V-shaped trench, a buried electric cable, or an above-ground steel cable. GPS may be used in the future as the guidance mechanism. Some guidance systems can follow a gentle curve; however, the travel speed may not be consistent.

Some linear systems are towable so they can irrigate more than one field. They also can be fixed on one end and pivoted wet or dry to move them to a second side of a field.

Many sprinkler packages are available with sprinklers and mounting configurations similar to those on center pivots. Sprinklers on linear moves are the same size throughout the length of the system. The water application rate (unlike the rate with a center pivot) is the same from one end to the other. If, however, the system is rotated wet at the end of a field, the outside edge of the semicircle will not receive the same amount of water as the area near the pivot point.

Pipe size on linear systems depends on the system's capacity and lateral length. Tower spacing (span length) can vary from approximately 100 to 200 feet depending on the terrain. One or more tower spacings may be used on the same system. A center-feed system can use smaller diameter lateral pipe than an end-feed system of the same capacity and length due to the lower friction loss in the center-feed. Calculated friction losses are similar to those in solid-set and hand-move laterals rather than those with center pivots (see Chapter 7).

For a ditch-feed system, a concrete-lined ditch presents fewer problems than an unlined ditch. A 1:1 side slope is required for most ditches. The minimum bottom width is 12 inches for a concrete-lined ditch and 30 inches for an unlined ditch. The minimum ditch depth should be 30 to 36 inches with a minimum water depth of 20

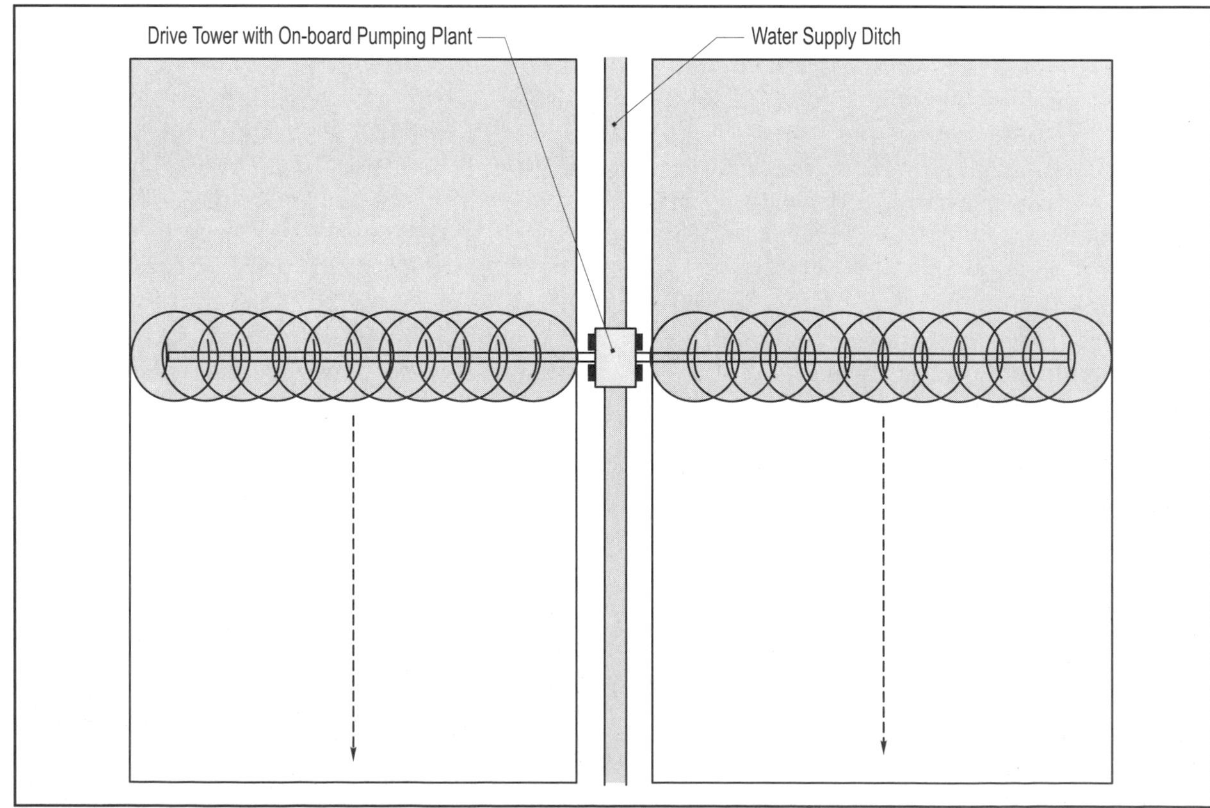

Figure 4-7. Typical field layout for a linear-move sprinkler irrigation system.

inches. Sloping ditches require a movable dike or water control devices to maintain the proper level of water.

For flexible-hose systems, select a hose diameter that minimizes friction losses. Linear systems use hoses similar to those used in a traveler. With flexible hoses, the length traveled will be about twice the length of the hose. Hoses may be any reasonable length depending on system capacity. Larger systems may use two hoses. Water is supplied to the hoses with buried pipe and appropriately sized risers. If the hose is too long, friction loss can be excessive, and the drive tower may have difficulty pulling the hose. For those reasons, a shorter hose may be used with risers spaced more closely.

Operation

Operating a linear system is similar to operating a center pivot. When the system completes a pass across the field, the machine must return to the starting point. This system also has some of the same management issues as a center-pivot system so the travel scheme should be adjusted to minimize wheel tracks and uneven watering. Irrigations should be scheduled to avoid exceeding the allowable soil water deficit and to prevent over-watering that can leach nutrients.

One way to operate a linear system is to apply less water on each pass and operate the system back and forth as needed to meet water demand, irrigating both ways. A second method of operation is to irrigate in one direction, and run the unit back dry to the starting point to be ready for the next irrigation period.

The best method may be to increase the travel speed as the system approaches the end of the field. When the system is reversed, the travel speed should be slower for the first part of the field and then increased to near maximum speed so the system is returned to the starting point as rapidly as possible. The total application depth should be equal for the entire field. Thus, the travel speeds must be selected so the forward and reverse cycles apply the same depth of water across the field. For long fields, it may be desirable to limit the water applied to the main portion of the field on the reverse cycle so the system can return to the starting point more quickly.

The alignment controls and safety shutoffs for linear systems are similar to those for center pivots. Usually, the two outside towers control the travel speed of the linear system. They travel at a preset speed, and an alignment system on each tower keeps each tower in line. The guidance system overrides the preset speed to alter the speed of one end tower versus the other as necessary to maintain the alignment along the guidance device. Linear-move systems can provide uniform application of chemicals because of high distribution uniformity and speed control.

Maintenance

Maintenance on control systems and the motor used to run the onboard generator and booster pump (when used) should follow manufacturer's recommendations to minimize system failures and damage to components. Hydraulic control systems powered by water must be drained in preparation for winter. This includes individual nozzle shutoffs, and automatic end-of-field or end-of-hose shutoff systems.

Travelers

Traveler systems use a large volume sprinkler with a nozzle between 0.50 and 2 inches in diameter operating at pressures of 60 to 120 psi to distribute water over a wetted diameter of 150 to 550 feet. Figure 4-8 shows a typical traveler system. All travelers require high pressure, thus using more

Figure 4-8. Typical traveler irrigation system.

energy for pumping than other systems. The system is pulled across the field using either a semi-rigid polyethylene hose or a cable. Travelers can irrigate between 0.2 and 0.8 acres for every 100 feet of travel length.

A traveler irrigation system consists of four components:

1. A large-volume gun sprinkler.
2. A cart for the sprinkler.
3. A reel for the hose or cable.
4. A drive system.

A traveler can be used on a variety of crops, field sizes, and shapes. It works best on long, level fields with straight rows, but it can be used on slopes up to 10%. With additional layout time, it can be used on terraced and contoured rows, but earth anchors and cable releases are required to handle the cable. A prepared travel lane is recommended. This travel lane should be at least 6 feet wide, and may need to be up to 16 feet wide depending on the type and diameter of hose. The travel lane should be grassed, and all abrasive materials should be removed to increase hose life.

Travelers deliver water with large droplets that may damage some plants, especially if pressure is too low. In addition, the large water droplets can destroy the structure of surface soil particles, causing a surface seal to develop. A surface seal could reduce the soil infiltration rate significantly and cause runoff.

Liquid manure or lagoon effluent can be applied with most travelers if they have a power unit to drive the machine. Water drive units may work for manure if the effluent has no solids to clog the drive mechanism. Generally, nonagitated manure effluent from lagoons can be handled with water turbine machines. Poultry, beef, dairy, and swine pit manure have enough large and fibrous solids to cause problems on water-driven machines. Check with a dealer for recommendations on using the unit for land application of manure. See Chapter 9 for more information.

Types of Travelers

Travelers are available in two basic configurations: cable-tow (Figure 4-9) or hose-tow (Figure 4-10). Hose-tow units may be more adaptable to irregular fields than the cable-tow units because cable-tow units must have the entire hose laid out.

The cable-tow (soft-hose) traveler has a large-volume sprinkler mounted on a cart to which a soft hose and cable are connected. The cable is anchored at the end of the traveler run and winds up on a cart-mounted winch. The winch is powered by either an auxiliary engine or water turbine. The hose is flexible, ranging in size from 2 1/2 to 5 inches in diameter and is fabricated of vulcanized rubber reinforced with multiple layers of synthetic threads woven into the rubber. This type of gun generally is set up on a rectangular 80 acres (1,320 feet x 2,630 feet).

The hose-tow (hard-hose) traveler has a large-volume sprinkler mounted on a cart, and a trailer-mounted hose reel at the end of the traveler run. The hose supplies water to the sprinkler and also pulls the sprinkler through the field. The hose reel is stationary during an irrigation run. The reel is powered by a water turbine or auxiliary engine to wind the hose. Any length of hose can be unrolled to irrigate varying lane lengths. Hose-tow travelers are available in many sizes ranging from less than 25 gpm to more than 500 gpm; 1 1/2 to 5 inches in diameter; and 500 to 1,000 feet long.

Design

To design a traveler system, determine the system flow rate required to irrigate the field area considering the soil types and the crops to be irrigated. Assemble a summary table (similar to Table 4-2) of soil types and slopes present in the field. See Chapter 2 for an explanation of how to estimate system capacities for various crops and soil types.

Conduct a field survey to establish the dimensions and elevation differences within the field. Use this to establish the location of mainline pipes and maximum elevation difference between field positions and pump.

System Hydraulics

Select the appropriate hose size based on the flow rate, Table 4-3, and the manufacturer's recommendations. Estimate the

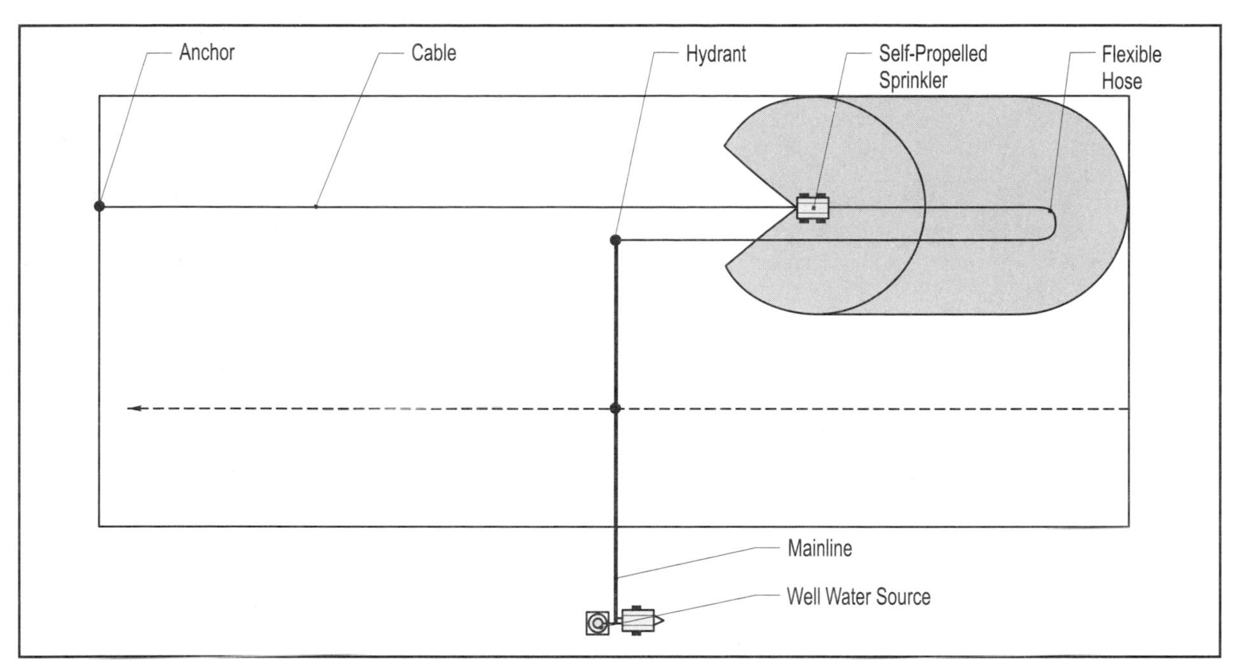

Figure 4-9. Typical field layout for a cable-tow traveler.

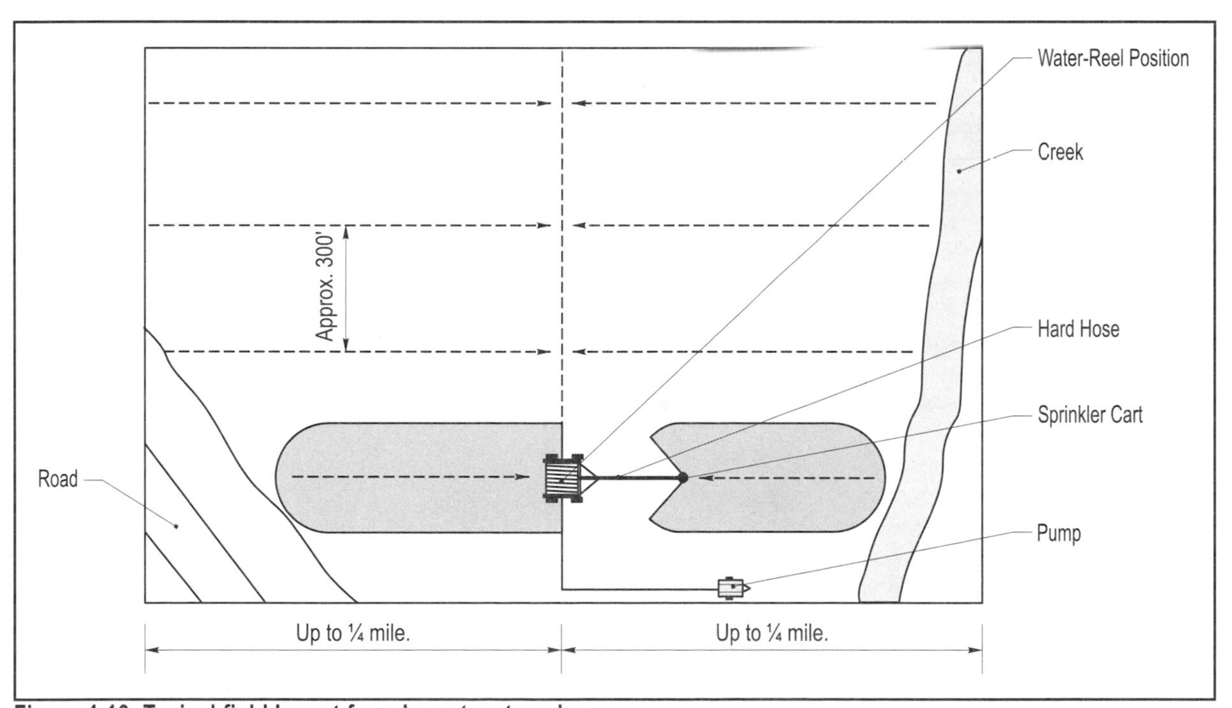

Figure 4-10. Typical field layout for a hose-tow traveler.

pressure loss in the hose using Tables 4-4 and 4-5, and add elevation change and the friction loss through the drive mechanism based on manufacturer's data. Operating pressure at the inlet to the traveler or traveler hose must be higher than the pressure required for the volume gun. For example, a hard-hose traveler discharging 400 gpm through 1,000 feet of 4-inch hose would have nearly 40 psi of friction loss in the hose. Thus, the inlet pressure required could be 50 to 60 psi higher than the sprinkler pressure. Choose the shortest hose length that will allow irrigating the field as laid out.

Table 4-3. Recommended flexible hose sizes for travelers.

Irrigated area (ac)	Flow rate (gpm)	Inside diameter (in)	Soft hose length (ft)[a]	Hard hose length (ft)[a]
less than 20	Up to 150	2 to 3	330 to 660	200 to 400
20 to 40	150 to 300	3 to 4	330 to 660	400 to 600
40 to 100	250 to 600	4 to 5	660	600 to 1200
60 to 120	400 to 750	4 to 5	660	600 to 1200

[a] Longer hose lengths are available, but are often impractical due to friction loss or other operating limitations.

Table 4-4. Estimated pressure loss due to friction for 100 feet of soft hose (psi).

	Nominal inside diameter (in)				
Flow (gpm)	2.5	3.0	4.0	4.5	5.0
100	1.6				
150	3.4	1.4			
200	5.6	2.4			
250		3.6	0.9		
300		5.1	1.3	0.6	
400			2.3	1.3	
500			3.5	2.1	1.1
600			4.9	2.7	1.6
700				3.6	2.1
800				4.6	2.7
900					3.4
1,000					4.2

Table 4-5. Estimated pressure loss due to friction for 100 feet of hard hose (psi).

	Nominal inside diameter (in)				
Flow (gpm)	2.5	3.0	4.0	4.5	5.0
100	2.8				
150	6	2.5			
200	10.2	4.2			
250		6.2	1.8		
300		8.8	2.4	1.2	
400			3.7	2.1	
500			5.8	3.2	1.9
600			8	4.5	2.7
700				6	3.6
800				7.7	4.6
900					5.9
1,000					7

Pressure losses due to friction for hard hose vary due to differences in inside diameter. These values should be used as an indcation of the relative friction losses only. To obtain actual friction loss values, contact the manufacturer.

Normally, sprinkler pressure is 70 to 100 psi. Select the nozzle size best suited to accommodate the desired flow rate with an operating pressure near the middle of the published pressure range. This will allow the system to function well should the nozzle pressure fluctuate due to changes in field elevation or changing pump hydraulics. Chapter 6 presents flow rate and pressure information for volume guns.

Based on the system flow rate and the length of the supply line, select a mainline pipe size that is economical and yet hydraulically appropriate for the installation. Estimate the total pressure loss through the mainline by calculating the friction loss and adding elevation change along the pipeline. Be sure to include losses for all pipeline fittings (tees, elbows, valves, etc.).

Information to estimate the friction loss in the mainline is presented in Chapter 7. Chapter 9 discusses these issues as they relate to for animal manure. For most rectangular fields, lay the supply line across the middle of the field to minimize the length of the travel lanes.

Field Layout

The field should be divided into rectangular areas delineated by nearly equally spaced travel lanes or alleys. If the field width is not evenly divisible by the desired lane spacing, reduce the lane spacing to create equal spacing. Because a traveler is continuously moving, the normal spacing between alleys is approximately 50 to 70% of the sprinkler's wetted diameter. See Chapter 6 for specific information on volume gun sprinklers.

The exact spacing depends on the combination of sprinkler performance, field dimensions, and prevailing wind conditions. Reduce spacing for higher wind speeds. Do not underestimate the effect wind has on the water distribution pattern of volume guns on travelers. Lanes spaced too far apart could lead to reduced yield resulting from under-irrigated strips between the lanes.

If adjoining lands should not be irrigated, position the first lane a distance equal to the wetted radius from the edge of the field to prevent watering the land that should not be irrigated. Some of the field will not be irrigated fully, but this may be necessary to be in accordance with laws or regulations.

Operation

The principles of operation of the cable-tow traveler and the hose-tow traveler are similar. In either case, position the traveler unit in a travel lane. Based on the wetted diameter, position the sprinkler 150 to 300 feet from the end of the field (depending on whether or not adjoining lands should be watered). Start the pump, and allow water application to continue for long enough to apply the desired depth of water to the beginning end of the field before engaging the drive mechanism. The stationary time is determined by the average water application rate of the sprinkler.

Set the travel speed to deliver the desired depth of water based on the sprinkler flow rate, lane spacing (Chapter 6), and desired set time. Common set times are 10 to 11 hours to allow making two sets per day. Time the machine while it travels down the lane to verify the accuracy of the travel speed throughout the run.

The sprinkler should be stopped half the wetted diameter from the end of the field so the water pattern does not reach outside the target area onto adjoining fields or roadways. Complete the irrigation by watering for the same period of time as was used for the startup cycle.

Cable Tow

With a cable-tow system, as seen in Figure 4-11, set the sprinkler cart and winch at the start of a travel lane. Uncoil the cable to the opposite end of the travel lane, and attach it to an anchor (earth anchor, tractor, truck, or a large tree). Be sure the anchor can withstand the pull exerted by the machine. For example, a 4 1/2-inch diameter hose 660 feet long could exert more than 6,000 pounds of pull. Unroll the hose, and connect it to a hydrant and the sprinkler unit leaving adequate hose looped behind the machine so the hose won't kink.

After completing the run, disconnect the hose from the hydrant, purge water from the hose, reel the hose onto the reel, and move the traveler to the next lane. Repeat the process of laying and anchoring the cable and laying out the hose and connecting it to the supply line. The process of moving and setting up the unit can be accomplished in about one hour. A tractor moves the machine.

Figure 4-11. Typical cable-tow traveler.

Hose Tow

The hose-tow traveler, as seen in Figure 4-12, operates like the cable-tow except that the hose pulls the sprinkler cart through the field.

Set the traveler unit at the end of a travel lane. Use a tractor to pull the sprinkler cart and attached hose to the start of the travel lane. This will uncoil the hose from the reel. Drop the stabilizers on the hose reel to support the machine so the hose reel will not tip over. Connect the supply line, and start the pump. Follow similar field boundary setbacks and stationary watering times as with the cable-tow traveler.

Once the sprinkler cart has reached the end of a lane, the pump is shut down, the supply line is disconnected, and the traveler is moved to the next lane using a tractor.

On some machines, the hose reel is mounted on a turntable. With these models, a lane on both sides of a center alley or road can be irrigated without moving the reel. With other machines it is necessary to turn the hose reel unit 180° to irrigate on both sides of a center alley. One person should be able to move the unit in 30 to 45 minutes.

Choose a tractor size that can pull the hose and sprinkler cart through the field. A large tractor may be needed if the hose is full of water, the soil provides little traction, or the field has steep slopes. If possible, purge water from the hose before pulling it out.

Most of the sprinkler carts have adjustable widths. As riser height increases, widen the width of the sprinkler cart, or add weight to stabilize it so it does not overturn.

The hose and sprinkler cart travel best in a straight line, but due to the thick wall and heavy weight of the hose when it is full of water, it will follow some contour. Ridges may aid in encouraging the hose to follow a contour. Experience with operating the machine will indicate the maximum contour that can be handled.

Solid Set

A solid-set system has sprinklers installed to irrigate an entire field. The system has a complete buried pipe system or above-ground pipe system that is placed in the field throughout the growing season. Buried laterals are generally PVC plastic, while above-ground laterals can be either aluminum or special PVC pipe formulated to resist degradation from ultraviolet light. Because of the high cost, solid-set systems normally are used on high-value crops. Many systems are designed to be multipurpose, supplying irrigation, frost protection, crop cooling, or chemical application. Figure 4-13 shows an above-ground configuration for a solid-set irrigation system.

Design

Solid-set system design depends on how the system will be used. A system designed strictly for irrigation where only a portion of the system is operated at one time requires a smaller pump, power unit, and supply line than a system designed for frost protection (the entire system watering at same time). For frost protection, the capacity of the

Figure 4-12. Typical hose-tow traveler.

Figure 4-13. Typical solid-set sprinkler irrigation system.

system should be able to provide a rate of at least 0.1 inch per hour. A local irrigation specialist can provide detailed recommendations for specific geographic areas.

An existing portable aluminum pipe system can be converted to solid-set by adding more laterals and sprinklers. Usually the pump, power unit, and much of the existing lateral line can be used. Additional equipment such as other lateral lines, lower volume sprinklers, and a larger pump and motor may be needed. A controller and automatic valves are needed if the solid set system is to be automated.

Solid-set systems typically use low output single- or double-nozzle impact sprinklers to give application rates of 0.20 inch per hour or less. Design should optimize lateral line sizes to minimize friction loss while, at the same time, minimizing costs. Pipeline size should be selected to limit friction loss to less than 20% of the sprinkler design pressure. For example, for a 50 psi sprinkler, limit the pressure difference between the first and last sprinkler on the line to 10 psi (0.2 x 50). Many times it may be possible to lay out the mainlines and laterals so that friction losses are equalized throughout the system. See Chapter 7 for a discussion of friction loss.

Select sprinklers, sprinkler spacing, and lateral spacing based on soils, topography, and wind conditions. Chapter 6 provides more detail about selecting sprinklers.

Most systems are designed by selecting a sprinkler spacing along the lateral (S_L) and the spacing between lateral (S_M) lines. Spacing may conform to crop row or tree spacing in orchards. Typical field layouts range from square to rectangular with S_L x S_M values ranging from 30 x 30 feet to 80 x 80 feet. Figure 4-14 shows a typical layout for a solid-set system.

Operation

By design, the labor requirements for a solid-set sprinkler system can be low, depending on the type of system and whether it is used on an annual or perennial crop. Most solid-set systems do not irrigate the entire crop field and are sequenced from one set to another throughout the irrigation cycle. An automated system requires less labor than a manual system. A solid-set system with manual valves will require setting valves between each set.

The pipe and sprinklers in above-ground systems must be installed and removed each growing season. Buried pipelines must be drained unless they are

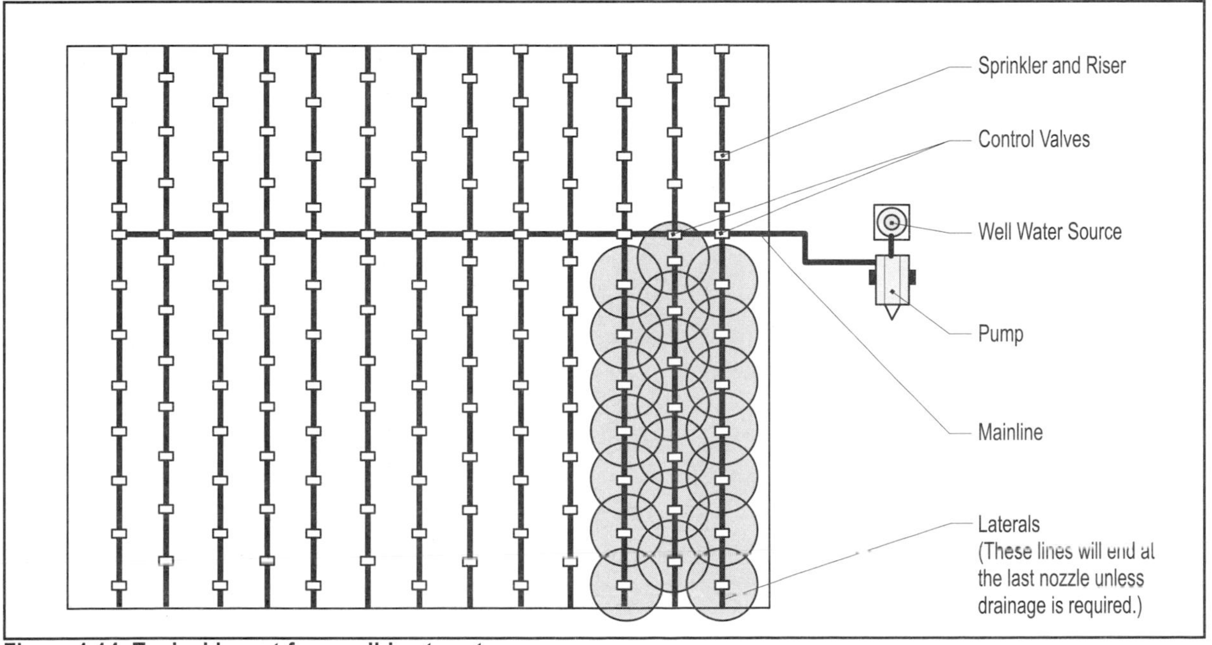

Figure 4-14. Typical layout for a solid-set system.

below the frost line.

Hand Move

Hand-move systems, like the one shown in Figure 4-15, are normally the lowest cost sprinkler systems. They are common on small irregularly-shaped fields. With hand-move systems, labor requirements are high, often requiring up to one worker hour per acre for each move with small sprinklers. Hand-move systems often are designed for one to three moves per day. Hand-move pipe is difficult to use on sizable areas or with tall crops such as corn.

Figure 4-15. Typical hand-move sprinkler irrigation system.

Design

The design of a hand-move system, depicted in Figure 4-16, is essentially the same as that of a solid-set system. Select sprinklers, sprinkler spacing, and lateral spacing based on soils, topography, and wind conditions. Select a sprinkler spacing along the lateral (S_L) and the spacing between lateral (S_M) lines. Chapter 6 covers sprinkler selection in more detail.

Design the system so that the desired water depth is applied in a 7-, 11- or 23-hour set. Planning 3, 2, or 1 sets per day, with at least an hour to move the pipe, optimizes labor. Since the system can irrigate with a variety of sprinkler choices and spacings, it can be designed to meet almost any field requirement, including soil intake rate and topography.

Limit friction loss in the lateral to 20% of sprinkler design operating pressure. For example, for a 50 psi sprinkler, limit the pressure difference between the first and last sprinkler on the line to 10 psi (0.2 x 50).

Couple laterals to the mainline (supply line) by tee-valves in the mainline spaced at the desired lateral interval. Select pipes, fittings, and latches for compatible size, fit,

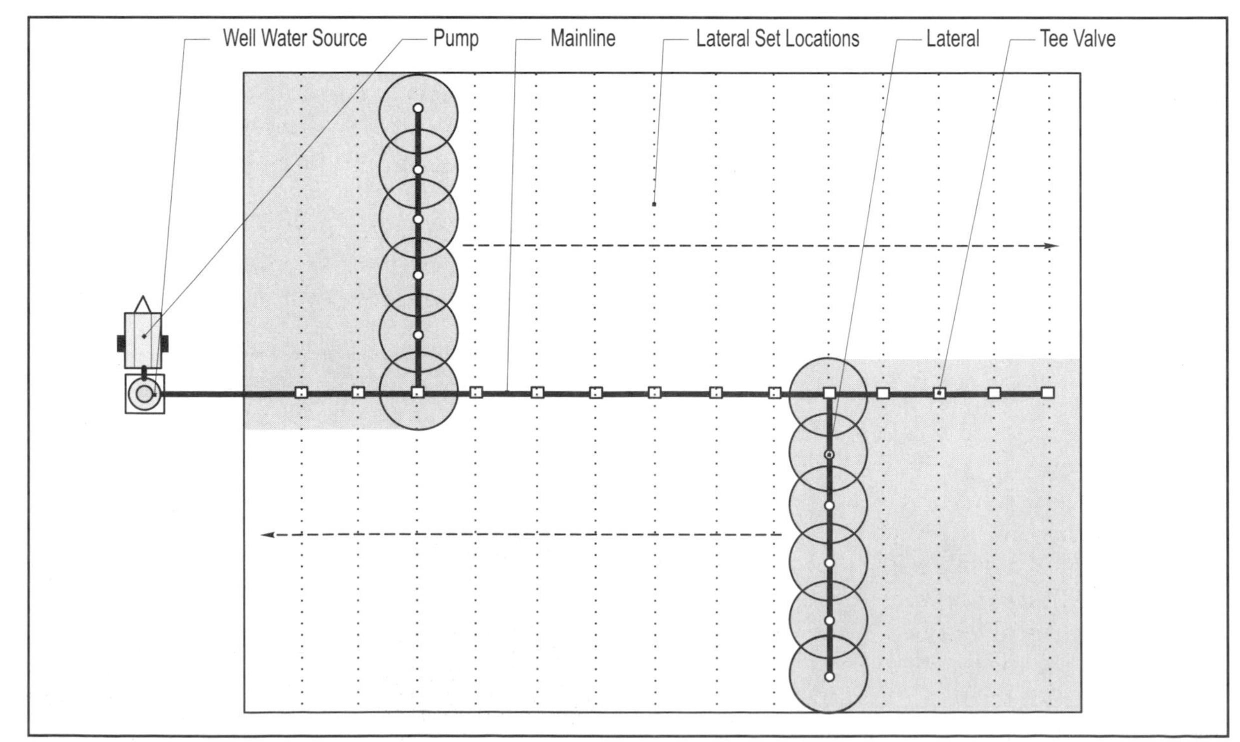

Figure 4-16. Typical layout for a hand-move sprinkler irrigation system.

and function.

Operation

The sprinklers on one or more laterals are operated at one time. This is considered one set. Labor and pumping time can be optimized if two sets of laterals are available. Water is delivered to one set while the second set is being moved to the next location.

Before operation, it is good practice to flush the lines to prevent plugging of sprinkler nozzles with debris and small animals. Based on the desired depth of application, determine the set time using sprinkler discharge and spacing data (Chapter 6). Operate the system until the desired water depth is applied. Shut off the water when a set is done and drain the laterals. Then hand move them to the next location.

When the lateral line reaches the end of the field, disassemble it and move it back to the first set position, or move it across the mainline to the original location of a second lateral. Lateral lines are returned to the starting position of the irrigation cycle so the area that was watered first will again be first in the following cycle.

Larger hand-move systems can be designed so that one or more lateral are being moved at any one time. A management scheme can be devised so the pump is operated continuously even while some laterals are being moved. This removes the difficulty of restarting the pump for each set.

Side Roll

The side-roll (or wheel-move) system, shown in Figure 4-17, is designed to be used on low-growing crops such as vegetables, potatoes, soybeans, forage crops, sod, and cereal grains (wheat and barley). It can be used on any soil suitable for sprinkler irrigation. However, the side-roll system is best suited to relatively level fields that are square or rectangular. Rolling topography makes alignment difficult.

The side-roll system is meant to be used on a single field or on side-by-side fields where the lateral can be moved directly from one field to the other field. It cannot be moved parallel to the pipe without putting dolly wheels or skids under each wheel of the system or disassembling the machine, which is time consuming. With additional effort, sections of pipe can be added to or removed from either or both ends to irrigate trapezoidal areas.

Design

The lateral line is usually 4- or 5-inch diameter aluminum pipe mounted so the pipe serves as the axle for a number of wheels. The wheels are spaced 30 to 40 feet apart with the sprinklers midway between the wheels. Wheels are available in diameters ranging from 4 to 10 feet to provide clearance for different crop heights. It is recommended that self-leveling sprinklers be used so the lateral can be stopped at each desired lateral position without concern for sprinkler orientation.

Sprinkler selection for side-roll systems is similar to that for hand-move systems but must conform to equipment limitations (Chapter 6). There is less flexibility of choosing sprinkler spacing (S_L) and lateral spacing (S_M). While certain lower pressure sprinklers can be used, sprinklers are typically medium to high pressure.

Select sprinklers to conform to the field conditions and desired operation. For example, a side-roll with sprinklers spaced every 40 feet and having 60 feet between sets would need a carefully selected nozzle size to achieve the desired application depth and set/time combination. As with hand move, set times of 7, 11, or 23 hours are often selected to optimize labor.

The lateral line size on a side-roll system, as for other set-move systems, should

Figure 4-17. Typical side-roll irrigation system.

maintain friction loss in the lateral less than 20% of the recommended sprinkler operating pressure. A mainline is positioned perpendicular to the travel direction and has appropriately spaced risers. Friction losses can be reduced by attaching to the mainline near the center of the lateral rather than the end (see Chapter 7 for friction loss in laterals).

Operation

The operation of the side-roll lateral is similar to that of the hand-move system, except the lateral line is moved by rolling the wheels. The principle of operation is to water the set, disconnect from the mainline, let the system drain, move the lateral by rolling it to the next set, then reconnect to the mainline. Normally, a engine mounted on a driver unit moves the lateral. A telescoping aluminum pipe or short flexible hose simplifies connecting to the lateral and mainline hydrant.

Figure 4-18 depicts a typical layout for a side-roll system. The distance between lateral sets (S_M) is some multiple of the distance that can be obtained by a given number of wheel rotations, typically 3 to 5 rotations. This distance is commonly 50 to 80 feet. The distance between lateral sets must be selected to match the type and size of the sprinklers. Improve uniformity by staggering the position of each of the sets to fall midway between sets of the previous irrigation.

When the lateral finishes the last set of the field, move it back to the starting point or to an adjacent field.

Sprinkler System Management

Each sprinkler system has advantages and disadvantages that best suit it for certain conditions, and management of the irrigation system should take advantage of the inherent features of the chosen sprinkler system. Design of the system uses the best information available to achieve the best results. Poor management can, however, accentuate undesirable features and result in unsatisfactory operation. No system can achieve the desired

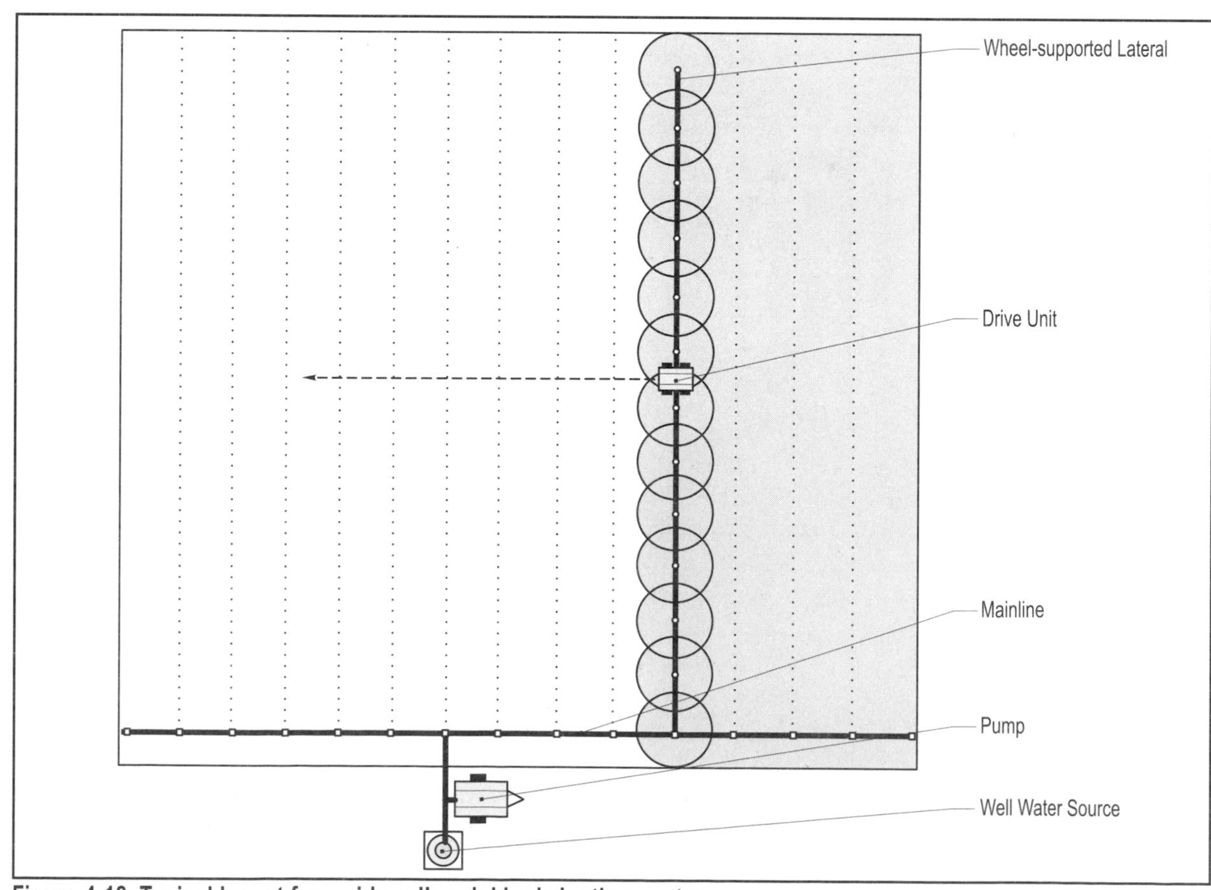

Figure 4-18. Typical layout for a side-roll sprinkler irrigation system.

objectives if the operational management defeats the design of the system. Likewise, no level of management can correct faults in a design that renders the sprinkler system unsuitable for the site.

The design of the sprinkler system should minimize the potential for negative outcomes such as runoff, deep percolation, excessive evaporation, under-application, and excessive energy use. It is likely, however, that design cannot match a sprinkler system perfectly to all parts of a field. Fine-tuning of the operation always is required to assure that the system functions as intended. Often modifications in operating procedure are needed to achieve high efficiency and uniformity. For example, if runoff occurs in areas of a center pivot system, it may be necessary to change tillage practices, add furrow dikes, or reduce the application depth.

It is good practice not to fill the soil profile completely with water at each application. Leaving some capacity to store rain fall often reduces the total seasonal water required and the potential for leaching of water and nutrients below the root zone. Because of limited soil water-holding capacity, some very sandy soils may not allow the flexibility of reserving soil moisture storage for rainfall.

Some areas with low-capacity water supplies may preclude full irrigation. Deficit irrigation may change the management scheme of the irrigation system. Maximizing the application efficiency may dominate the operation at the same time that uniformity concerns are reduced because all water will be used regardless of where it falls.

Whenever possible, operate the sprinkler system during favorable periods such as when winds speeds are low and evaporation potential is minimal. This is especially important for high-pressure sprinkler systems such as travelers or high-pressure center-pivots. Most sprinkler irrigation designs, however, require full-time operation during periods when water use is high and rainfall is lacking.

For center-pivot irrigation, use rotational times other than multiples of 24 hours (1 day) to minimize the effect of irrigating a particular area of the field at the same time each rotation. Also for center-pivots avoid light, frequent irrigations that keep the canopy wet and have higher evaporation, unless the desired operational scheme is for cooling the crop.

LEPA (low energy precision application) and in-canopy spray systems require a complete management system to achieve success. Farming in circles for center pivots and in-row furrow diking are required to maintain the high application efficiency that is possible with these systems.

For set-move systems, such as hand-move and side-roll, a good management practice is to stagger successive sets midway between the previous sets. Irrigating with the lateral between the previous sets will help to average out uneven distribution from sprinkler patterns and wind effects.

Sprinklers that malfunction along with normal wear and tear on the nozzles cause the system water application uniformity to decrease with time. Systems that pump sand or have poor quality water will exhibit the most severe declines in water application uniformity over time. A good management practice is to walk beside the system while it is operating to verify that each sprinkler is functioning. Some nozzles may be partially plugged with debris. Walking beside the system closely enough to view the operation of each sprinkler or nozzle will help identify these problems.

Monitor and record operating pressure during the season and from year to year to spot potential problems early. Lower than normal pressure or sand in the trap indicate well and pump problems.

Water hammer could result in undue wear on the pipeline and system. Watch the system when it is filling. If the system lurches and sways up and down, the symptoms may indicate water hammer. Systems with long delivery pipelines are most susceptible to water hammer problems. Installation of a throttling valve at the pump may help reduce the impact of water hammer problems for the system.

Chapter 5

Sprinkler Characteristics

Chapter 5 Contents

The desire for energy conservation has driven sprinkler technology toward lower operating pressures. The sprinkler industry has developed a multitude of sprinkler design innovations over the last 20 years to achieve that objective. Most of the recent developments and improvements in sprinkler performance have been with center pivot applications, largely in sprinkler distribution packages.

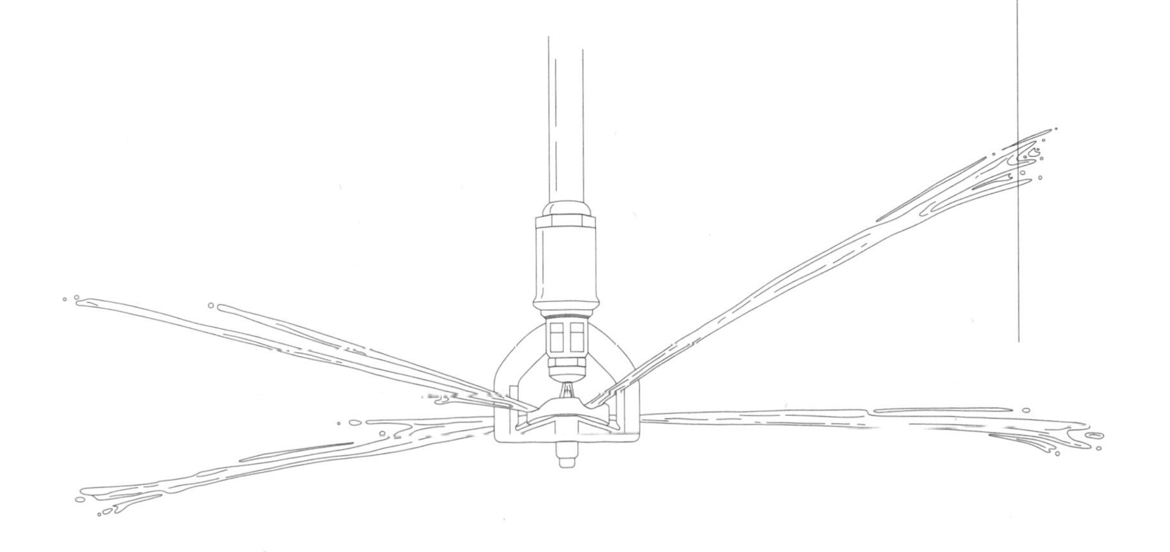

Sprinkler Performance Characteristics

The goal of any sprinkler system is to apply the desired amount of irrigation water to the crop's root zone as efficiently and uniformly as possible. Listed below are factors that determine performance characteristics such as droplet size, wetted diameter, rotational speed, and the application rate. These performance characteristics, in turn, affect the selection of a sprinkler system and how well a sprinkler system will perform on a given crop field.

- Sprinkler type: Impact, spray, fixed/rotating, bubbler, gun
- Nozzle type and size: Circular, non-standard, tapered, pressure compensating
- Configuration: Trajectory angle, mounting location and height, spacing/overlap
- Operating pressure

Figure 5-1 illustrates common types of sprinklers. For any sprinkler type, there are often a variety of options and configurations, including trajectory angle, nozzles, spray plates, etc. Table 5-1 provides options and operating characteristics for several types of sprinklers.

Nozzles

Sprinklers use some type of nozzle to control the discharge of the water. The nozzle also may have a device to spread the water into a desired pattern or to control water droplet size. Flow through a nozzle is governed by the principles of orifice flow based on the nozzle size and operating pressure. In general terms, the discharge (gallons per minute or gpm) is determined by the cross-sectional area (size) of the nozzle and the square root of the pressure. Equation 5-1 is used to predict discharge of a round, straight bore nozzle.

Equation 5-1. Nozzle Discharge.

$$q = c_d \cdot (29.83) \cdot d^2 \cdot p^{0.5}$$

Where:

q = Nozzle discharge (gpm)
c_d = Discharge coefficient based on nozzle type and configuration (between 0.95 and 1.00)
d = Nozzle diameter (inches)
p = Pressure (psi)

Most nozzles have circular openings (Figure 5-2). Nozzles may incorporate tapered entrances or have straightening vanes in their design. Other nozzle openings may have rectangular or triangular shapes that are used to achieve the desired water application patterns and droplet sizes.

Table 5-1. Sprinkler options and operating characteristics.

Type	Material	Trajectory angle	Droplet size	Sprinkler pressure (psi)[a]	Wetted diameter (ft)
Bubbler	Plastic	Down	Large	5 to 20	1
180° spray	Brass or plastic	-10° to 10	Small	10 to 30	10 to 20
360° spray					
Smooth	Plastic	0°	Small	10 to 30	20 to 40
Fine groove	Plastic	-5° to 10°	Small to medium	10 to 30	30 to 50
Coarse groove	Plastic	0° to 15°	Medium to large	15 to 40	40 to 60
Combination	Plastic	Combination	Small to medium	10 to 40	40 to 50
Rotating	Plastic	-5° to 30°	Medium to large	15 to 45	40 to 70
Impact					
Low pressure	Brass or plastic	0° to 15°	Medium to large	25 to 50	50 to 70
High pressure	Brass or plastic	10° to 27°	Medium to large	40 to 80	60 to 100
Volume gun	Various	18° to 27°	Medium to large	40 to 100	100 to 550

[a]Pressure listed is for the sprinkler device. Operating pressure for the system must be higher to compensate for friction lossesand elevation differences within the field.

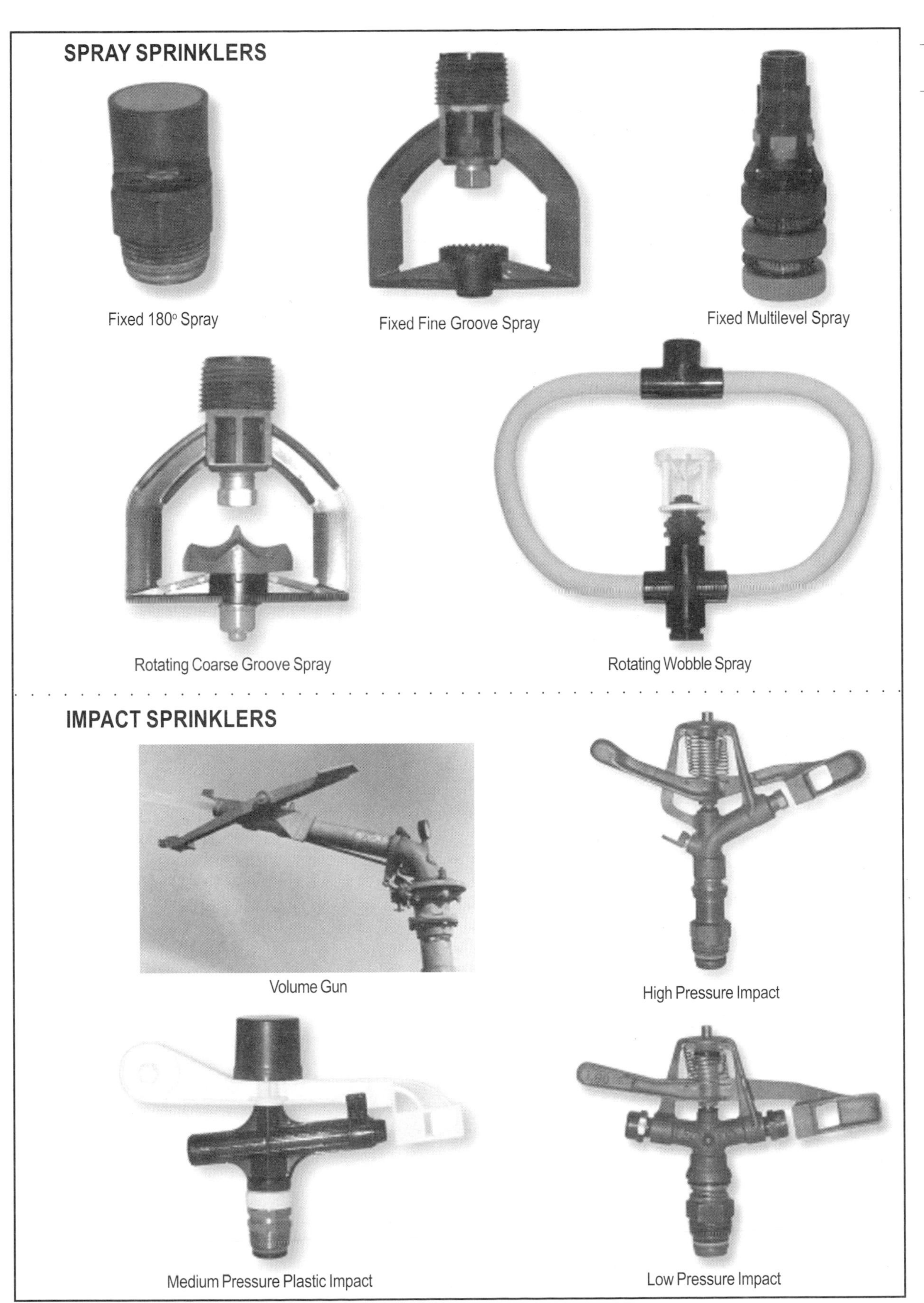

Figure 5-1. Examples of common sprinkler types.

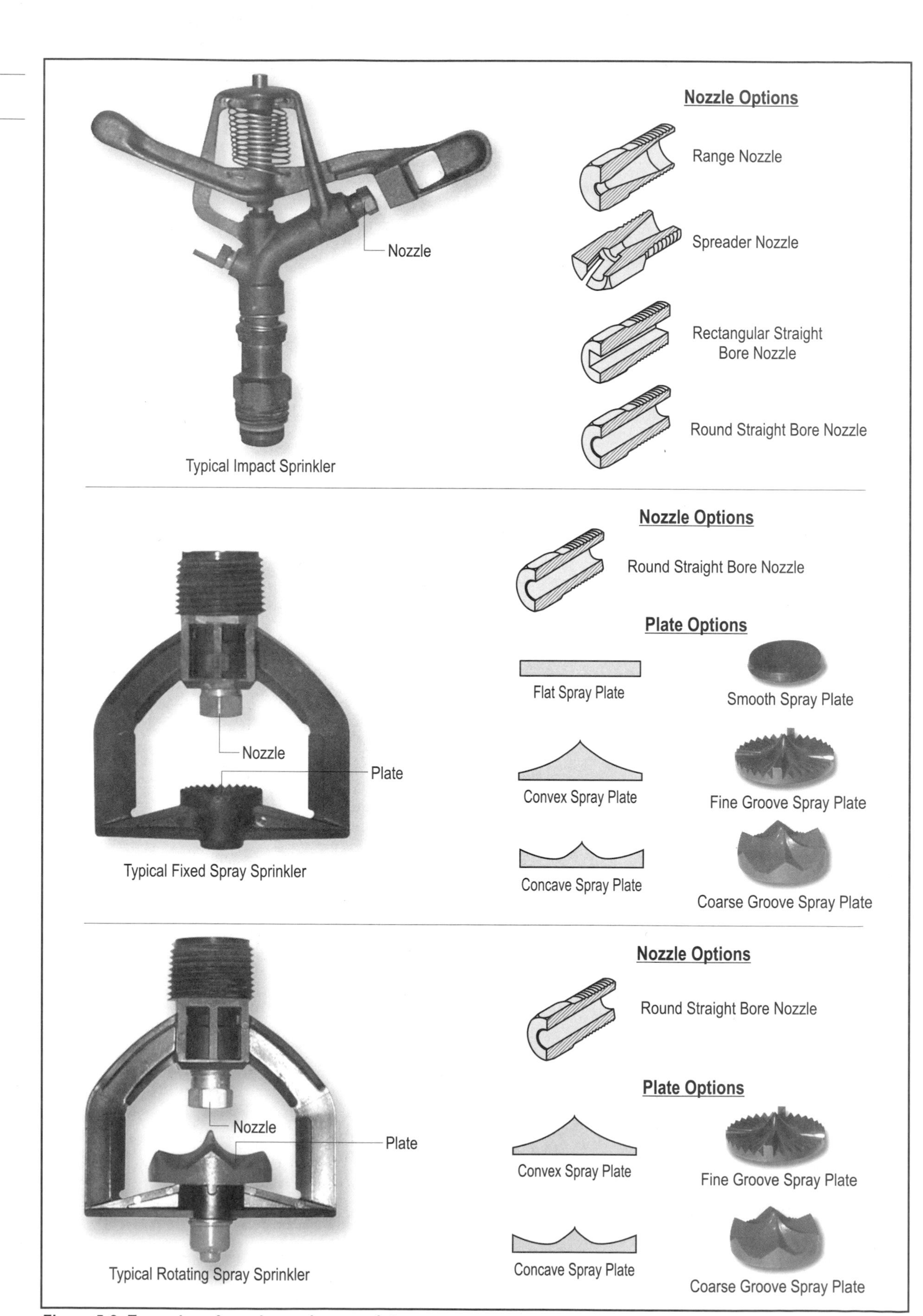

Figure 5-2. Examples of nozzles and spray plates.

Traditional nozzle size/pressure relationships may no longer apply when non-standard (non-circular) nozzles are used. Some nozzles will even compensate for variations in pressure.

Spray heads more commonly use spray plate geometries, where the stream of water from the nozzle opening is directed onto the spray plate, to achieve desired patterns or droplet sizes. The possible configurations of nozzles and spray plates are almost unlimited.

Nozzle Pressure

Pressure not only governs the output of a nozzle, but also the droplet sizes and distance that the sprinkler throws water. It is important that sprinklers be operated within the manufacturer's recommended pressure range to achieve effective wetted coverage, optimum droplet size, and uniformity. For a given nozzle, operating at the high end of the recommended pressure range results in maximum wetted coverage. Excessive pressure, however, decreases water droplet sizes, making them more susceptible to wind drift effects and reducing wetted coverage. Figure 5-3 illustrates these concepts.

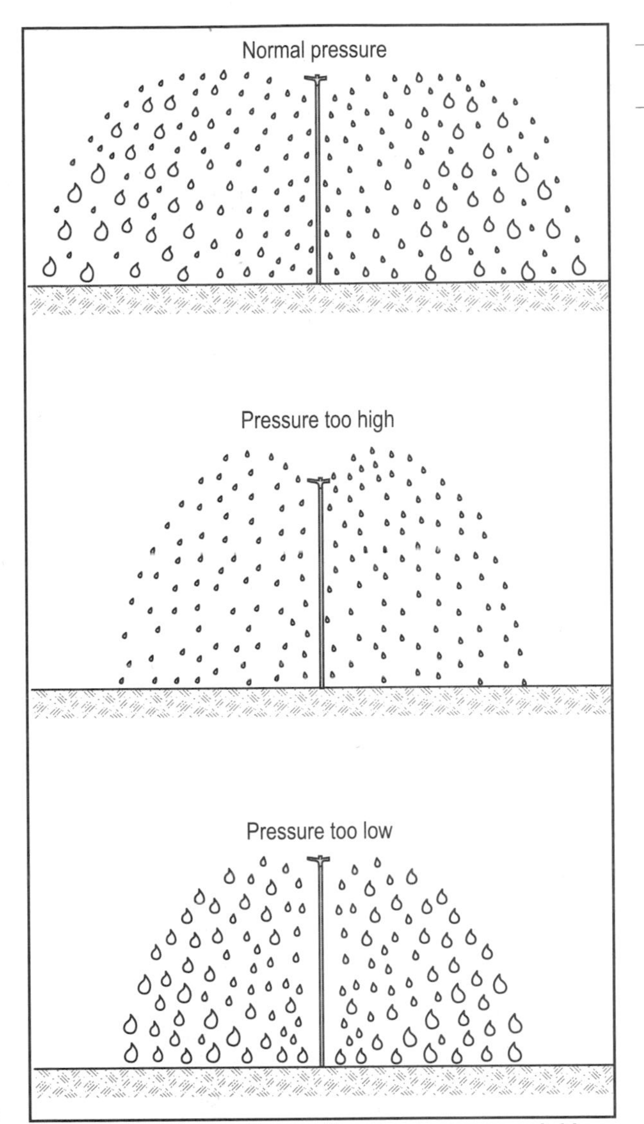

Figure 5-3. Effects of operating pressure on sprinkler distribution.

Information provided by sprinkler manufacturers gives output (gallons per minute) versus pressure for sprinklers and nozzle configurations. A limitation is that only a specific number of nozzle size combinations; for example, 1/8 inch, 9/64 inch, 5/32 inch; can be selected to obtain the desired output at a given pressure.

In other words, because nozzle sizes are limited, there are a series of step-like outputs for any given pressure. Table 5-2 shows example nozzle outputs. Some sprinkler manufacturers use numbered or color-coded nozzles that may be sized in either inches or millimeters. Many nozzle sizes are coded based on fractions of the total diameter, so size #11 could be 11/64 inch or 11/128 inch.

Trajectory Angle

Trajectory angle generally varies from 0° to 30° with respect to horizontal. Some sprinkler heads have multiple trajectories designed to provide uniform distribution under specific operating conditions. For example, a large stream may be directed at a higher angle to get more distance and coverage while a smaller stream is directed at a lower angle to fill the wetted pattern. When spray heads on center pivots are mounted on drop tubes, the trajectory angle of the water from the spray head is directed to spray at an upward angle. In-canopy or bubbler applications (see later sections) often direct the water spray downward.

Higher angles tend to throw the water a greater distance but are more subject to drift due to wind (Figure 5-4). A larger stream, for example are produced by a coarse spray plate, also will be less subject to wind.

Table 5-2. Example nozzle outputs for straight, round-shaped nozzles (gpm).
Based on the equation $q = (c_d)29.83d^2p^{0.5}$. Table assumes a discharge coefficient (c_d) equal to 0.95.

Diameter, d	Pressures, p (psi)															
(in)	5	10	15	20	25	30	35	40	45	50	55	60	65	70	75	80
1/16	0.248	0.350	0.429	0.495	0.554	0.606	0.655	0.700	0.743	0.783						
5/64	0.387	0.547	0.670	0.774	0.865	0.947	1.023	1.094	1.160	1.223						
3/32	0.557	0.788	0.965	1.114	1.245	1.364	1.474	1.575	1.671	1.761	1.847	1.929				
7/64	0.758	1.072	1.313	1.516	1.695	1.857	2.006	2.144	2.274	2.397	2.514	2.626				
1/8	0.990	1.400	1.715	1.980	2.214	2.425	2.620	2.801	2.970	3.131	3.284	3.430	3.577	3.712		
9/64	1.253	1.772	2.171	2.506	2.802	3.070	3.315	3.544	3.759	3.963	4.156	4.341	4.527	4.697		
5/32	1.547	2.188	2.680	3.094	3.459	3.790	4.093	4.376	4.641	4.892	5.131	5.359	5.578	5.789	5.992	6.188
11/64	1.872	2.647	3.242	3.744	4.186	4.585	4.953	5.295	5.616	5.920	6.209	6.485	6.749	7.004	7.250	7.488
3/16	2.228	3.151	3.859	4.456	4.982	5.457	5.894	6.301	6.683	7.045	7.389	7.717	8.032	8.336	8.628	8.911
13/64			4.529	5.229	5.846	6.404	6.918	7.395	7.844	8.268	8.672	9.057	9.427	9.783	10.13	10.46
7/32			5.252	6.065	6.780	7.428	8.023	8.577	9.097	9.589	10.06	10.50	10.93	11.35	11.74	12.13
15/64			6.029	6.962	7.784	8.527	9.210	9.846	10.44	11.01	11.54	12.06	12.55	13.02	13.48	13.92
1/4			6.860	7.921	8.856	9.701	10.48	11.20	11.88	12.52	13.14	13.72	14.28	14.82	15.34	15.84
9/32					11.21	12.28	13.26	14.18	15.04	15.85	16.62	17.36	18.07	18.76	19.41	20.05
5/16					13.84	15.16	16.37	17.50	18.57	19.57	20.52	21.44	22.31	23.15	23.97	24.75
11/32					16.74	18.34	19.81	21.18	22.46	23.68	24.83	25.94	27.00	28.02	29.00	29.95
3/8					19.93	21.83	23.58	25.20	26.73	28.18	29.56	30.87	32.13	33.34	34.51	35.64
13/32					23.39	25.62	27.67	29.58	31.37	33.07	34.69	36.23	37.71	39.13	40.50	41.83
7/16					27.12	29.71	32.09	34.31	36.39	38.36	40.23	42.02	43.73	45.38	46.98	48.52
15/32					31.13	34.11	36.84	39.38	41.77	44.03	46.18	48.23	50.20	52.10	53.93	55.70
1/2					35.42	38.81	41.91	44.81	47.53	50.10	52.54	54.88	57.12	59.28	61.36	63.37
17/32					39.99	43.81	47.32	50.58	53.65	56.56	59.32	61.95	64.48	66.92	69.27	71.54
9/16					44.83	49.11	53.05	56.71	60.15	63.40	66.50	69.46	72.29	75.02	77.65	80.20
5/8					55.35	60.63	65.49	70.01	74.26	78.28	82.10	85.75	89.25	92.62	95.87	99.01

Gun sprinkler nozzles, d(in)	25	30	35	40	45	50	55	60	65	70	75	80	85	90	95	100	105	110	115	120
0.5	35.42	38.81	41.91	44.81	47.53	50.10	52.54	54.88	57.12	59.28	61.36	63.37	65.32	67.21	69.05	70.85	72.60	74.31	75.98	77.61
0.6	51.01	55.88	60.36	64.52	68.44	72.14	75.66	79.03	82.25	85.36	88.35	91.25	94.06	96.79	99.44	102.0	104.5	107.0	109.4	111.8
0.7	69.43	76.06	82.15	87.82	93.15	98.19	103.0	107.6	112.0	116.2	120.3	124.2	128.0	131.7	135.3	138.9	142.3	145.6	148.9	152.1
0.8	90.69	99.34	107.3	114.7	121.7	128.2	134.5	140.5	146.2	151.7	157.1	162.2	167.2	172.1	176.8	181.4	185.9	190.2	194.5	198.7
0.9			135.8	145.2	154.0	162.3	170.2	177.8	185.1	192.1	198.8	205.3	211.6	217.8	223.7	229.5	235.2	240.8	246.2	251.5
1.0			167.7	179.2	190.1	200.4	210.2	219.5	228.5	237.1	245.4	253.5	261.3	268.9	276.2	283.4	290.4	297.2	303.9	310.4
1.1			202.9	216.9	230.0	242.5	254.3	265.6	276.5	286.9	297.0	306.7	316.1	325.3	334.2	342.9	351.4	359.6	367.7	375.6
1.2			241.4	258.1	273.8	288.6	302.6	316.1	329.0	341.4	353.4	365.0	376.2	387.1	397.8	408.1	418.2	428.0	437.6	447.0
1.3					321.3	338.7	355.2	371.0	386.1	400.7	414.8	428.4	441.6	454.4	466.8	478.9	490.8	502.3	513.6	524.
1.4					372.6	392.8	411.9	430.3	447.8	464.7	481.0	496.8	512.1	526.9	541.4	555.5	569.2	582.6	595.7	608.5
1.5					427.7	450.9	472.9	493.9	514.1	533.5	552.2	570.3	587.9	604.9	621.5	637.6	653.4	668.8	683.8	698.5
1.6					486.7	513.0	538.0	562.0	584.9	607.0	628.3	648.9	668.9	688.3	707.1	725.5	743.4	760.9	778.0	794.7
1.7								634.4	660.3	685.2	709.3	732.5	755.1	777.0	798.3	819.0	839.2	859.0	878.3	897.2
1.8								711.2	740.3	768.2	795.2	821.3	846.5	871.1	894.9	918.2	940.9	963.0	984.7	1006
1.9								792.5	824.8	855.9	886.0	915.0	943.2	970.5	997.1	1023	1048	1073	1097	1121
2.0								878.1	913.9	948.4	981.7	1014	1045	1075	1105	1134	1162	1189	1216	1242

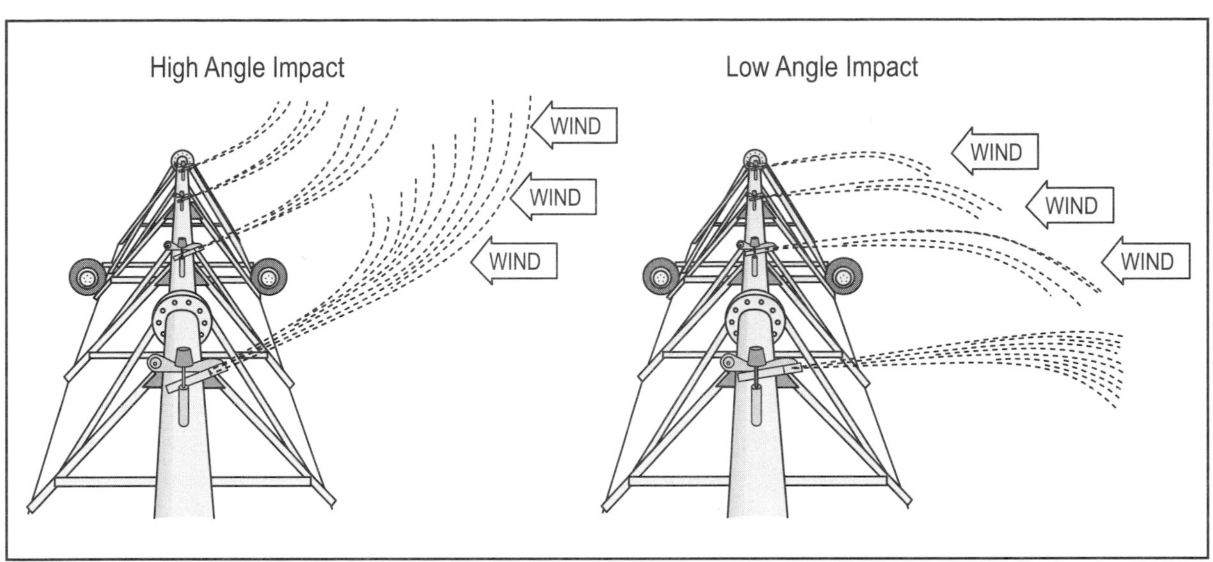

Figure 5-4. Effects of wind and sprinkler trajectory angle on drift.

Wetted Diameter

Each sprinkler head can distribute the water over a given area, referred to as the wetted diameter (Table 5-3). Water application patterns (see a later section) are very important when matching a certain sprinkler type to the field. Larger wetted diameters distribute the water over a larger area, thereby making it easier to apply the water at a rate less than the soil infiltration rate. Smaller wetted diameters concentrate the water application on a smaller area and have higher average application rates that can more easily exceed the soil infiltration rate. There is a tradeoff between the wetted diameter and the pressure required. Large wetted radii are possible only by using higher pressures.

Table 5-3. Examples of wetted diameters of impact sprinklers with standard nozzles.
Pressure of 60 psi and trajectory angle of 27°.

Diameter (ft)	Capacity (gpm)
70	2
80	5
110	10
140	20
180	50
250	100
320	200
420	500

Higher trajectory angles increase the wetted diameter (see Figure 5-5). The height of sprinklers above the soil or canopy surface also affects wetted diameter. Increasing wetted diameter by either method exposes the spray to greater potential for wind drift. Others have spray plates that rotate and use larger grooves to get bigger droplets for an increased wetted diameter.

For moving systems, a larger wetted diameter increases the length of time that the crop canopy is wet. This increases the opportunity for canopy evaporation losses, thus reducing application efficiencies. These losses would be more significant in regions where evaporation potential is high, but may not be important in areas that are more humid.

Droplet Size

Droplet size increases with larger nozzles and/or lower pressures. Table 5-4 summarizes general characteristics of sprinkler droplets.

Table 5-4. Droplet size characteristics related to potential interactions.

Potential	Smaller droplets	Larger droplets
Wind drift	high	low
Evaporation	high	low
Infiltration reduction (soil sealing)	low	high

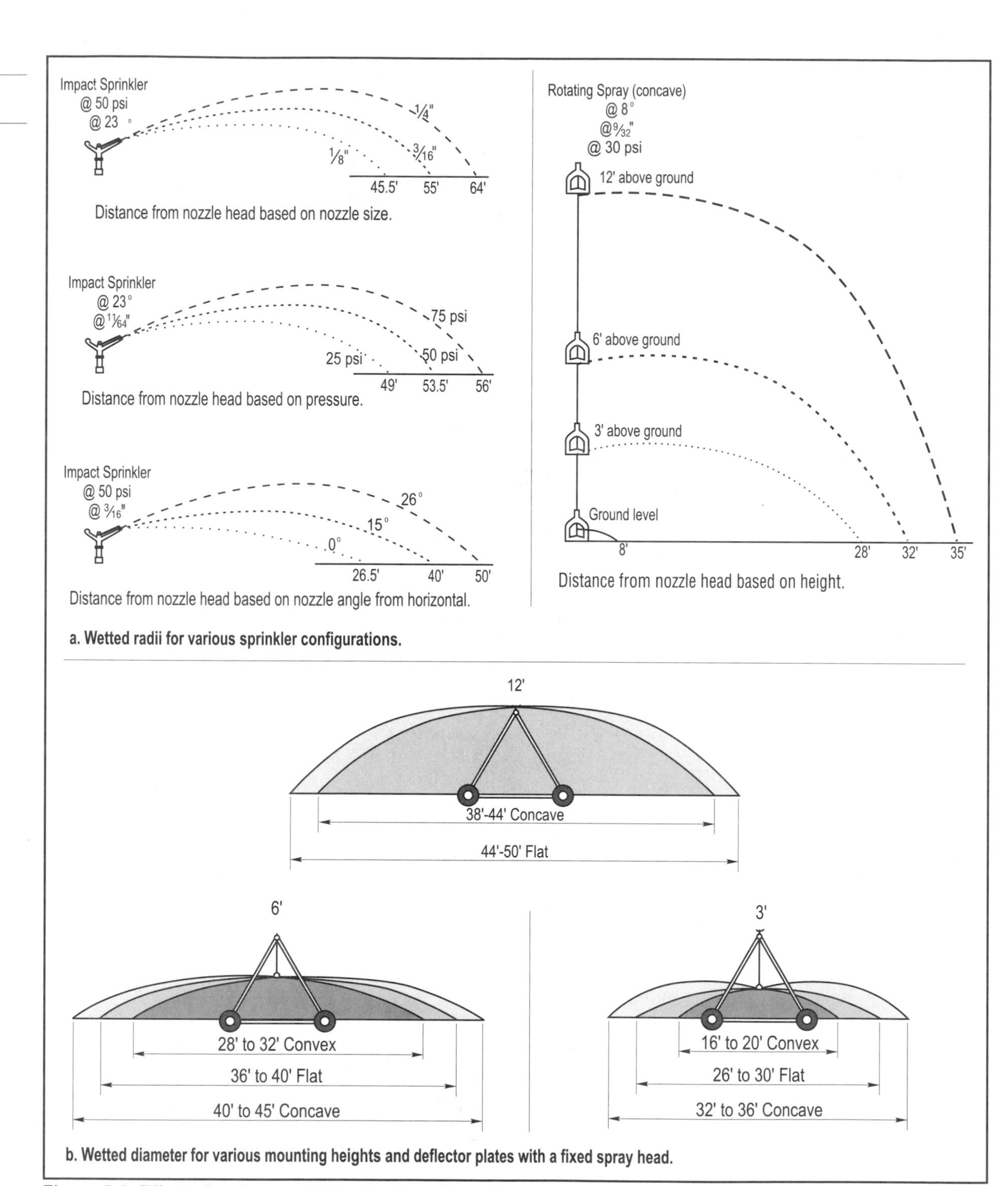

Figure 5-5. Effect of various sprinkler head configurations on coverage.

Large droplets resist wind drift better than the small droplets, but may have detrimental effects on the crops or the soil surface sealing. Impact sprinklers often use various nozzle configurations (Figure 5-2) to control droplet sizes. Spray heads use various spray plate configurations to control the droplet size. Smooth plates will have smaller droplets; fine grooves will have medium droplets, and coarse grooves will have larger droplets. Some spray heads have a combination of fine and coarse grooves to obtain a mix of

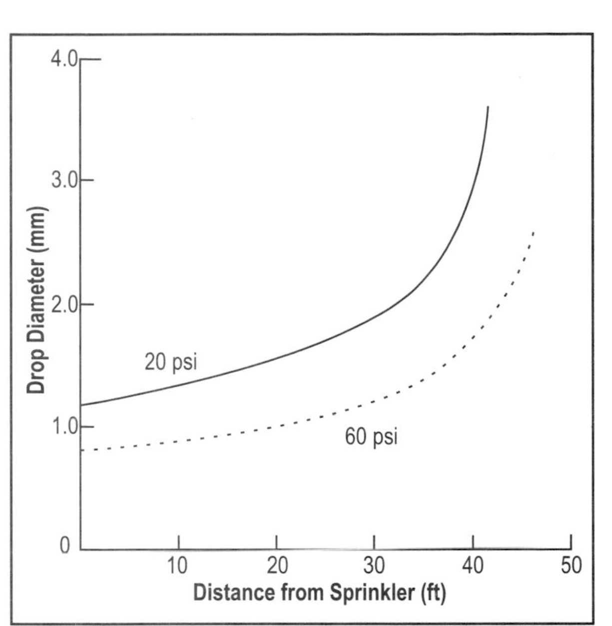

Figure 5-6. Example droplet size at various distances from a sprinkler. Based on a standard 5/32-inch nozzle.

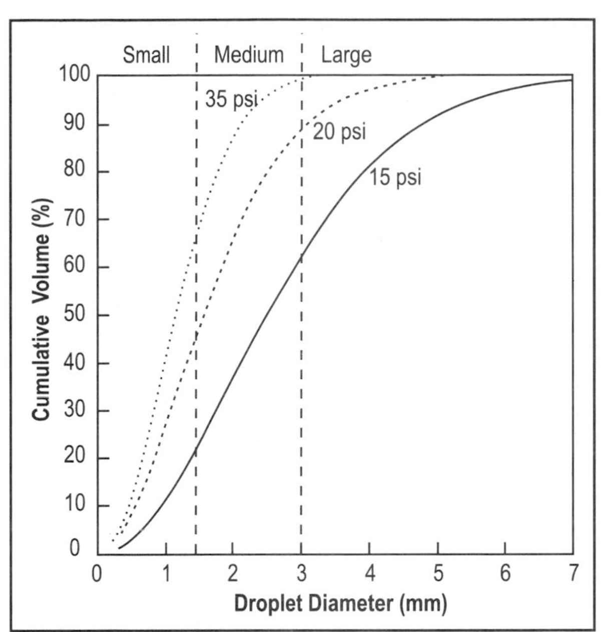

Figure 5-7. Example droplet size distribution for an impact sprinkler at 35, 20, and 15 psi pressure.

droplet sizes. Others have spray plates that rotate and use larger grooves to get bigger droplets for larger wetted diameters.

Figure 5-6 shows droplet sizes at various distances away from a typical sprinkler set at two different pressures. Figure 5-7 shows typical total droplet distribution for an impact sprinkler operated at three different pressures. Avoid using sprinklers that produce small droplets in climates where the temperature is high and the humidity is low. Evaporation potential may not be as important in more humid areas where only a few inches of irrigation water are applied per year. Wind drift potential is an important consideration where winds are high or with systems used for chemigation.

Sprinkler selection should consider the interactions of droplet size on the soils and crop to be grown at the site. Large drops striking the soil surface can reduce the infiltration rate of the soil. Crop residue cover and the crop canopy cover reduce the surface sealing effect of irrigation. Large droplets also can affect delicate crops and crops that are sensitive during germination and emergence. Figure 5-8 gives estimates of the infiltration rate reduction for three soil types and the various drop sizes.

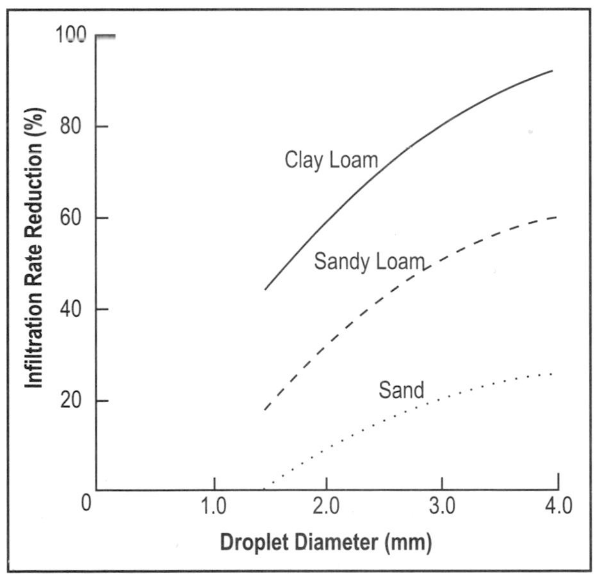

Figure 5-8. Relation of infiltration rate reduction to sprinkling three different soils at an application rate of approximately 0.5 inches per hour.

Application Rates

Application rates can be viewed from several perspectives:

- Instantaneous application rates.
- Average application rates.
- System application rates.
- Peak application rates.

Each needs to be considered when selecting sprinklers.

The instantaneous application rate indicates how fast the water is applied to an area at any point in time. The average application rate indicates how fast the water is applied on the average over the wetted diameter. The average system application rate indicates how fast the water is applied during the time the irrigation system irrigates one area of the field. The peak application rate often is used to indicate maximum system application rate. Table 6-7 in Chapter 6 is an example of system application rates for a different solid-set system.

Impact sprinklers have the highest instantaneous application rate because they have only one or two water streams coming from the sprinkler. Water is concentrated as it falls over a narrow section of the wetted diameter, whereas, many types of spray heads disperse the water over a larger portion of their wetted diameter at one time. With the impact sprinkler, however, the water applied has until the sprinkler comes around again to infiltrate. Spray heads have a lower but more constant application rate over the wetted area.

Sprinklers that have the largest wetted diameter will have the lowest average application rate. Hence, impact sprinklers or rotating spray sprinklers will have lower average application rates than spray heads.

The system application rate is the application rate of the combined sprinklers on the system for a given area. For example, because of overlap, one or more sprinkler application patterns may be contributing to the application rate at one time. Very high instantaneous application rates are possible if three or four impact sprinkler streams were falling on the same point at one time. The average system application rate generally is used for system design to match field conditions.

The peak application rate is the highest application rate during the time water is applied. For example, the average application rate might be 1.0 inch per hour for a center pivot, while the peak application rate could be 2 to 8 inches per hour.

Application Patterns

Sprinklers apply water over a surface area based on the wetted diameter of the sprinkler. Sprinklers do not apply equal amounts of water over the wetted area. Figure 5-9 shows distribution patterns for common sprinklers under various operating conditions. Each sprinkler and nozzle combination has an optimum operating pressure range that assures proper operation of the sprinkler and water distribution. When selecting a sprinkler package, use detailed information from the manufacturer, in addition to company literature, for specific sprinkler characteristics and operating conditions.

Figure 5-9, shows several general relationships between sprinkler patterns and application rates. Higher pressures tend to have a uniform distribution and larger wetted diameter. At lower pressures, nonstandard nozzle shapes on impact sprinklers can improve sprinkler distribution and decrease water droplet size. Higher trajectory angles give better distribution and larger wetted diameter. Sprays using coarse, grooved plates have larger wetted coverage than do smooth spray plate deflectors.

Wind Effects

Wind has a pronounced effect on sprinkler irrigation. Even if a sprinkler with perfect distribution could be designed, the results can be very different when used under field conditions. Figure 5-10 illustrates how wind affects the wetted coverage of a typical sprinkler. Not only does the wind shift the water downwind from the sprinkler, but it also shrinks the diameter perpendicular to the wind direction. It is common under high wind conditions to have all water deposited downwind from the sprinkler. By shrinking the coverage area, the wind causes higher application rates, which increases the potential for surface runoff. Wind also can carry smaller droplets entirely out of the intended coverage area.

Sprinkler Spacing and Overlap

Application patterns govern the required overlap between sprinklers to get optimum uniformity. For solid-set and set-type sprinklers, overlap must be maintained not only between sprinklers, but also between lateral settings. Figure 5-11a illustrates overlap

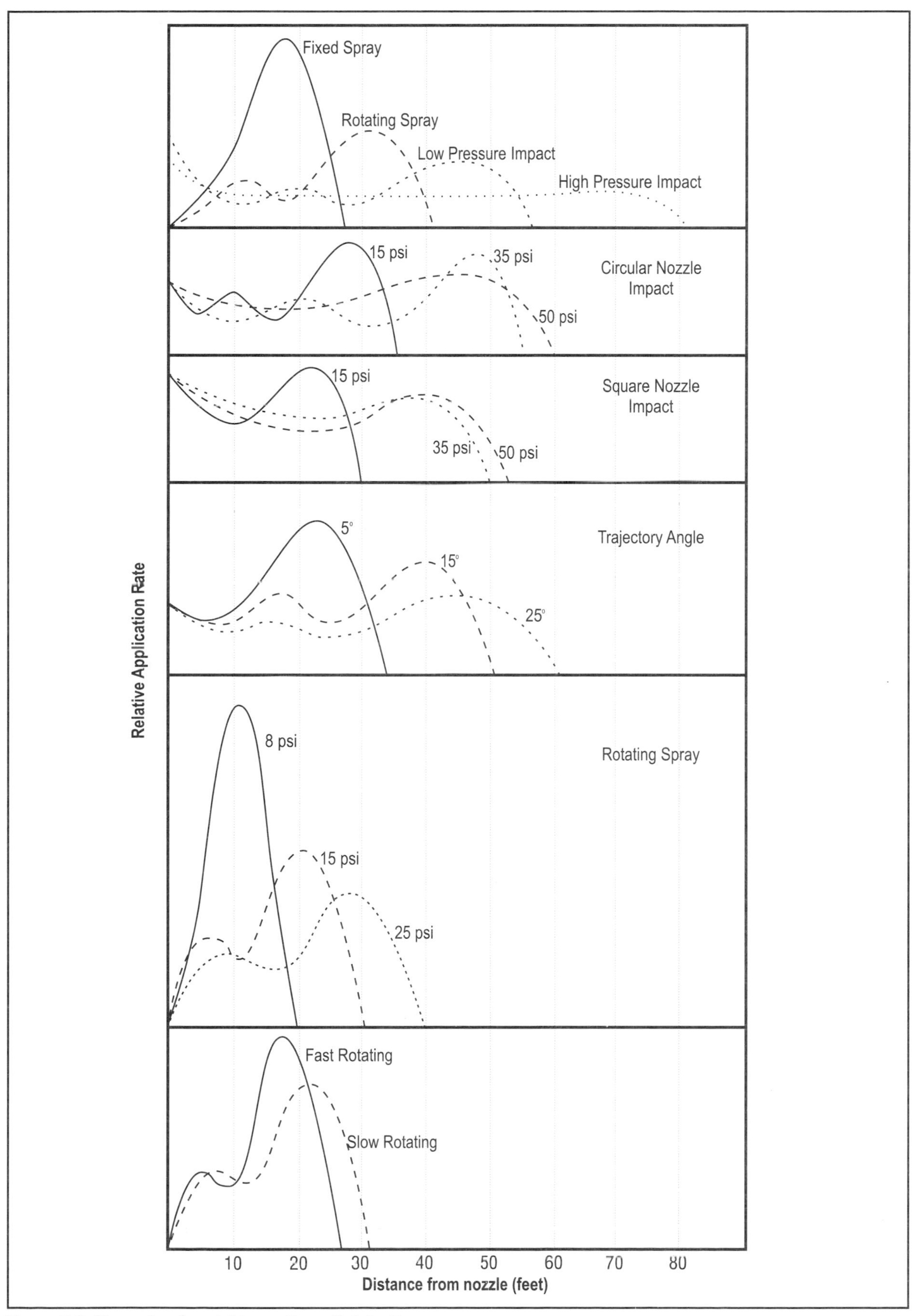

Figure 5-9. Example of sprinkler pattern distribution.

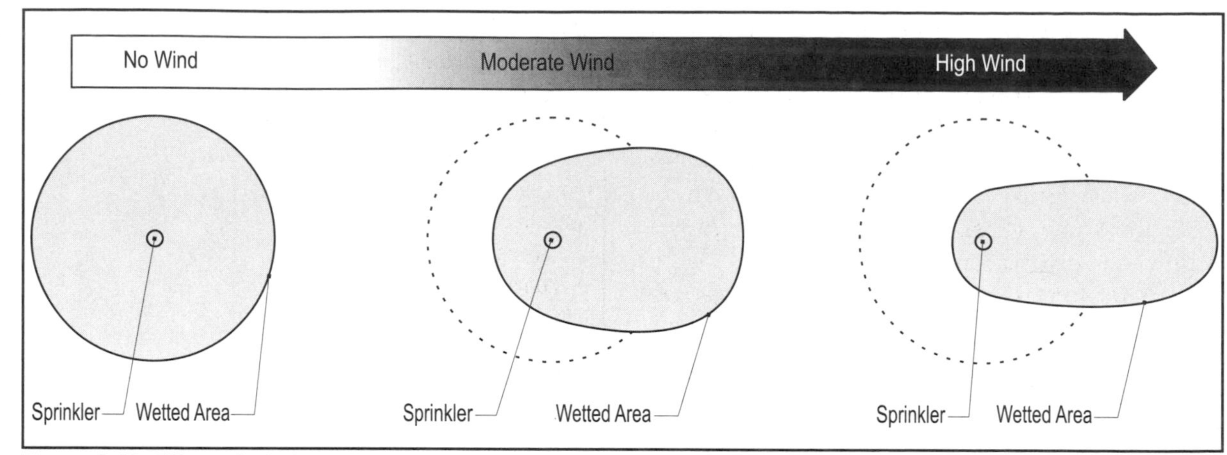

Figure 5-10. Effect of wind on shape and distribution of a sprinkler's wetted area.

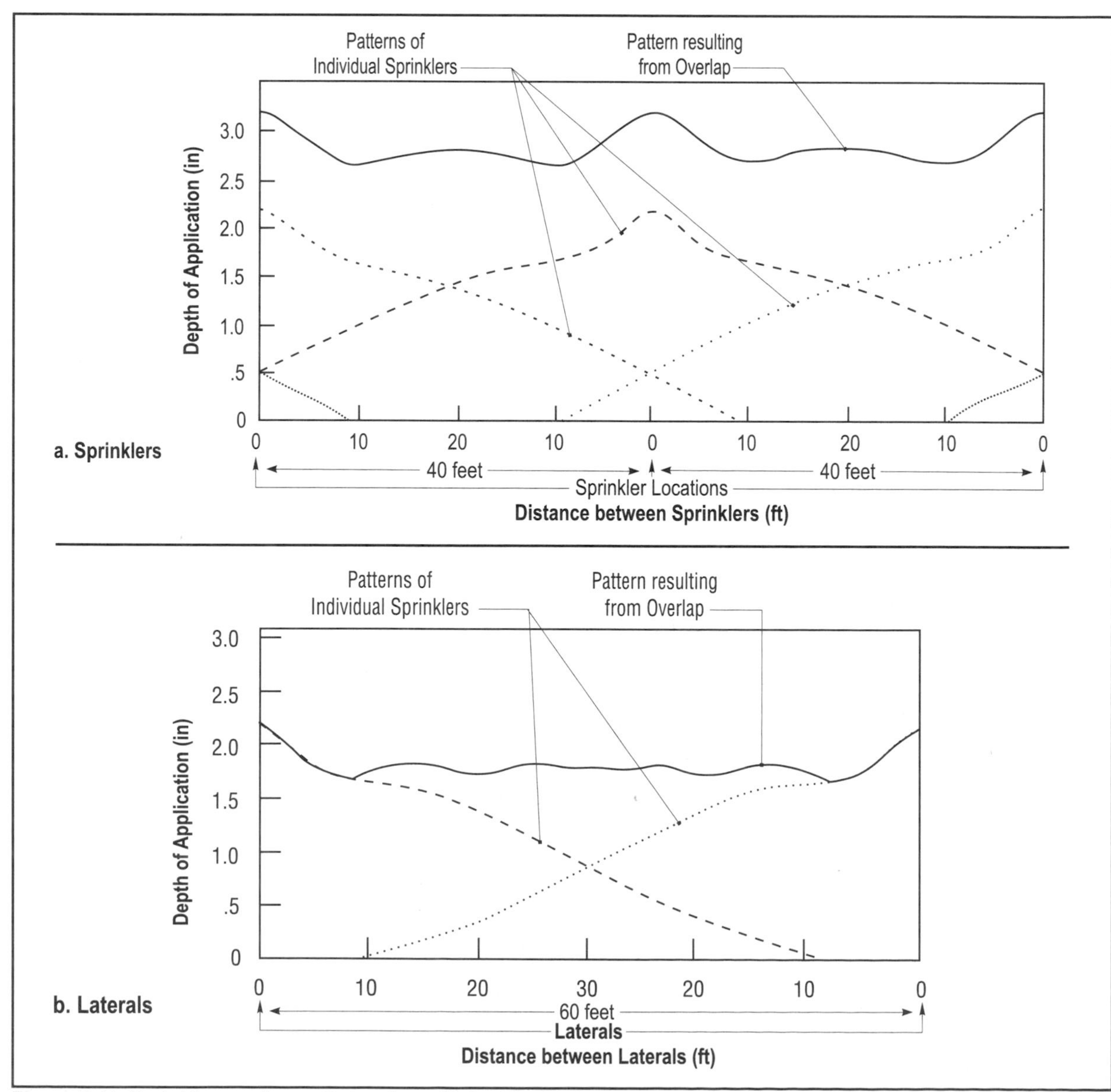

Figure 5-11. Sprinkler overlap patterns for a 40 foot by 60 foot set-type sprinkler system.

along the lateral, and Figure 5-11b illustrates interactions of overlap between laterals. For travelers, only the overlap between travel lanes is important because the gun moves down the lane during a set. Center pivots and linear-move units travel across the field in a near continuous fashion; thus, the design requires overlap along the lateral.

Sprinkler spacing for center pivots generally provides more than 100% of overlap. This, together with the movement around the field, can give high uniformity. Spacing for many of the newer sprinklers, however, is dictated by more than just overlap. For example, the sprinklers have unique sprinkler patterns (see Figure 5-9) that can cause severe non-uniformity if the spacing results in high and low portions of the pattern falling at the same location on the lateral.

The spacing of low pressure spray heads along a center pivot lateral is often best when the spacing is less than 20 to 25% of the wetted diameter. Final selection, however, is determined by the actual pattern distribution for that sprinkler and operating conditions. Figure 5-12 illustrates how improper spacing for a specific sprinkler can have a negative impact on the water distribution. Table 5-5 gives recommended maximum sprinkler spacing on center pivots where the sprinklers are placed close to the top of the crop canopy.

Special cases of spray head spacing for center pivots and linear-move machines exist when in-canopy drop tubes are used. Bubbler systems are designed to operate within a crop canopy to minimize the wetted diameter. Using spray head packages not designed for in-canopy use results in the crop canopy intercepting the water trajectory and at the same time concentrating water application on a smaller area.

Spray head spacing with in-canopy drop tubes spaced greater than twice the row spacing will reduce the application uniformity. Because spray head spacing affects equipment costs, narrow spacing with long drop tubes can add substantially to the sprinkler package cost.

Other factors that affect sprinkler spacing include sprinkler type, nozzle height, and direction of rows. Consult your Cooperative Extension service for recommendations before selecting sprinkler spacing for in-canopy packages.

Table 5-5. Recommended maximum sprinkler spacings for some selected above-canopy sprinklers.

Plate type	Sprinkler pressure (psi)			
	10	15	20	30
Fixed spray plate	6 ft	6 ft	8 ft	10 ft
Rotating plate (slow)	8 ft	10 ft	12 ft	14 ft
Wobbling plate (low angle)	12 ft	14 ft	14 ft	16 ft
Wobbling plate (high angle)	14 ft	16 ft	16 ft	18 ft

Table 5-6. Maximum travel lane spacing for travelers at various wind speeds.

Average wind speed (mph)	Maximum lane spacing (% of wetted diameter)
0 to 5	70 to 80
5 to 10	60 to 65
10 to 15	50 to 55
over 15	45 to 50

Table 5-6 gives the overlap between travel lanes for travelers under various wind conditions. Practical system design should allow extra down time so the system does not have to be operated under very high winds that could reduce uniformity.

For set-type systems, overlapping coverage of the sprinklers in rectangular or triangular configurations gives acceptable distribution of water as long as the correct sprinkler is chosen and then operated properly. A general rule is that overlap for high-pressure impact sprinklers should be 50 to 60% of the no-wind wetted diameter. Stated another way, sprinkler spacing should be about 40 to 50% of the wetted diameter. Under high wind conditions, the spacing should be reduced.

For low pressure sprinklers, overlap depends on the particular sprinkler (Figure 5-9) along with the actual operating conditions. These sprinklers may require more overlap than the 50 to 60% used for impact sprinklers.

Matching Sprinklers to Field Conditions

Sprinkler irrigation must match the water application with field conditions. Consider the following important site factors

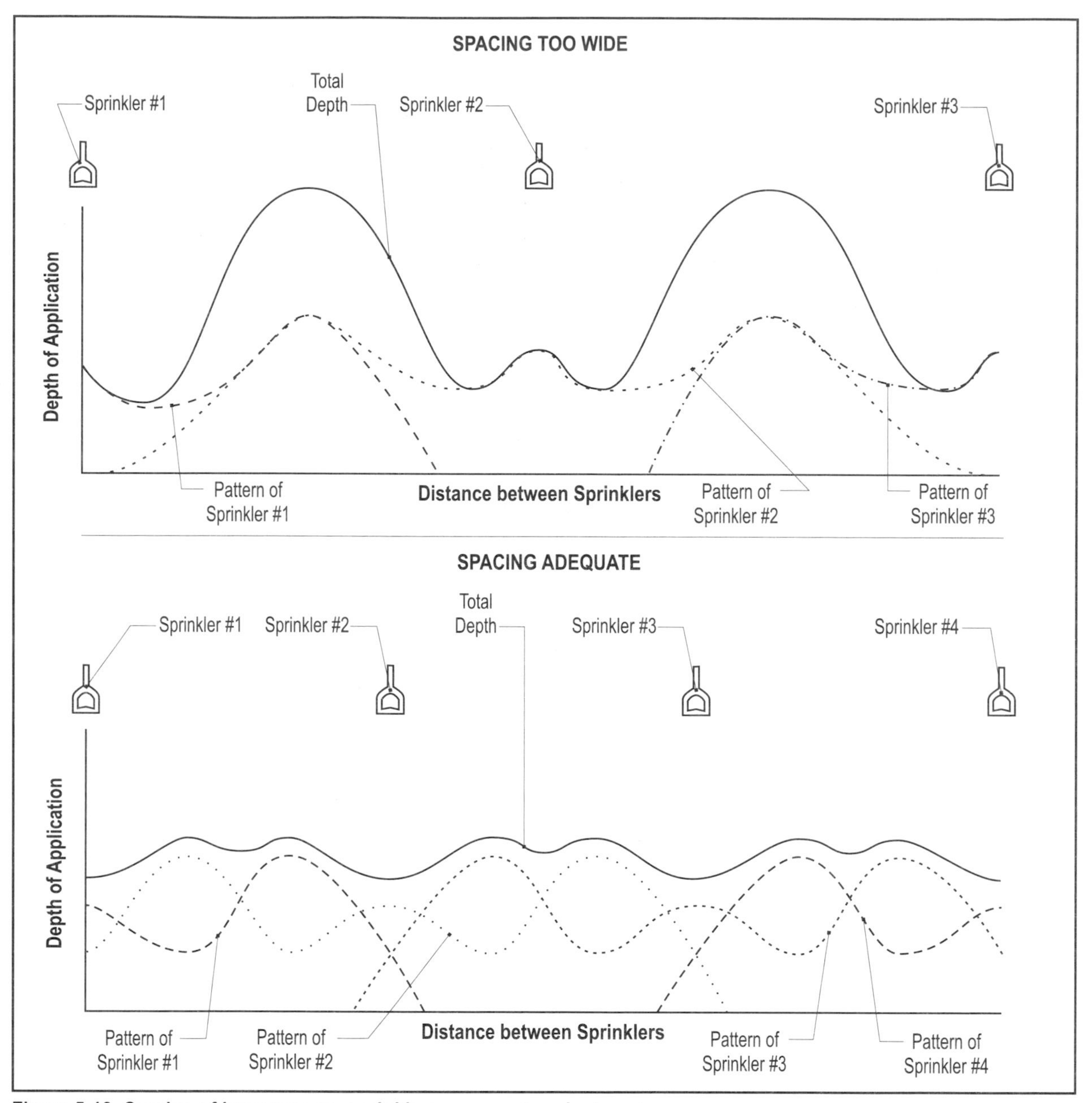

Figure 5-12. Overlap of low pressure sprinklers on a center-pivot system.

when selecting sprinklers:

- Soil infiltration rate and surface storage.
- Field slope.
- Topography (elevation changes across the field).
- Climate and growing season conditions.
- Prevailing wind conditions.
- Crop type and canopy during irrigation.
- Tillage practices.
- Anticipated management practices.

The above factors contribute to the overall application uniformity and efficiency of the irrigation system. Some factors such as soils, topography, and climate cannot be changed while others can be modified. For example, tillage can change surface storage and crop residue coverage. Modifying the design or operation of the system may amplify a limitation due to one or more factors.

Field Slope

Sloping terrain contributes to the potential for runoff. Water application rates must be lower than the infiltration rate on sloping terrain. Water is considered to be runoff if it moves over the soil surface within the field or off the field. Runoff can move soil and/or chemicals from the field and contribute to water quality problems. Water movement within the field can result in erosion and wheel tracking problems. It can result in lack of soil moisture in some areas and an excess in others, resulting in excessive deep percolation and leaching of chemicals. Yield loss and increased pumping costs are additional problems associated with runoff from sloping soils. Table 5-7 gives sprinkler application rate reductions based on slope for use in designing set-type sprinklers.

Table 5-7. Suggested sprinkler application rate reductions based on slope.

Field slope	Application rate reduction
0% to 5%	0%
6% to 8%	20%
9% to 12%	40%
13% to 20%	60%
over 20%	75%

Runoff control must be considered both in the design of the sprinkler system and in subsequent management during operation. What may work for one operation or field may not work well on a neighboring field. It is very important for an operator to routinely monitor conditions in the field and adjust the sprinkler irrigation applications to limit any irrigation runoff.

Topography

Elevation changes in a field can greatly alter the water distribution uniformity of irrigation systems, especially moving systems such as center pivots and travelers. This is particularly true for low pressure sprinkler packages. Elevation is directly proportional to water pressure (Chapter 7). For example, an elevation change of 10 feet equates to a change in pressure of 4.3 psi. Since each sprinkler has an orifice through which water is discharged, altering the pressure supplied to that orifice changes the sprinkler output (see earlier discussion).

Often topography, as depicted in Figure 5-13, is the primary reason for pressure differences in the field. If the field slopes uphill, the sprinklers located at the highest elevation will be distributing less water, and those at the inlet will be distributing more water than the design indicates. The water distribution is inversely proportional to the field elevation. The changes in pressure due to elevation can contribute to a much greater relative change in sprinkler discharge with low pressure than with high pressure.

Table 5-8 illustrates how small changes in elevation can create dramatic changes in sprinkler output for a system that moves across the field, especially at low pressures. Rises of more than 30 feet generally require pressure regulation even for high pressure center pivot systems. Sprinkler output changes of more than 10% across the field should be avoided. A general rule is to require pressure regulators on moving laterals whenever the elevation change in the field is greater than one half of the lateral end pressure in psi. For example, if the desired end pressure on a pivot is 20 psi, then pressure regulators should be used if elevation change in the field is greater than 10 feet.

Because pressure regulators cannot add pressure, sufficient pressure must be delivered to the system to supply operating pressure to the highest elevation in the field. Pressure regulators installed ahead of sprinklers control excess pressure at points of lower elevation. Additionally, there is pressure loss through the regulator itself, which must be added to the total operating pressure for the system. For this reason and because of cost, regulators should be used only when needed.

Differences in elevation also need to be considered in the design of set-type irrigation systems. This is accomplished by choosing the right combinations of lateral layout and sprinkler sizes to achieve uniform application. As with moving systems, set-type systems should have pressure regulators when elevation results in more than a 10% variation in sprinkler flow rate.

System Application Rate

The system application rate is the application rate of the combined sprinklers on the

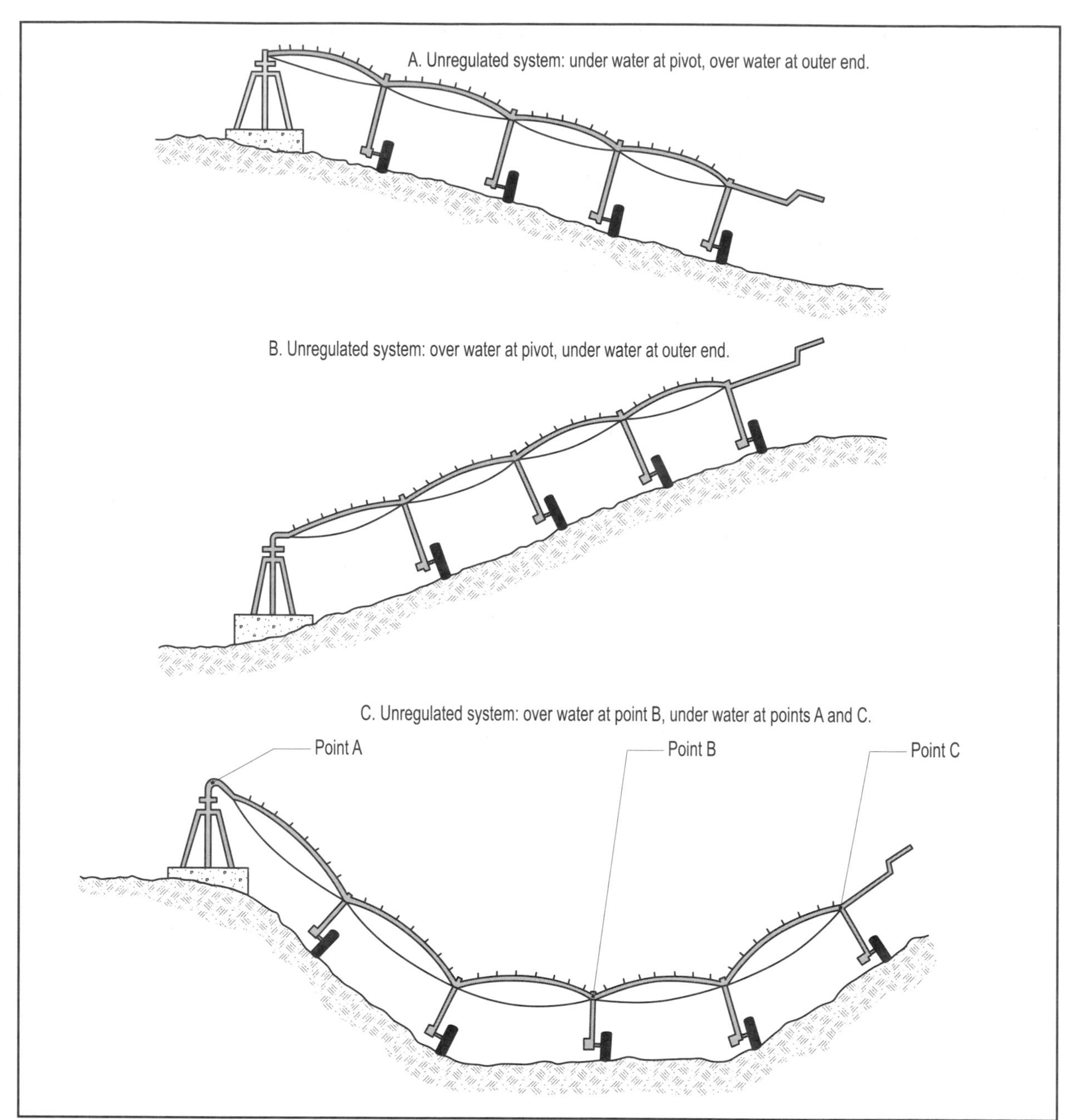

Figure 5-13. Water application characteristics for unregulated center pivots that result from operating on various terrain.

Table 5-8. Reduction in sprinkler output from elevation rise in the field. Shaded areas indicate systems with minimal problems.

Sprinkler pressure (psi)[a]	Field elevation rise (ft)							
	5	10	15	20	25	30	35	40
	Approximate reduction in sprinkler output (% decrease)							
15	7	16	25	35	47	63		
25	4	9	14	19	25	31	37	45
35	3	7	10	13	17	21	25	29
45	2	5	7	10	13	16	18	22
55	2	4	6	8	10	13	15	17
65	2	4	5	7	9	11	12	14
75	1	3	4	6	7	9	11	12

[a] Sprinkler pressure at the end of the lateral.

irrigation system for a given area. Overlap of the individual sprinkler application rates of several sprinklers contributes to the system application rate. The average system application rate is generally used for system design to match field conditions. Equation 5.2 provides an estimate of the average application rate for set-type sprinklers. Equation 5.3 estimates peak application rates for center pivot systems. Equation 5.4 estimates peak application rate for a linear- move system. Equation 5.5 gives the average application rate for a traveler system.

Equation 5-2. Average Application Rate for Set Systems.

$$R_{set} = \frac{(96.3) \cdot Q_{set}}{S_M \cdot S_L}$$

Where:

R_{set} = Average application rate for set systems (inches per hour)
Q_{set} = Sprinkler discharge (gpm)
S_M = Lateral spacing (feet)
S_L = Sprinkler spacing (feet)

Equation 5-3. Peak Application Rate for a Center Pivot.

$$R_{cp} = \frac{(122.6) \cdot Q_{cp}}{L \cdot r}$$

Where:

R_{cp} = Peak application rate for center pivot (inches per hour)
Q_{cp} = Total discharge for center pivot (gpm)
L = Length of pivot lateral (feet)
r = Wetted radius of sprinkler coverage (feet)

Equation 5-4. Peak Application Rate for a Lateral Move.

$$R_{LM} = \frac{(61.3) \cdot Q_{LM}}{L \cdot r}$$

Where:

R_{LM} = Peak application rate for linear move (inches per hour)
Q_{LM} = Total discharge of linear (gpm)
L = Length of linear (feet)
r = Wetted radius of sprinkler coverage (feet)

Equation 5-5. Average Application Rate for a Traveler.

$$R_T = \frac{(122.6) \cdot Q_T}{D^2 \cdot F}$$

Where:

R_T = Average application rate for traveler (inches per hour)
Q_T = Traveler discharge (gpm)
D = Wetted diameter for sprinkler (feet)
F = Fraction of circle wetted

Application rates are significant for the outer sections of center pivots. Figure 5-14 illustrates application rates for typical sprinkler packages on a center pivot. As the center pivot approaches a point in the field, the application rate steadily increases until the system is over the point, then gradually decreases as it moves past. Sprinkler package design should attempt to have the system application rate less than the infiltration rate of the soils irrigated.

Soil Infiltration Rate

Properly designed sprinkler irrigation systems limit the sprinkler application rate to less than the soil infiltration rate, except for in-canopy bubblers (Table 5-9). Soil texture, soil structure, soil compaction, organic matter, surface roughness and sealing, moisture content, and tillage practices affect infiltration rates. Because of these factors, the infiltration rate changes continuously and depends on conditions at the time of water application. The sprinkler system must be designed for the worst-case conditions, not the best case.

Table 5-9. Peak application rates for some typical center-pivot sprinklers.

System	Application rate, (in/hr)	Operation
Impact (high pressure)	1 to 2	Above canopy
Impact (low pressure)	1.5 to 2.5	Above canopy
Rotating spray	2 to 3	Above canopy
Fixed spray	3 to 6	Above canopy
Bubbler	40 to 100	In canopy

Sprinkler package selection should consider the effect of water droplet impact on infiltration rate. The energy of water droplets striking the soil surface can cause crusting or sealing of the surface layer.

In general, droplet size is less important where irrigation occurs mainly during periods when there is full crop canopy coverage. Larger water droplets have more energy than smaller droplets. Sprinklers with smaller droplets often will have a smaller wetted diameter thereby resulting in a higher potential for runoff. Table 5-10 gives maximum application rates for design of stationary or set-type systems. Table 5-11 provides guidelines for reducing application rates when the soil is bare versus covered.

Application Uniformity

Water application uniformity is a measure of the evenness of water distribution over the entire irrigated area. Irrigation systems should apply the water uniformly in sufficient quantities to meet the crop water needs without over watering or generating runoff.

In a practical sense, achieving perfect uniformity is not possible. However, the goal of a well designed and operated irrigation system is to apply the water as uniformly as possible. Table 5-12 gives some typical uniformities for irrigation systems.

To apply water uniformly requires that the correct sprinklers/nozzles be installed at the correct spacing along the lateral relative to the crop canopy, that the pumping plant

Table 5-10. Maximum application rates for design of stationary or set-type systems (inches per hour).

	Application amount (in)						
Soil texture	0.25	0.5	0.75	1.0	1.25	1.5	2.0
Sand	6.0	6.0	6.0	6.0	6.0	6.0	6.0
Loamy sand	6.0	6.0	4.8	4.2	3.9	3.6	3.3
Sandy loam	4.9	3.0	2.3	2.0	1.8	1.7	1.5
Loam	3.1	1.7	1.2	1.0	0.8	0.7	0.6
Silt loam	2.7	1.5	1.0	0.8	0.7	0.6	0.5
Sandy clay loam	1.7	1.0	0.7	0.6	0.5	0.4	0.4
Clay loam	1.3	0.7	0.5	0.4	0.3	0.3	0.2
Silty clay loam	1.1	0.6	0.4	0.3	0.3	0.2	0.2
Sandy clay	0.6	0.3	0.2	0.2	0.2	0.1	0.1
Silty clay	0.8	0.4	0.3	0.2	0.2	0.2	0.1
Clay	0.4	0.2	0.1	0.1	0.1	0.1	0.1

Note: This table is for infiltration rate for full cover conditions and initial moisture content at 50% of the available water capacity. Field capacity of sand through sandy loam is assumed to be at 1/10 bar.

Table 5-11. Recommended water application rates for bare and covered soil for stationary or set-move systems (inches per hour).

Soil characteristics	Covered	Bare
Clay; very poorly drained	0.3	0.15
Silty surface; poorly drained, clay and claypan subsoil	0.4	0.25
Medium textured surface soil; moderate to imperfectly drained profile	0.5	0.30
Silt loam, loam, and very sandy loam, well to moderately well drained	0.6	0.40
Loamy sand, sandy loam, or peat; well drained	0.9	0.60

Table 5-12. Ranges of typical application uniformities for sprinkler irrigation.

Type of irrigation	Uniformity
Center pivot	0.85 to 0.92
Linear-move	0.85 to 0.95
Traveling gun	0.75 to 0.85
Side roll	0.70 to 0.80
Solid-set	0.70 to 0.90
Set-move	0.70 to 0.80

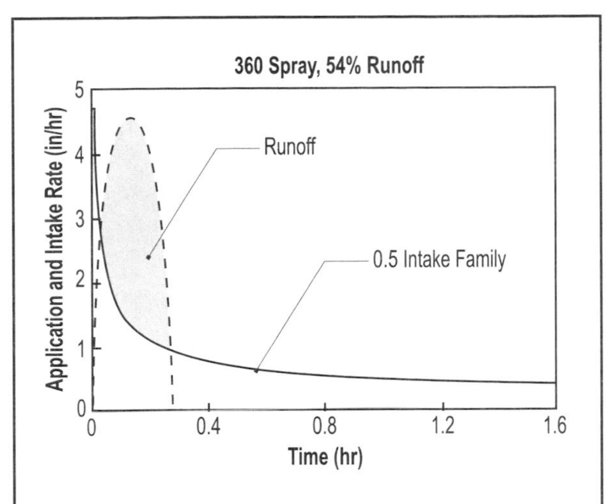

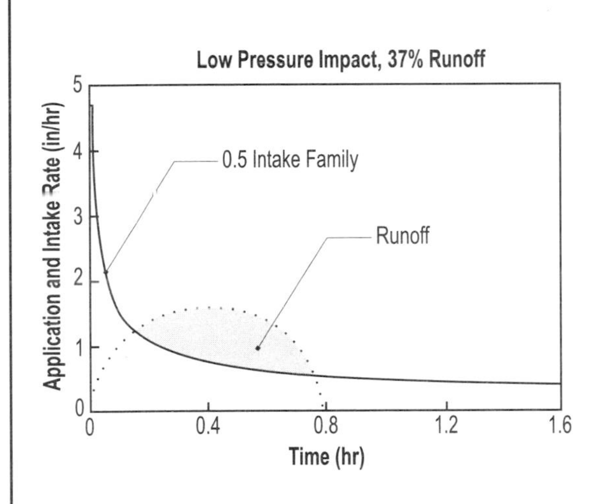

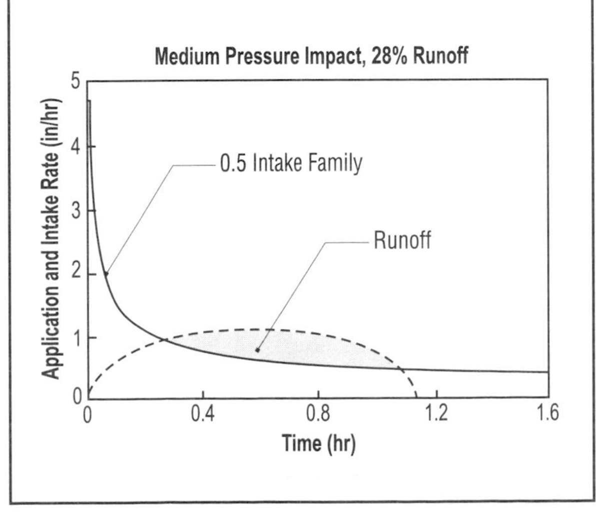

Figure 5-14. Potential runoff for a silt loam soil without surface storage receiving 1.1 inch water application from a center pivot.

deliver water at the appropriate pressure and flow rate, and that the system is not operated under adverse weather conditions. Another aspect of water application uniformity is the uniformity of infiltration. Water applied uniformly can be overshadowed by surface runoff problems. Thus, the system designer must consider how the sprinkler package will match the field conditions.

Factors that govern uniformity include the sprinkler distribution pattern, local weather conditions, and how the sprinkler system is operated. As discussed earlier, sprinkler water distribution is influenced by sprinkler type, nozzle configuration, operating pressure, trajectory angle, mounting configuration, and overlap.

Uniformity of water application generally increases with a decrease in sprinkler spacing, assuming that the operating conditions of the sprinkler do not change. Narrowing the spacing results in more overlap of the water application patterns of individual sprinklers. A narrow spacing makes it more difficult for wind to alter the overall water application uniformity of the system. Some sprinkler devices, however, have individual application patterns that may influence the choices of sprinkler spacing (see Figures 5-9 and 5-12). For example, if the spacing caused the peaks in the individual sprinkler patterns to overlap, uniformity would be compromised.

Spray heads placed in the canopy need to be spaced close together (Table 5-5). Figure 5-15 illustrates inadequate irrigation when spray heads were spaced 12.5 feet apart. In some cases where corn is tall, spray heads close to the ground may perform better than those higher in the canopy.

Uniform water application may not mean the irrigation system is highly efficient because water can be uniformly overapplied. High irrigation efficiency, however, requires uniform water application over the entire application area. Nonuniformity that contributes to over-irrigation of all or parts of the field is undesirable.

Coefficient of uniformity (CU) is a term used to describe how uniformly water is applied over a given area. It is determined

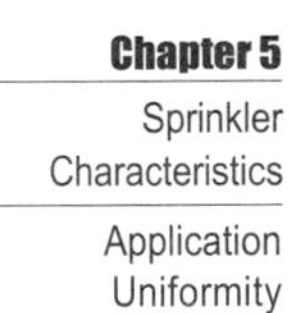

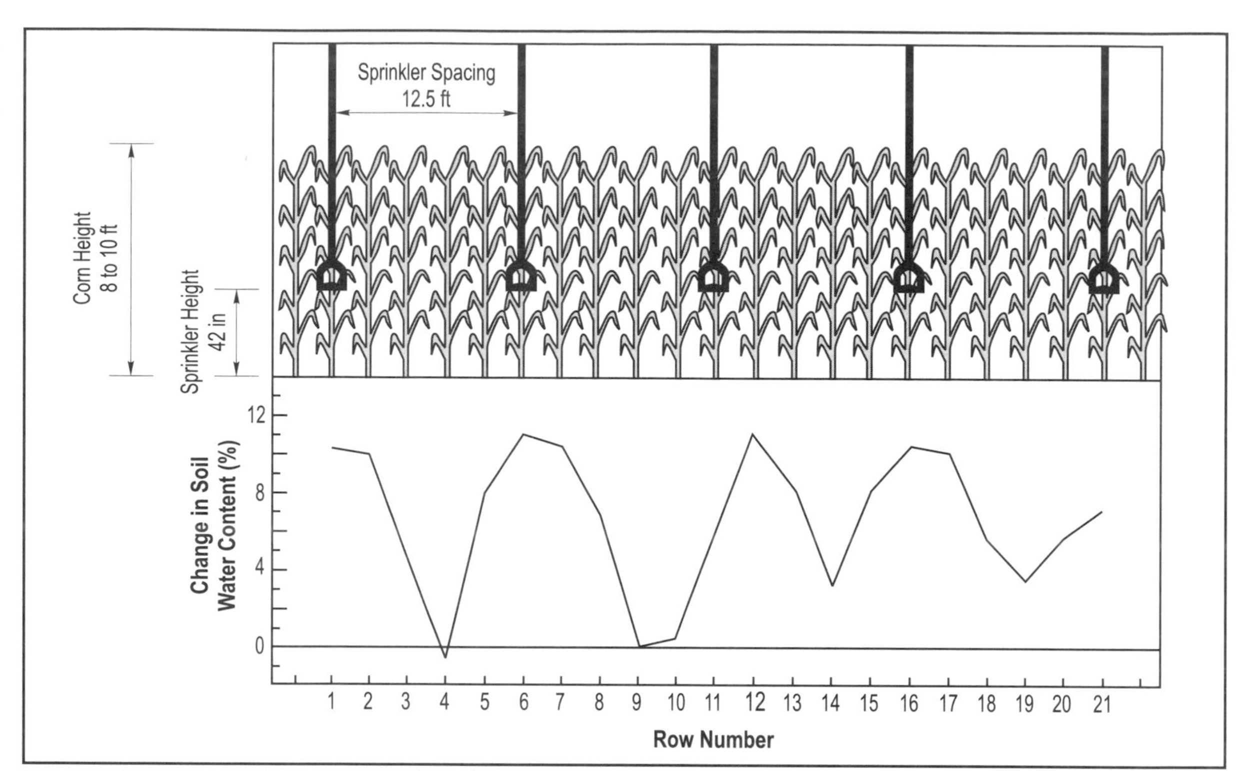

Figure 5-15. Example of change in soil moisture content after irrigation with fast-rotating spray heads. The heads are 42 inches above the ground surface and 12.5 feet apart.

by calculating the relative deviation of average water applied versus the average application depth. Distribution uniformity (DU) is another common term used to describe water application uniformity where the low quarter of the water applied is compared to the average water applied. Testing of water application uniformity is discussed in Chapter 6.

Application Efficiency

Application efficiency is the fraction of water that is stored in the soil and available for use by the crop (net irrigation) versus the total water applied (gross irrigation). Application efficiency results from a combination of the system characteristics, the operating conditions such as weather, and how the system is managed. Figure 5-16 is a simplified diagram of sprinkler irrigation application efficiency. Losses subtract water that is applied from the total water that the plant could use. High efficiency application strives to minimize all the losses.

Efficient irrigation applies water to the soil/crop system with minimum losses. The concept of efficiency is very complex and difficult to quantify even with rigorous research setups. Often it is oversimplified either due to lack of knowledge or to promote a certain concept or product. Fallacies relative to sprinkler losses abound. For example, air evaporation losses are often grossly overestimated while the maximum of those losses seldom exceeds 3 to 5%. Because of the interactions between losses, minimizing one loss or portion of a loss may contribute to increased losses elsewhere. For example, reducing air evaporation or wind drift loss by placing the sprinkler in the crop canopy may result in increased application rates and increased runoff while reducing only one component of the losses.

Because local weather conditions vary daily and during each day, the total of all components of water losses during sprinkler irrigation can be in the range of 3 to 60% of the water applied. Obviously, few fields experience the maximum level of loss for all factors. Irrigating during conditions of low air temperature, high humidity, low wind velocity, and under cloudy skies would maximize the amount of water reaching the soil surface.

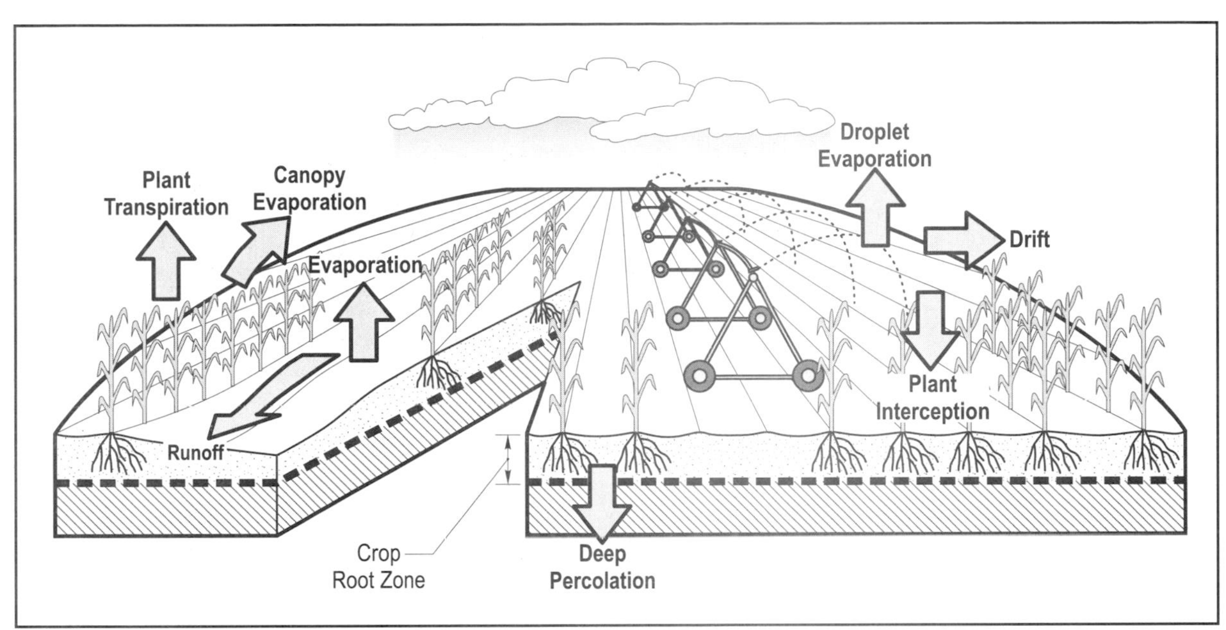

Figure 5-16. Sprinkler irrigation water balance.

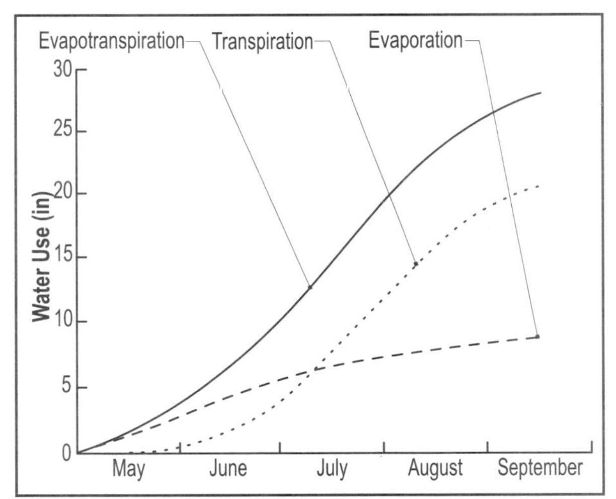

Figure 5-17. Relative portions of evaporation and transpiration comprised in ET.

Properly matched sprinkler packages to the soils would limit runoff losses. Crop residue management or limiting the area to which water is applied, with bubblers for example, may minimize canopy evaporation losses.

Evapotranspiration (ET) is the combination of transpiration from the crop and evaporation of the water from the soil or plant surfaces. Figure 5-17 shows the relative portions of evaporation and transpiration during the growing season for row crops.

The potential water losses during water application from sprinkler irrigation fit into three general categories—soil, air, and canopy losses. These combine to reduce the amount of water available for plant use. Soil losses can be further divided into losses due to runoff, and deep percolation. Air losses include both droplet evaporation and wind drift. Canopy losses are canopy evaporation and plant interception. Table 5-13 illustrates some sprinkler water losses for various sprinkler packages with center pivots.

Air losses refer to the water that evaporates between the sprinkler head and the soil or plant canopy surface. Air losses depend on local weather conditions. In-air evaporation is

Table 5-13. Example sprinkler water losses for center pivots based on 1-inch application depth.

Water loss component	Impact sprinkler water loss	In-canopy spray head loss	LEPA water loss
Air evaporation and drift	0.03 in	0.01 in	0.00 in
Net canopy evaporation	0.08 in	0.03 in	0.00 in
Plant interception	0.04 in	0.04 in	0.00 in
Evaporation from soil	negligible	negligible	0.02 in
Total water loss	0.15 in	0.08 in	0.02 in
Application efficiency	85%	92%	98%

typically in the 1 to 3% range, even though it is often perceived as higher. Evaporation is in direct relation to the water temperature and the surface area of the water droplets. Small water droplets represent greater surface area and hence, greater air evaporation loss. To reduce air losses, the water should spend little time in the air. This can be accomplished by directing the water stream downward rather than up into the air; however, this decreases wetted diameter. Smaller wetted diameters result in higher water application rates creating higher potential for runoff.

Wind drift often is a much greater problem than is evaporation. High trajectory angles and small droplets combined with high winds may move water away from the intended irrigated area. In some cases, wind drift can move a considerable amount of water out of the area.

Canopy losses are direct evaporation of water that is intercepted by the plant foliage. Canopy evaporation losses can range from 0 to 10% depending on the local weather conditions. Canopy evaporation cools the plant and reduces transpiration. However, current theory is that canopy evaporation exceeds the reduction in transpiration, thus, the net difference is toward the evaporation loss side.

Canopy evaporation occurs for the length of time water is on the leaf surface. Large wetted diameters give longer irrigation times. For example, a high pressure impact sprinkler would apply 1 inch in about 80 minutes, while in-canopy sprays would apply 1 inch in less than 8 minutes. High pressure impact sprinkler packages have the greatest canopy losses. In-canopy packages irrigate for short time periods due to plant interception of the spray pattern. However, they normally wet the entire canopy except when the drop tubes hang well into the canopy.

Runoff loss is the water that reaches the soil but does not infiltrate where it falls. Runoff losses could range from 0 to 60% or more of the water applied. Runoff water either is redistributed to other portions of the field (usually low-lying areas) or leaves the field boundaries. The amount of runoff loss depends on the match of the sprinkler package with the soil and slope conditions. Sprinkler packages with high application rates, matched with soils with steep slopes and low infiltration rates will produce maximum runoff if soil surface storage is not provided. Conversely, low water application rates, low slopes, and high infiltration rates produce little runoff.

Deep percolation loss is water infiltrated in excess of the soil water-holding capacity. It is water that passes through the soil profile and does not contribute to plant growth. Deep percolation losses can be minimized with sprinkler irrigation systems. Water should not be applied unless the soil is able to hold it. If this strategy is followed, deep percolation will occur only after an excessive rainfall or one that follows the irrigation event. The way to minimize the potential for deep percolation loss is to reserve a portion of the soil's water-holding capacity for rainfall; this is called deficit irrigation.

An important loss that may be overlooked is conveyance loss. Although conveyance losses do not contribute directly to sprinkler inefficiency, they can be significant when compared to other losses. Conveyance losses include leaks from pipelines, valves, boots, low pressure drains, and gaskets that do not seal under pressure.

Field research with low pressure sprinklers has shown mixed results relative to improving water use efficiencies. Several studies have concluded that sprinkler packages must be designed to fit the soil and environmental conditions. Low Energy Precision Application (LEPA) systems may show very high application efficiency but result in low uniformity if not designed to contain runoff. Other spray head packages placed in or under the crop canopy have resulted in water runoff problems with center pivots when not used as designed. Runoff within the field may contribute to localized deep percolation.

Chapter 6

Sprinkler Selection and Management

Chapter 6 Contents

The ultimate goal when selecting the sprinkler type to be used on an irrigation system is to achieve efficient and uniform water application. However, obtaining the desired performance from different sprinkler types depends on more than just selecting a sprinkler. When used properly, most sprinklers will provide uniform application of water. Using these same sprinklers, however, in the wrong situation can result in poor water distribution, runoff, or reduced efficiency.

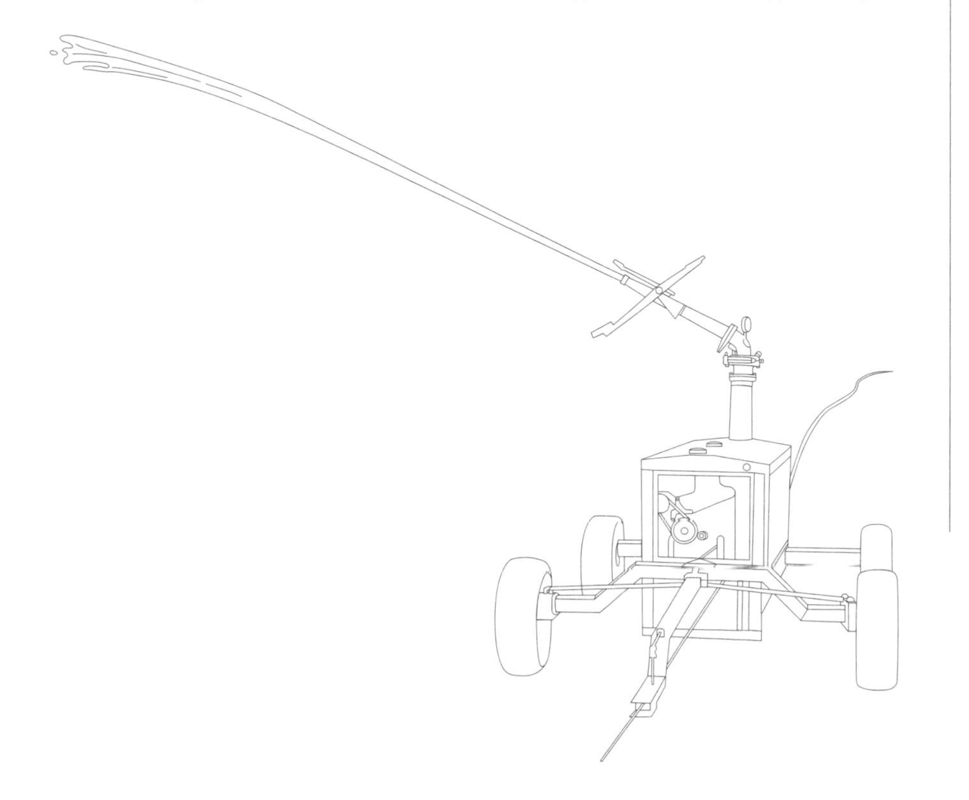

Sprinkler Selection

A wide variety of irrigation sprinkler types is available for agricultural use, but the type of irrigation system to be installed governs the choice of sprinklers for a particular application. For example, a traveler would use only a volume gun sprinkler.

Figure 5-1 (page 67) illustrates some of the sprinkler types. Figures 5-1 and 5-2 (page 67 and 68) also show various sprinkler options and how they may affect sprinkler patterns. Table 6-1 lists general sprinkler types and their suitability for various irrigation systems. Table 6-2 provides a summary of sprinkler characteristics that influence selection for each application.

Many of the characteristics are interrelated, thus choosing a sprinkler with a desirable trait in one aspect can result in an undesirable trait that may counteract the expected outcome. For example, a low-angle sprinkler will have less wind drift, but at the same time will have a smaller wetted diameter, thus a higher application rate that could result in runoff.

Table 6-2. Sprinkler design characteristics that influence selection.

Design characteristic	Result
Lower pressure	Saves energy
Lower pressure	Smaller wetted diameter
Lower pressure	Larger droplets
Lower angle	Smaller wetted diameter
Lower angle	Less wind drift
Larger nozzle	Higher flow rate
Larger nozzle	Larger droplets
Larger nozzle	Less plugging
Smaller wetted diameter	Higher application rate
Larger wetted diameter	Lower application rate
Higher flow rate	Higher application rate
Larger droplets	Higher potential surface sealing

Center Pivot and Linear Move Selection

Center pivots are the most common agricultural sprinkler system available today.

Table 6-3 gives sprinkler technology adaptability for center pivots with general suitability and constraints for various sprinkler packages. It is important to note that specific applications can amplify constraints or enhance suitability based on individual characteristics at a given operating pressure, including wetted diameter, drop size distribution, and application pattern. Many of the alternatives also require specific system approaches involving application of the hardware and subsequent management.

Sprinkler orientation options have increased dramatically in recent years. Not only is there a choice of spacing of the sprinkler along the center pivot lateral, but also the sprinklers can be mounted on top of or below the lateral in a variety of configurations. Traditionally, sprinklers were mounted on

Table 6-1. General sprinkler types used for agricultural irrigation systems.

Type	Sprinkler pressure (psi[a])	Orientation (with respect to vertical)	Suitability for systems
Spray			
Fixed	10 to 40	Both up and down	Center pivot, linear move
Rotating	15 to 45	Both up and down	Center pivot, linear move, some set-type
Bubbler	5 to 20	Down	Center pivot, linear move
Impact			
Circular nozzle	40 to 80	Up	Center pivot, linear move, set-type
Nonstandard nozzle	25 to 50	Up	Center pivot, linear move, some set-type
Volume gun	40 to 100	Up	Traveler, set-type, some center pivots, end guns
Gear-driven	25 to 80	Up	Set-type, center pivot, linear move

[a]Pressure listed is for the sprinkler device. Operating pressure for the system must be higher to compensate for friction lossesand elevation differences within the field.

Table 6-3. Various sprinkler packages and wetted diameters for center pivots.

Sprinkler system	Best suited to	Recommended constraints
Bubbler and in-canopy sprays	Converting from surface irrigation Low capacity water supply High loss environments High energy cost areas	Level fields, less than 1% slope Narrow spacing, less than 6 ft Furrow diking Farming in a circle Low application depths
Above-canopy sprays	Converting from high pressure High energy cost areas	Low slopes, less than 5% Narrow to medium spacing, less than 20 ft Tillage system benefit Low application depths
Low pressure impacts, less than 40 psi at sprinkler	Converting from high pressure	Moderate to low slopes, less than 10% Narrow to medium spacing, less than 25 ft Tillage system benefit
Medium to high pressure impacts, greater than 40 psi at sprinkler	Low evaporation loss environment Low seasonal application Moderate application depths Low energy cost areas	Moderate slopes, less than 15% Wide spacing, less than 40 ft Best to operate during low wind conditions

top of the center-pivot laterals. Over the years, a greater share of the installations have used drop tubes (drops) with low pressure spray heads placed below the truss, just above the crop canopy, and at various levels into the canopy. Figure 6-1 is an illustration of mounting heights of sprinklers on a center pivot.

The bubbler mode package often is referred to as LEPA (Low Energy Precision Application). LEPA is the practice of placing the bubbler sprinkler near the soil surface (less than 24 inches) and in many cases, using a drag sock to drop the water onto the soil surface.

With in-canopy sprays, sprays other than bubblers are placed within the crop canopy. In-canopy sprays have been referred to as LESA (Low Elevation Spray Application) and MESA (Mid Elevation Spray Application).

While applications in the crop canopy can reduce atmospheric losses, in-crop canopy application patterns are distorted, and runoff/water movement within the field potentially can be greater because of the higher application rates from reduced wetted diameters. For these reasons, in-canopy operation of sprinklers is not recommended.

Figure 6-2 is a chart showing the normal ranges of operating pressures for center-pivot systems for various sprinkler packages. The operating pressure shown is pivot point pressure assuming about 10 psi pressure loss for the pivot lateral. In situations with higher pressure losses because of longer than normal laterals or smaller pipe size, adjust the operating pressure ranges upward on the chart. Adjust the pivot pressure by adding friction loss in the lateral (see Chapter 7) to the recommended sprinkler pressure.

Most of the center-pivot sprinkler packages are suitable for installation on linear-move systems. Linear-moves, however, have constant sprinkler discharge along the lateral, while discharge from sprinklers on center pivots increases from the pivot point to the outer end.

High Pressure Systems

High pressure systems (55 to 80 psi) typically are equipped with impact sprinklers or volume guns mounted on top of the lateral pipe. End gun booster pumps seldom are needed with high pressure systems. Few new high pressure systems are installed today because of the desire to reduce pumping costs and the availability of lower pressure alternatives.

In general, high pressure sprinkler

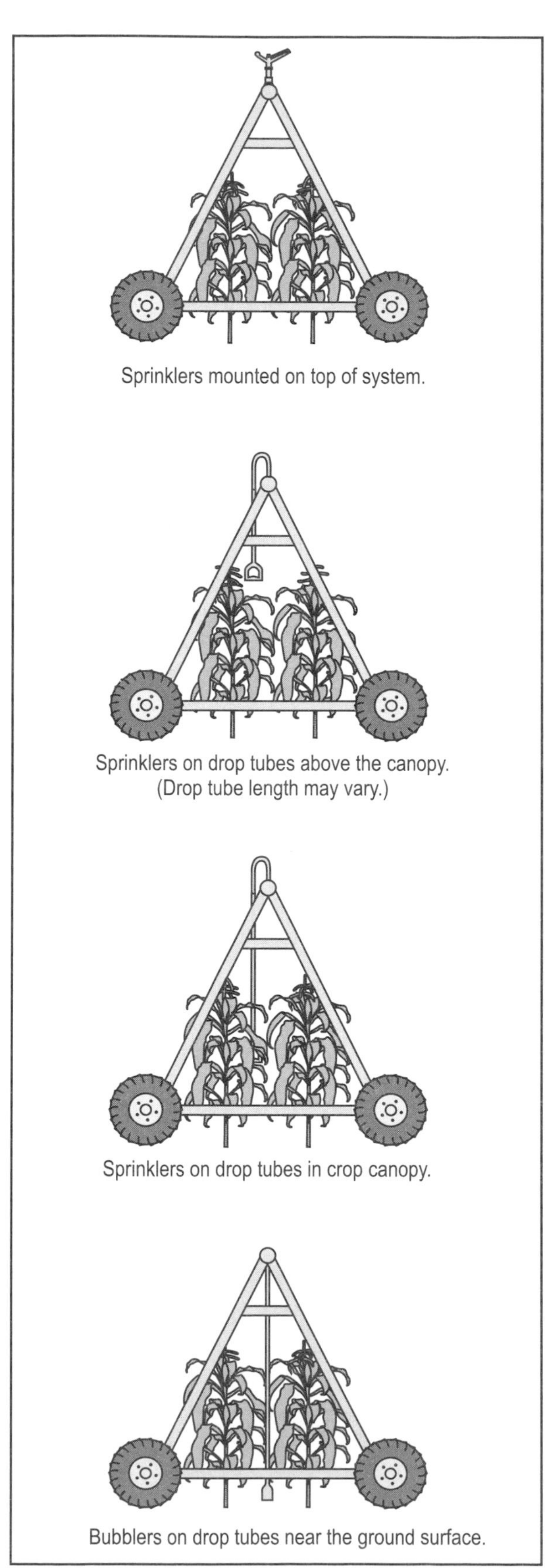

Figure 6-1. Mounting heights of sprinklers on a center pivot.

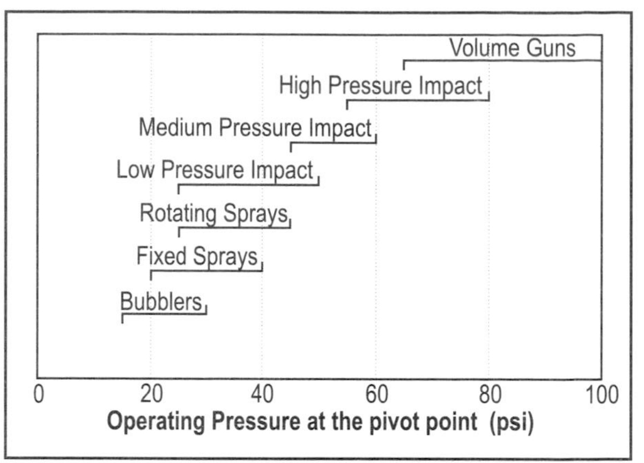

Figure 6-2. Normal ranges of operating pressures at the pivot point for center-pivot systems.

packages are better suited to fields that have steep slopes, up to 15%, and soils with lower infiltration rates. These sprinkler packages have large wetted diameters that result in low water system application rates. However, the water stream from an individual sprinkler is concentrated where it strikes the soil surface. At those points, the sprinkler delivers water to the soil at rates that may be greater than the soil infiltration rate for some soil textures.

These two factors combine to break down soil clods and move small clay particles into the spaces between larger sand and silt particles. Soil textures with a high proportion of clay at the surface are susceptible to surface crusting that can greatly reduce the soil infiltration rate.

Medium Pressure Systems

Medium pressure systems (45 to 60 psi) can have either medium pressure impact or spray sprinklers, often mounted on top of the lateral pipe. Spray heads should be chosen with coarse grooves to minimize small water droplets. It may be necessary to use pressure regulators on sprinklers near the pivot point to prevent sprinkler pressures that are above the recommended ranges. An end gun may require a booster pump.

Low Pressure Systems

Low pressure systems (25 to 50 psi) are equipped with spray heads or low pressure impact sprinklers. In general, low pressure systems are restricted to fields with low slopes and high infiltration rate soils. Sprays distribute the water from the nozzle using a spray plate to break the stream into smaller streams. These streams usually have smaller water droplets and a lower amount of energy striking the soil surface, thus reducing the opportunity for surface sealing.

Impact sprinklers are mounted on top of the lateral, while low pressure sprays may be installed in a variety of positions relative to the pivot lateral. Common placement includes on top of the lateral, on drop tubes below the trusses, and on drop tubes to some distance above the soil surface within the crop canopy. Each change in mounting position can result in a different water application pattern, a different water application efficiency, and a different potential for surface runoff. Booster pumps are generally required when end guns are used.

Low pressure sprinkler packages have a narrower wetted diameter compared to high pressure sprinklers. Wetted diameters are even smaller when sprays are operated within a crop canopy (in-canopy spray) because the water pattern will be intercepted by the stem and leaves of the plant. Overall application efficiencies of in-canopy sprays are not significantly different from above-canopy packages, thus it is recommended that spray packages be mounted above the level of the tallest crop canopy.

The bubbler system (LEPA) can achieve water application efficiencies of more than 95% due to low evaporation losses from wetting only a portion of the soil surface. The extremely narrow water application pattern (less than 2 feet) of the bubbler system results in a water application rate that is well above the infiltration rate of all soils. Thus, runoff will occur unless soil surface storage such as furrow diking is provided.

Another factor that may be overlooked in sprinkler selection is the effect of the start-stop cycle of the drive units of many center pivots. This factor will have little influence on uniformity of a medium or high pressure sprinkler package (wetted diameters greater than 50 feet). However, with some low pressure spray heads and all in-canopy packages, water application uniformity can be reduced with high speed settings. This is because the tower travels 5 to 15 feet per minute while moving, then stands stationary for 10 to 50 seconds of each minute.

Pressure Regulators

The flow rate leaving a sprinkler nozzle is determined by the pressure supplied to the base of the device. Pressure at a sprinkler is controlled by the pressure at the pivot point, the friction loss in the pivot pipeline, and the elevation changes across the field. Thus, although the drive units on center pivots are well suited to rolling terrain, differences in elevation along the pivot lateral will change the amount of water delivered by each sprinkler. For uniform water application, the pressure supplied to each nozzle or sprinkler should remain nearly constant throughout the entire irrigation cycle.

Pressure regulators or flow control nozzles can help ensure uniform water application if changes in elevation are large and should be considered for low pressure systems (see Chapter 5). In general, pressure regulating devices are justified if differences in elevation result in a flow variation greater than 10% among sprinklers or nozzles.

LEPA and In-canopy Sprays

LEPA (Low Energy Precision Application) Figure 6-3, is a sprinkler and cropping system used on center pivots and linear-moves. It involves discharging water from bubbler nozzles on drop tubes placed near the soil

Figure 6-3. Low Energy Precision Application (LEPA) system.

surface with a cropping system that includes crop rows parallel with the direction of travel and reservoir tillage (furrow diking). This would mean farming in a circle with the center pivots. LEPA has a very low pressure requirement.

LEPA was developed as an alternative to flood irrigation in locales where water supplies were becoming scarce. LEPA is best suited when converting from flood irrigation, with low capacity water supplies, and where high application losses result from wind and evaporation. LEPA should be selected for fields with less than 1% slope. Where the prescribed cropping system is used, high application efficiencies are possible (Table 5-13, page 85).

In-canopy sprays should not be confused with LEPA even though they may appear similar. In-canopy spray systems use spray heads mounted on drops located within the crop canopy during a majority of the cropping season. They could be located from 2 feet above the ground to near the top of the canopy. Like LEPA, practices for in-canopy systems include farming with rows parallel to the direction of travel and furrow diking.

Because of the widely varying applications of in-canopy sprays, application efficiencies and uniformities can vary considerably. Placing the nozzles near the soil surface may reduce canopy losses, but will increase the application rates and create higher runoff potential. In-canopy sprays near the top of the canopy may have little advantage over above-canopy sprinklers except for reduction of air losses (but the wetted diameter also is reduced).

In-canopy sprinklers will have small improvements over above-canopy sprinklers unless the rows run the same way as the direction of travel because sprinklers and drops will drag over the canopy as the system moves across the field. Sprinkler spacing also must be closer (see Table 5-5 and Figure 5-12).

Converting Center Pivot Sprinkler Packages

Converting to a new sprinkler package from an older center pivot may offer one or more advantages, including using new sprinkler technology, replacing a poor design on the original package, reducing energy requirements, or simply replacing worn out sprinklers.

Most new sprinkler packages can offer reduced pressure and, therefore, lower power requirements. When changing to a new sprinkler package, consider equipment modifications, economic costs and benefits, and any changes in management needed to make the new system work properly. Make sure the new sprinkler package conforms to the goals and needs of the overall irrigation system (see Chapter 2). It would be unwise to simply select a new sprinkler package that merely saves energy if it is not suited to the field or management scheme.

When thinking about new equipment, consider the following factors. Select a sprinkler that is suited to the field, crop, and operating conditions. New holes may need to be cut in the lateral pipeline to accommodate sprinklers if the old spacings do not match the new package. In addition, drop tubes and pressure regulators may be desired. An end-gun booster pump may be needed with the typical lower pressure packages. The main pumping plant supplying pressure for the system must be modified. Simply using a throttling valve at the pump to achieve lower pressure at the system is not acceptable.

All costs for the new sprinkler package must be weighed against potential energy savings. Economic feasibility is normally considered on a time frame of three to five years for system upgrades. Additional considerations must include the costs of any potential negative impacts such as reduced yields from non-uniformity or loss of anticipated savings due to system inefficiency.

Traveler Sprinkler Selection

Travelers have a single sprinkler mounted on a cart that travels along lanes spaced across the field. A volume gun sprinkler with adequate capacity is selected to meet the design capacity for the field irrigated (Chapter 2 and Table 5-2 in Chapter 5). Traveler system design is covered in more detail in Chapter 4. Options may include part-circle and slow-reverse features. Most other options will relate to the field layout and machine features such as the soft hose versus hard hose.

Sprinkler selection for a traveler is straightforward because only one sprinkler is used. Different manufacturers rate their traveler machines for various capacities, but they generally fall within the range given in Table 6-4. Using gun sprinklers with high flow rates may result in excessive friction loss in the hose, which must be overcome by the pump (Tables 4-4 and 4-5). The design also must specify the lane spacing required to provide adequate sprinkler pattern overlap under the prevailing wind conditions (Table 5-6).

The desired travel speed depends on the sprinkler, lane spacing, and depth of water needed. Use Table 6-5 to determine the travel speed for different sprinkler flow rates, lane spacing, and application depths.

All volume gun sprinklers must be operated within the correct range of operating pressures based on the manufacturer's recommendations. Table 5-2 gives sprinkler discharges at various pressures for gun-type sprinklers. Droplet sizes from volume gun

Table 6-4. Typical volume gun performance for travelers.

System capacity (gpm)	Operating pressure (psi)	Wetted diameter (ft)	Full circle application rate (in/hr)
50 to 200	50 to 80	200 to 330	0.18 to 0.26
120 to 500	70 to 100	275 to 440	0.22 to 0.36
300 to 1,000	80 to 110	375 to 580	0.30 to 0.42

Table 6-5. Average applied water depth by traveling sprinklers.

Discharge rate (gpm)	Lane Spacing (ft)	Travel speed (in/min)						
		6	12	18	24	36	48	60
50	105	1.50 in	0.80 in	0.50 in	0.40 in			
	155	1.00 in	0.50 in	0.30 in	0.30 in			
100	130	2.50 in	1.20 in	0.80 in	0.60 in			
	190	1.70 in	0.80 in	0.60 in	0.40 in			
150	145	3.30 in	1.70 in	1.10 in	0.80 in	0.60 in	0.40 in	0.30 in
	215	2.20 in	1.10 in	0.70 in	0.60 in	0.40 in	0.30 in	0.20 in
200	160		2.00 in	1.30 in	1.00 in	0.70 in	0.50 in	0.40 in
	235		1.40 in	0.90 in	0.70 in	0.50 in	0.30 in	0.30 in
250	175		2.30 in	1.50 in	1.10 in	0.80 in	0.60 in	0.50 in
	260		1.50 in	1.00 in	0.80 in	0.50 in	0.40 in	0.30 in
300	180		2.70 in	1.80 in	1.30 in	0.90 in	0.70 in	0.50 in
	265		1.80 in	1.20 in	0.90 in	0.60 in	0.50 in	0.40 in
400	200		3.20 in	2.10 in	1.60 in	1.10 in	0.80 in	0.60 in
	300		2.10 in	1.40 in	1.10 in	0.70 in	0.50 in	0.40 in
500	225		3.60 in	2.40 in	1.80 in	1.20 in	0.90 in	0.70 in
	330		2.40 in	1.60 in	1.20 in	0.80 in	0.60 in	0.50 in
600	240			2.70 in	2.00 in	1.30 in	1.00 in	0.80 in
	365			1.80 in	1.30 in	0.90 in	0.70 in	0.50 in
700	250			3.00 in	2.20 in	1.50 in	1.10 in	0.90 in
	375			2.00 in	1.50 in	1.00 in	0.80 in	0.60 in
800	260			3.30 in	2.50 in	1.60 in	1.20 in	1.00 in
	385			2.20 in	1.70 in	1.10 in	0.80 in	0.70 in
900	300				2.40 in	1.60 in	1.20 in	0.96 in
	360				2.00 in	1.30 in	1.00 in	0.80 in
1,000	330				2.40 in	1.60 in	1.20 in	0.97 in
	400				2.00 in	1.30 in	1.00 in	0.80 in

$$\text{Average Applied Depth} = \frac{19.25 \cdot (\text{Discharge Rate})}{(\text{Lane Spacing}) \cdot (\text{Travel Speed})}$$

sprinklers can be large, and care should be used to prevent soil surface sealing and crop damage to sensitive crops when irrigating. Instantaneous application rates of volume gun sprinklers are quite high as the stream passes over a given portion of the wetted coverage. Wind also reduces the wetted diameter of the sprinklers, resulting in even higher application rates. Care needs to be taken to ensure that the application rate does not result in runoff.

Many manufacturers recommend operating sprinklers in a part-circle mode with a small, pie-shaped section in front of the traveler left unirrigated so the machine travels on a dry lane. However, the sprinkler also may be operated in a full-circle mode. Check the owner's manual for recommendations on how the system should be operated.

Set-type Sprinkler Selection

Historically, medium to high pressure impact sprinklers have been the dominant sprinkler type used in set-type systems like side-roll and solid-set systems. Most set systems use low capacity single- or double-nozzle impact sprinklers giving application rates of 0.20 inches per hour or less. Table 6-6 gives recommended pressure ranges for impact sprinklers used with set-type systems.

Most solid-set systems are designed by selecting a sprinkler spacing along the lateral (S_L) and the spacing between laterals along the mainline (S_M). Figure 6-4 illustrates spacing between sprinklers and laterals. Spacing may conform to crop row or tree spacing in orchards. Typical field layouts range from square to rectangular with S_L x S_M values ranging from 30 feet x 30 feet to 80 feet x 80 feet. Table 6-7 provides average water application rates for a range of spacing between sprinklers and laterals at various sprinkler discharges. Table 6-8 gives examples of nozzle sizes and pressures to achieve various water application rates. The application rate is multiplied by the set time

Table 6-6. Pressure ranges for standard impact nozzles with set-type systems.

Nozzle size (in)	Pressure range (psi)
1/16	20 to 45
3/32	20 to 50
7/64	20 to 50
1/8	20 to 50
9/64	20 to 50
5/32	30 to 55
11/64	30 to 55
3/16	35 to 60
13/64	35 to 60
7/32	35 to 60
15/64	40 to 75
1/4	40 to 80
9/32	50 to 80
5/16	55 to 80
11/32	55 to 80
3/8	60 to 80

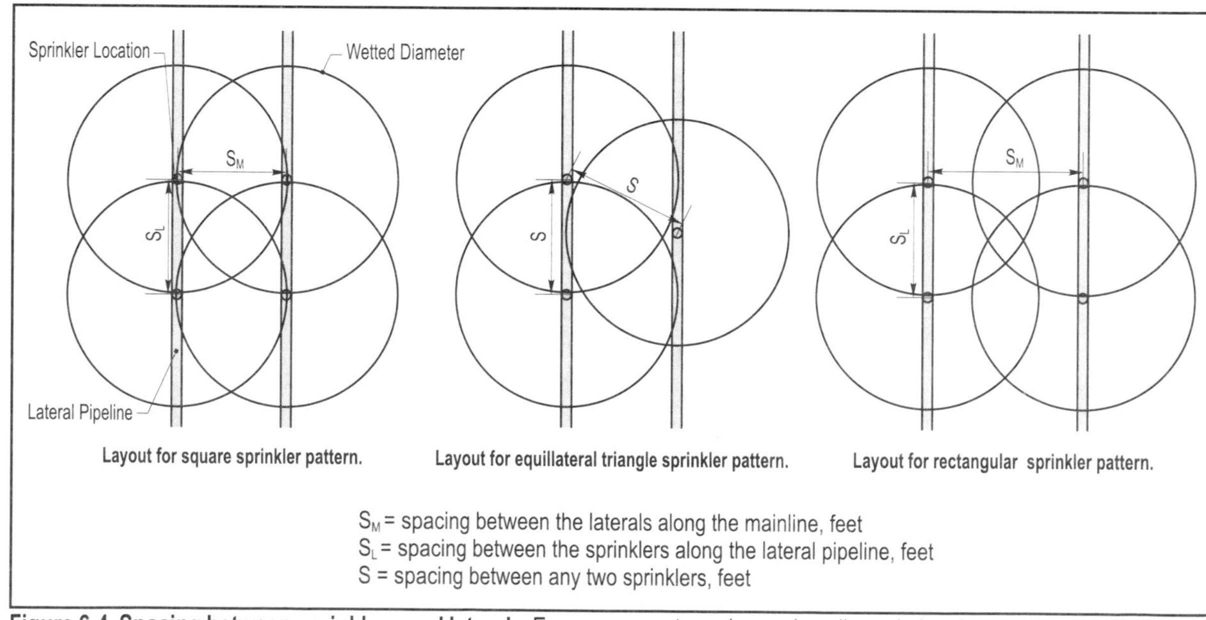

Figure 6-4. Spacing between sprinklers and laterals. For square, rectangular, and equilateral triangle sprinkler arrangements.

in hours and the application efficiency to arrive at an estimated net application rate.

The wetted diameters of many low pressure sprinklers require narrow spacing along and between the laterals. Distribution patterns (see Figures 5-9 and 5-11) and smaller wetted diameters also restrict the use of certain low pressure sprinklers, especially fixed sprays. Some applications, orchards for example, do not require uniform coverage over the entire area.

Low pressure impact sprinklers and rotating sprays are currently being used on solid-set and set-move systems. Spacing of sprinklers both along the linear and between laterals or lateral-move usually is less than for high pressure sprinklers. For each sprinkler type, there is an optimum pressure that should be used for any given sprinkler set spacing, for example, 20 x 30 or 30 x 40. Low pressure sprinklers often possess unique distribution patterns (Figure 5-9) that preclude using traditional spacing and overlap configurations. Use research information or manufacturer's recommendations to design the system including sprinkler and nozzle selection, spacing, and operating pressure. Side-roll systems could be configured with trailing laterals that provide closer spacing to offset the smaller wetted diameters at lower pressure.

It is very important to do sprinkler uniformity testing in the field when installing reduced pressure sprinkler packages on solid-set and set-move applications, as noted in the next section.

Table 6-7. Average water application rate (inches per hour).
For various sprinkler spacings with set irrigation systems with a specified flow rate per sprinkler.

Flow Rate	Sprinkler spacing (ft)											
(gpm)	20 x 20	30 x 30	30 x 40	40 x 40	40 x 60	60 x 60	60 x 80	80 x 80	120 x 120	180 x 180	240 x 240	300 x 300
2	0.48	0.21	0.16	0.12								
3	0.72	0.32	0.24	0.18	0.12							
4	0.96	0.43	0.32	0.24	0.16	0.11						
5	1.20	0.54	0.40	0.30	0.20	0.13	0.10					
6	1.44	0.64	0.48	0.36	0.24	0.16	0.12					
7	1.69	0.75	0.56	0.42	0.28	0.19	0.14					
8		0.86	0.64	0.48	0.32	0.21	0.16	0.12				
9		0.96	0.72	0.54	0.36	0.24	0.18	0.14				
10		1.07	0.80	0.60	0.40	0.27	0.20	0.15				
12		1.28	0.96	0.72	0.48	0.32	0.24	0.18				
15		1.61	1.20	0.90	0.60	0.40	0.30	0.23				
20			1.61	1.20	0.80	0.54	0.40	0.30				
25			2.01	1.50	1.00	0.67	0.50	0.38				
30			2.41	1.81	1.20	0.80	0.60	0.45				
35					1.40	0.94	0.70	0.53				
40					1.61	1.07	0.80	0.60	0.27			
45						1.20	0.90	0.68	0.30			
50						1.34	1.00	0.75	0.33			
60						1.61	1.20	0.90	0.40			
70						1.87	1.40	1.05	0.47			
80							1.61	1.20	0.54			
100								1.50	0.67	0.30		
150									1.00	0.45		
200									1.34	0.59	0.33	
300										0.89	0.50	0.32
400										1.19	0.67	0.43
500										1.49	0.84	0.54

Based on Equation 5-2.

Table 6-8. Example nozzle sizes and pressures for different application rates and sprinkler spacings.

Sprinkler spacing	Average water application rate (in/hr)						
	0.10	0.15	0.20	0.25	0.30	0.40	0.50
	Nozzle Size, in (Pressure, psi)						
30 x 30	5/64 (30)	3/32 (45)	7/64 (30)	1/8 (40)	1/8 (45)	9/64 (45)	5/32 (50)
30 x 40	3/32 (30)	3/32 (50)	7/64 (45)	1/8 (45)	9/64 (45)	5/32 (45)	11/64 (55)
40 x 40		1/8 (35)	9/64 (35)	5/32 (35)	11/64 (35)	3/16 (45)	13/64 (50)
30 x 40			9/64 (45)	5/32 (45)	11/64 (45)	3/16 (50)	13/64 (55)
40 x 60			5/32 (50	11/64 (50)	3/16 (50)	7/32 (50)	15/64 (65)
60 x 60				7/32 (50)	15/64 (55)	1/4 (70)	9/32 (70)

Evaluating System Performance

Irrigation systems can be evaluated for application efficiency and uniformity. Evaluation involves laying out a grid of catch cans in the area to be irrigated (see Figure 6-5). Measuring the amount of water caught in the cans during an irrigation event allows estimates of efficiency and uniformity to be made. It is important that enough catch cans are used to provide reasonable estimates of the actual net irrigation. Consult with an irrigation specialist for information on procedures for in-field testing of sprinkler irrigation systems.

Efficiency

All irrigation systems will have losses (see section on application efficiency). An estimate of the application efficiency can be calculated by averaging the amount of catch in the cans and dividing by the amount of water delivered to the system. An accurate flow rate is needed for determining the water delivered. Some application losses cannot be estimated by this procedure such as runoff and deep percolation.

Uniformity

No irrigation system can apply water precisely to all areas of the field. Hence, an estimate of uniformity is important to judge the performance of the system. Measured irrigation application with catch cans is compared to the average catch to arrive at a value for uniformity.

The two most common methods of expressing uniformity are the CU (coefficient of uniformity) and DU (distribution uniformity). CU calculates the average deviation of the catch compared to the average depth of the catch. For center-pivot evaluations, the deviations and averages are weighted for the relative area represented by each catch can. DU is calculated by dividing the average catch of the low one-quarter of the catch by the average catch. It also needs to be weighted when evaluating center pivots. Refer to ASAE S436, *Test Procedure for Determining the Uniformity of Water Distribution of Center Pivot, Corner Pivot, and Moving Lateral Irrigation Machines Equipped with Spray or Sprinkler Nozzles,* for more information on uniformity.

Figure 6-6 is an example showing a plot of application depth, and application distribution for a can test on a center pivot system. The system delivers 750 gpm to a 95-acre field. The system should apply 0.63 inches gross depth of application in 36 hours. The average catch can depth was 0.51 inches, thus the application efficiency would be about 80% ($0.51 \div 0.63 = 0.81 \sim 80\%$) if no runoff or deep percolation occurs. The CU was calculated as 0.92 and DU as 0.90. Uniformity would be acceptable for this example, but efficiency is somewhat low.

Most center pivots are designed to apply water with a coefficient of uniformity of 90% or greater. New center-pivot systems can have nearly 200 nozzles and 200 pressure-regulating valves installed on a single system. This means that there are more than 400 opportunities for a water distribution component to malfunction. For the system to deliver water uniformly, the first requirement is that each nozzle is installed in the correct location on the system. Even on new systems, check the nozzle installed with the computer printout showing the correct nozzle location.

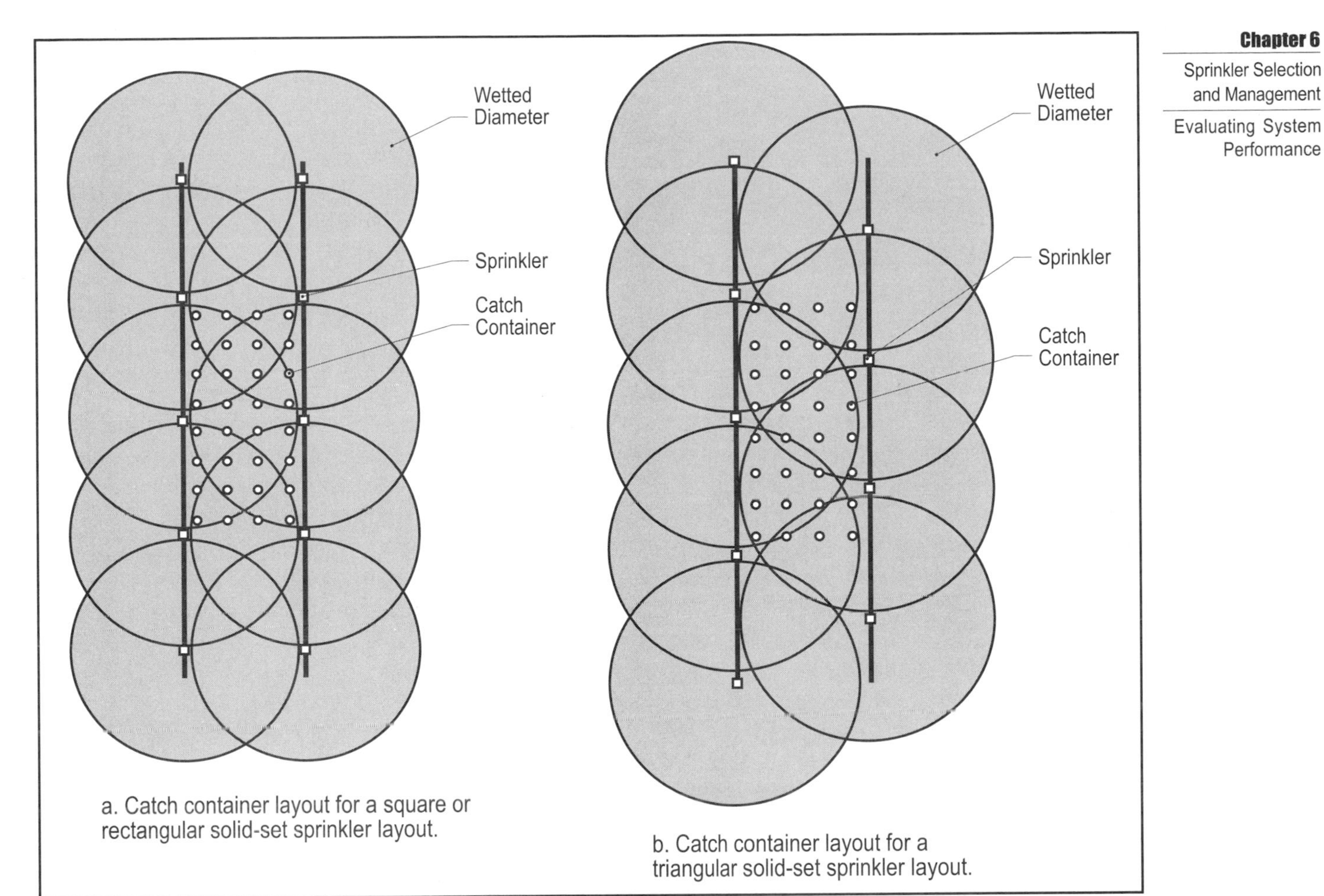

Figure 6-5. Catch container layouts for solid-set sprinkler layouts. Catch container should not be more than 10 feet apart. Run test until average depth. In the containers is at least one inch; longer tests will reduce errors in measuring small amounts.

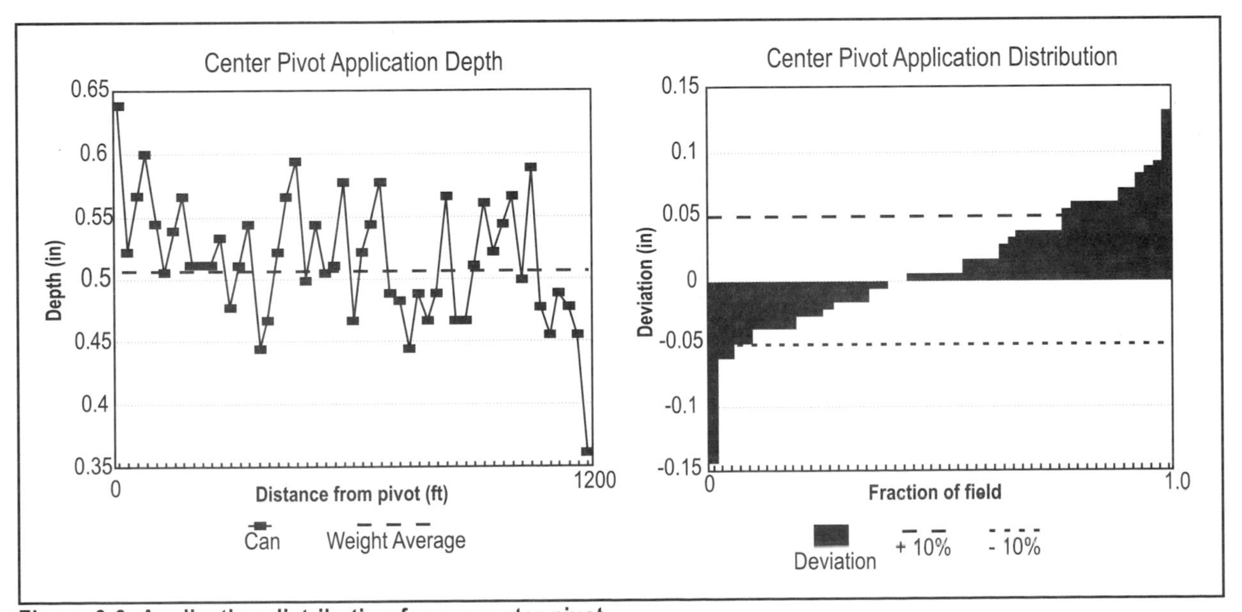

Figure 6-6. Application distribution from a center pivot.

Table 6-9. Example of runoff potential from a center-pivot irrigation system.

Soil intake family	Slope (%)	System capacity (gpm)	System length (ft)	Application depth (in)	Wetted diameter (ft)	Potential runoff (%)
Base system characteristics						
0.3	3 to 5	800	1340	1.2	40	26
Influence of soil intake family (soil texture) on runoff						
0.1	3 to 5	800	1340	1.2	40	44
0.5	3 to 5	800	1340	1.2	40	11
1.0	3 to 5	800	1340	1.2	40	0
Influence of slope on runoff						
0.3	0 to 1	800	1340	1.2	40	0
0.3	1 to 3	800	1340	1.2	40	8
0.3	greater than 5	800	1340	1.2	40	35
Influence of system capacity on runoff						
0.3	3 to 5	500	1340	1.2	40	14
0.3	3 to 5	700	1340	1.2	40	22
0.3	3 to 5	900	1340	1.2	40	29
Influence of application depth on runoff						
0.3	3 to 5	800	1340	0.6	40	3
0.3	3 to 5	800	1340	0.9	40	16
0.3	3 to 5	800	1340	1.5	40	33
Influence of wetted diameter on runoff						
0.3	3 to 5	800	1340	1.2	30	48
0.3	3 to 5	800	1340	1.2	60	15
0.3	3 to 5	800	1340	1.2	80	8
Influence of application depth and wetted diameter on runoff						
0.3	3 to 5	800	1340	0.6	60	0
0.3	3 to 5	800	1340	0.9	60	7
0.3	3 to 5	800	1340	1.5	60	22
0.3	3 to 5	800	1340	0.6	80	0
0.3	3 to 5	800	1340	0.9	80	2
0.3	3 to 5	800	1340	1.5	80	15

Source: Table generated using the CPNOZZLE computer model developed at the University of Nebraska.

Runoff Potential

With the center pivot, the rate of water application is the same regardless of the speed of rotation. If a potential for runoff exists or is encountered after installation, a number of management changes can be implemented. Increasing the rotation speed decreases the application depth, thus reducing the potential for runoff. For some nozzle packages, providing soil surface storage and planting the crop in a circle may be the only option for reducing the runoff potential.

For new installations, reducing the system capacity reduces the potential for runoff, but this alternative also reduces the net water that can be applied per day (see Chapter 2). Another option is to increase the wetted diameter of the sprinklers to eliminate runoff. Table 6-9 provides an estimate of the potential for runoff from sprinkler packages with various wetted diameters, soil intake families, application depths, and levels of soil surface storage.

Chapter 7

Pumps, Piping and Power Units

Chapter 7 Contents

Selection of the pump will depend on the amount of water required and the water source. The next step is to match the pump to a properly sized power unit. The power unit can be of several different types, depending on the proximity of the pump to available energy sources. Correct design of the pump, power unit, and pipelines is critical for low cost operation of the irrigation system. Proper design can also extend the life of the system by preventing excessive pressure losses.

Overview of Pump Selection

This chapter will discuss pumping plant design only for single, constant speed pump and power unit operations. If an irrigation project indicates a need for a multiple pumping plant or a variable speed pumping plant arrangement, consult with an experienced irrigation system engineer during the project's pre-planning and designing phases.

Most irrigation systems require a pump and a power unit, which must match the specific requirements of a particular water source, distribution pipeline, and sprinkler system. Before selecting an irrigation pumping plant, evaluate the following conditions:

- Water source.
- Pumping rate.
- Vertical suction lift.
- Length and size of suction pipe.
- Friction losses in suction pipe and fittings.
- Foot valve and strainer.
- Vertical lift on discharge side of pump.
- Size, length, and type of pipeline.
- Friction loss in the distribution pipeline.
- Pressure required at sprinkler system.
- Total head on pump.
- Available pumps, power sources, and performance curves.
- Type of automatic controls.

Pump Alternatives and Selection

Pumps commonly used for sprinkler irrigation systems include centrifugal, deep-well turbine, and submersible. Turbine and submersible pumps are actually special forms of a centrifugal pump, but they are normally called by their common industry names.

Centrifugal pumps are versatile and are generally used above ground where the vertical suction lift is less than 20 feet. Deep-well turbine pumps are best suited for groundwater wells or wet well (sump) installations pumping from surface waters. Submersible pumps are used for wells but usually in smaller horsepower sizes. Table 7-1 gives some general advantages and disadvantages to consider when selecting the most appropriate irrigation pump for a situation.

Table 7-1. Factors to consider in selecting an irrigation pump.

Pump type	Advantages	Disadvantages
Centrifugal	1. High efficiency over a range of operating conditions 2. Easy to install 3. Simple, economical and adaptable to many situations 4. Electrical, internal combustion engines or tractor power can be used 5. Does not overload with increased head 6. Vertical centrifugal pump may be submerged and not need priming	1. Suction lift is limited. Pump needs to be set less than 20 vertical feet from the pumping water surface 2. Priming required 3. Loss of prime can damage pump 4. If total head is much lower than design value, the motor may overload
Vertical turbine	1. Adapted for use in wells 2. Provides high total dynamic head and flow rates with high efficiency 3. Electric or internal combustion power can be used 4. Priming not needed 5. Can be used where water surface fluctuates	1. Difficult to install, inspect and repair 2. Higher initial cost than centrifugal pump 3. To maintain high efficiency, impeller clearance must be adjusted to specifications 4. Repair and maintenance is more expensive than centrifugal
Submersible	1. Can be used in deep wells 2. Priming not needed 3. Adaptable to misaligned wells 4. Easier to install than turbine 5. Smaller diameters are less expensive than vertical turbines	1. More expensive in larger sizes than deep well vertical turbines 2. Only electric power can be used 3. More susceptible to lightning 4. Water movement past motor is required

Centrifugal Pumps

Centrifugal pumps are used to pump from reservoirs, lakes, streams, and shallow wells. Some models also can be used as pressure booster pumps within irrigation pipelines. All centrifugal pumps must be completely filled with water or primed before they will operate. The suction line, as well as the pump, has to be filled with water and free of air. Airtight joints and connections on the suction pipe are essential.

Pumps can be primed by hand-operated vacuum pumps (Figure 7-1), use of a internal combustion engine vacuum, by motor-powered vacuum pumps, or by small water pumps (Figure 7-29) that fill the pump and suction pipe with water. A foot valve is commonly used near the entrance of the suction pipe to help maintain a prime after each shutdown for easy startup during the next irrigation event.

Centrifugal pumps are designed for either horizontal or vertical operation. Figure 7-1a depicts a horizontal installation. The pump has a vertical impeller connected to a horizontal shaft. A vertical centrifugal has a horizontal impeller connected to a vertical shaft. The impeller of both types is located inside a sealed housing called the volute case.

Horizontal centrifugals are most common for irrigation. They are generally less costly, require less maintenance, are easier to install, and are more accessible for inspection and maintenance than a vertical centrifugal. The horizontal pump's impeller shaft is typically powered by direct connection or v-belts to a horizontal electric motor or to a PTO shaft of a combustion engine.

Some centrifugal pumps are designed with special open or semi-open impellers to handle effluent waters with some solids content. Limitations on solid contents for different impeller types and other operating cautions are discussed in Chapter 9.

Vertical centrifugals may be mounted with the impeller under water, making priming unnecessary, but bearing maintenance usually is high in this type of installation. A vertical pump and power unit are typically mounted to a support frame over an open sump casing or on a floating pontoon for placement in a surface water body. Single-stage vertical centrifugals usually are limited to applications of 50 feet or less total pumping head, but some single and multi-impeller assemblies have been arranged to discharge at over 200 feet total dynamic head.

Operating Characteristics

Centrifugal pumps draw water into their impellers after they have been primed, so they must not be set any higher than the manufacturer's recommended maximum practical suction lift (MPSL) above the water surface while pumping (also called the total dynamic suction lift, TDSL). MPSL values can vary from less than 10 to about 20 feet. Impeller rotation throws water toward the outside of the impeller by centrifugal force and creates a lower pressure at the center of the impeller. This action develops a partial vacuum near the impeller inlet that causes the water to be forced into the impeller.

Operating a centrifugal pump at suction lifts greater than designed, or under conditions of high or fluctuating vacuum pressure, can result in pump damage by cavitation. Cavitation is caused by imploding air bubbles and water vapor that make a distinctive noise (like gravel flowing through the pump). Cavitation erodes the surface of the impeller and can eventually cause it to become deeply pitted.

The maximum suction lift can depend on the pump and the installation of the pump. Manufacturers determine and report the net positive suction head (NPSH) for pumps at sea level on each impeller performance curve, shown in Figure 7-9 (page 112). To find the maximum suction lift (head) at which a pump will operate, subtract the NPSH from 33 feet.

For example, if a pump's NPSH is 20 feet at a given temperature, speed, capacity, and discharge head, the pump has a maximum suction lift of 13 feet at sea level. Suction lift also is reduced by 1.2 feet for every 1,000 feet of elevation above sea level. Maximum suction lift includes not only the vertical suction lift, but also the friction losses in the intake pipe, elbows,

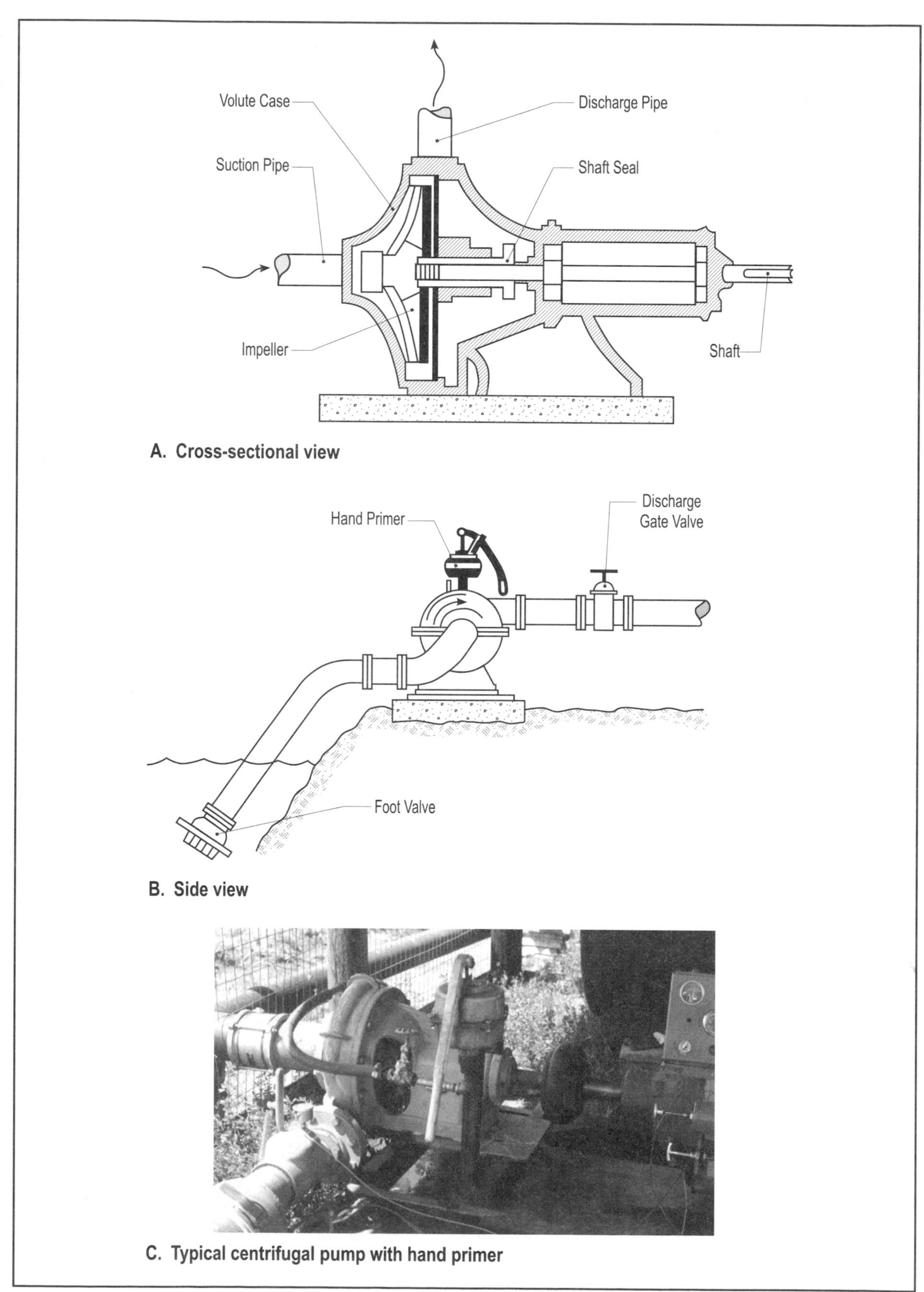

Figure 7-1. Horizontal centrifugal pump.

foot valve, and other fittings. Because of these losses, a pump rated for a maximum suction lift of 13 feet may effectively lift water only 8 to 10 feet.

Estimate the maximum practical suction lift (MPSL) with the following equation:

Equation 7-1. Maximum Practical Suction Lift.

$$MPSL = H_{atm} - (H_f + e_s + NPSH + f_s + ELT)$$

Where:

- $MPSL$ = Maximum practical suction lift, feet
- H_{atm} = Atmospheric pressure (H_{atm} = 33 feet of head at sea level)
- H_f = Friction losses on suction side of pump, feet
- e_s = Vapor pressure of water (about 0.5 feet at 50 F)
- $NPSH$ = Net positive suction head required by the pump at the capacity the pump will operate, feet (look on the pump's performance curve)
- f_s = Safety factor, feet (about 2.0 feet)
- ELT = 0.0012 times the site elevation above sea level

Installation

For a centrifugal pump to operate at its designed efficiency throughout its life, it should be correctly located, have a good foundation, and be properly aligned. When locating a pump, do the following:

- Provide easy access for inspection and maintenance.
- Protect it from the elements. House it in a permanent installation; allow headroom for servicing equipment.
- Protect it against floods.
- Locate it close to the water supply for a short and direct suction line.
- Shield rotating shafts, pulleys, and moving parts for personal safety.
- Use ground-fault outlets for electrically driven pumps, and properly wire, ground, and protect in conduit.

Make sure the power unit drive shaft or belt drive rotates in the same direction as the arrows on the pump casing. The pump and power unit should be secured to a concrete or steel base according to the manufacturer's recommendations and aligned to provide for proper operation of support bearings and drive shaft or v-belts. Unless a flexible shaft is used between the pump and power unit, avoid using anchor bolts.

Be sure the pump lines up with the power unit and piping. Do not force pipes into place with flange bolts, which may draw the pump out of alignment. Provide independent support for suction and discharge pipelines to keep strain off the pump casing. Use some type of flexible hose or connector on suction and discharge pipelines to isolate pump vibration.

Size suction pipes so that water velocity is less than 2 to 3 feet per second (fps). The intake pipe, especially with long intakes and high suction lifts, should have a uniform slope upward from the water source to the pump. Avoid high spots where air can collect and cause the pump to lose prime. Suspend the inlet end of the suction pipe above the bottom of a stream or pond, or lay it in a concrete or metal sump. On horizontal suction lines with a reducer, use the eccentric-type inlet, Figure 7-2, with the straight section on the upper side of the line and the tapered section on the bottom side.

Air that enters the suction pipe may cause the pump to lose prime, or the air may become trapped in the impeller, which reduces the output of the pump. If the water level is too low or the suction pipe inlet is not sufficiently submerged in the

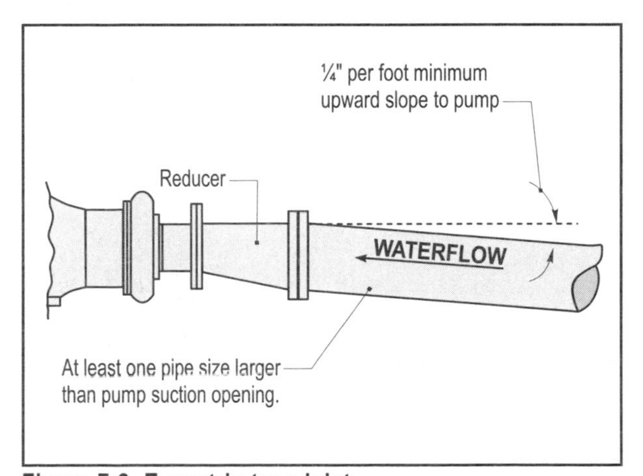

Figure 7-2. Eccentric-type inlet.

water, air can enter the suction pipe through a vortex or whirlpool. In shallow water, a mat or float with vertical straightening vanes will reduce the potential for the problems a vortex causes.

Use screens or strainers on the suction line if debris is present in the water supply. For large amounts of small debris in the water supply, place a screen around, and 2 to 3 feet from, the suction inlet for good protection and less clogging. Strainers are generally small and are fastened to the end of the suction pipe. They are satisfactory in relatively clear water but may collapse when coated with algae or debris. Consider using rotating screen units that are self-cleaning for use in water supplies with high amounts of debris.

Some centrifugal pumps are used in more than one location, increasing the difficulty of providing a proper foundation. Portable pump units usually are mounted on wheels or skids. For an adequate foundation, locate the unit on level, firm ground, and secure it in place to prevent shifting during operation.

Next to a river with moderately sloping banks, the centrifugal pump may be on skids, sloping timbers, or a track so it can be removed quickly during floods. Also use this method where the water level fluctuates enough to be out of range of suction lift if the pump is permanently installed. With steep banks, consider a foundation platform on piling, or place the pump on a floating barge or boat.

Most centrifugal pumps are typically primed manually by the operator. If the pumping plant is inaccessible, an independently powered priming system should be installed. If an automatic restart of the pumping plant is desired, automatic priming systems are available and described later in this section.

When water is pumped from a well or small sump, the water tends to rotate, which interferes with flow into the suction line. This is particularly true in cylindrical sumps or wells. Install baffles along the suction pipe at right angles to the water rotation, and select a pump suction line less than one-third the diameter of the well.

Falling water into a sump near the intake pipe also entrains air into the water. Extending the suction line to deeper water will minimize air entry into the pump.

A short elbow at the pump inlet also disturbs the water flow and may cause noisy operation, efficiency loss, and heavy end-thrusts, particularly with high-suction lift. Make any bend in the suction line a long sweep or long radius elbow; place it as far away from the pump as practical.

Deep-Well Turbine Pumps

Deep-well turbine pumps are used in cased wells or where the water surface is below the practical limits of a centrifugal pump. Priming is not a concern because the intake for the turbine pump is continuously under water. Turbine pumps also are used with some surface water supplies where the pumps are set in wet wells or sumps. Turbine pump efficiencies are comparable to or greater than those of most centrifugal pumps. Turbine pumps give long and dependable service if properly installed and maintained. Deep-well turbine pumps are

- Usable where the water surface fluctuates, as in reservoirs.
- Suitable for high heads and large discharges.
- Not likely to lose prime.
- Higher in initial cost than centrifugal pumps.
- More sensitive to adjustments than centrifugal pumps.
- Efficient over a narrower range of operating conditions than centrifugal pumps.
- Dependent on additional stages to produce greater pressures or heads.
- Difficult to install and repair.

A turbine pump, Figure 7-3, has three main parts:

(1) Discharge head assembly.
(2) Bowl assembly.
(3) Shaft and column assembly.

The discharge head assembly is made of cast iron and designed for installation on a foundation. The head supports the column and shaft assembly and the bowl assembly in the well. The discharge head includes a port for the water discharge and a mounting platform for an electric motor or a right-angle gear drive to connect a horizontal power shaft from an engine.

The bowl assembly contains an inlet and one or more bowl stages with impellers and is located below the pumping water level in the well. The shaft and column assembly

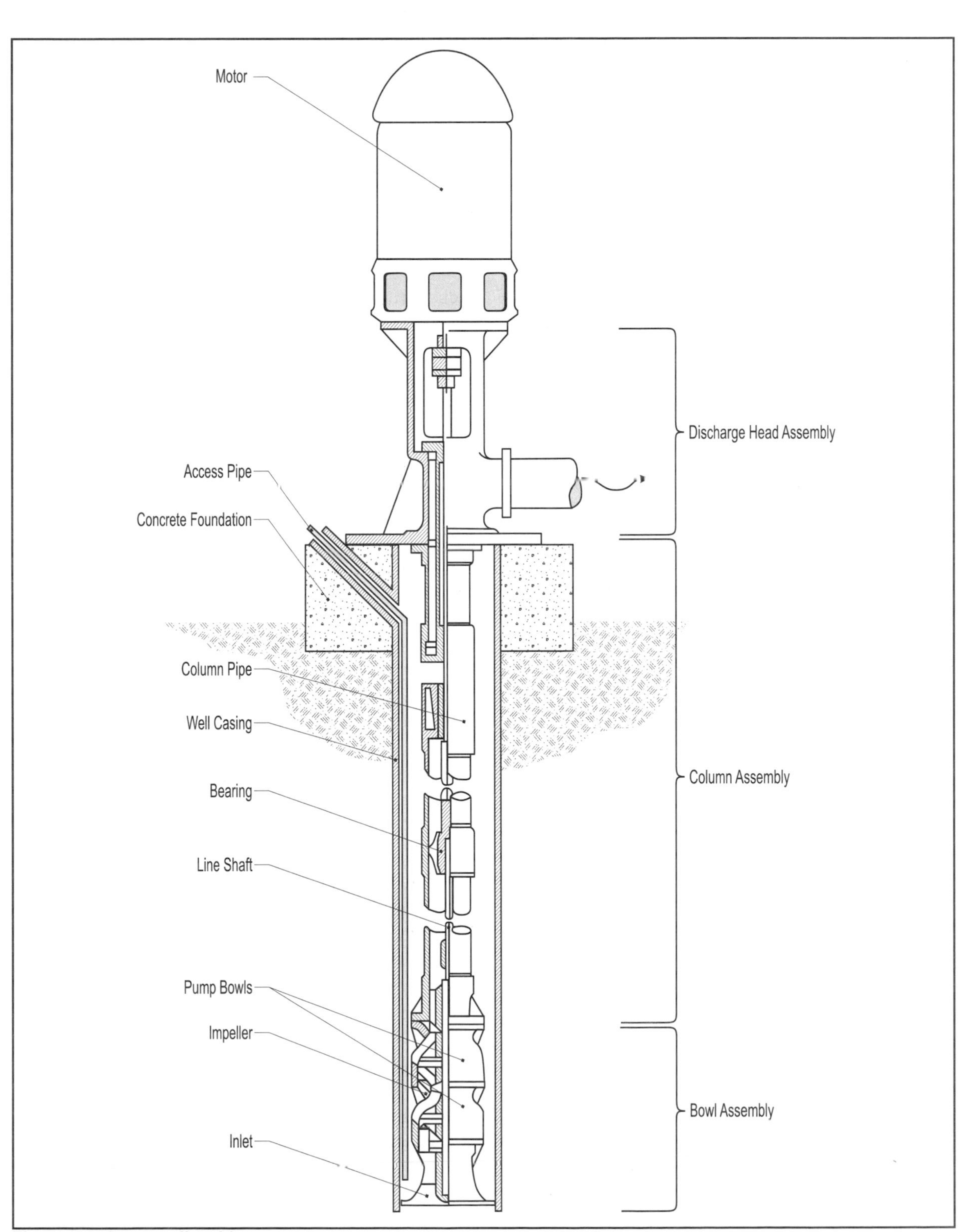

Figure 7-3. Deep-well turbine pump.

connects the head and bowl assemblies and is positioned inside the well casing. The line shaft carries the power, and the column carries the water upward. The line shaft on a turbine may be lubricated by oil or water.

An oil-lubricated pump has an enclosed shaft into which oil drips, lubricating the bearings; it keeps any sand out of the bearings. A water-lubricated pump has an open shaft; the bearings are lubricated by the pumped water. A water-lubricated pump is free of oil and is better for a domestic or municipal water supply.

Operating Characteristics

The impeller in a turbine pump operates similarly to an impeller in a centrifugal pump, but because the bowl and impeller diameter is limited in size, each impeller develops a relatively low pressure head. In typical applications, several impellers are installed in series, one impeller above the other, with each impeller having its own bowl to achieve the desired pressure head.

Each rotating impeller unit raises the operating head for a turbine pump a set amount at a given flow rate regardless of the number of stages. With this staging, a three-stage bowl assembly has three impellers attached to a common shaft through the separate bowls and develops three times the pressure head of a single-stage pump.

The operating characteristics of a turbine pump, like a centrifugal pump, depend mainly on bowl design, impeller diameter, and speed of rotation. Head capacity, efficiency, horsepower, and the rate of speed are similar to those for centrifugal pumps. Turbines, however, operate at a higher efficiency over a narrower range of rotational speeds than do centrifugal pumps.

Impellers in turbine pumps, as illustrated in Figure 7-4, are either semi-open or closed. A semi-open impeller has only the top side of the impeller vanes closed. The bottom side runs with a close clearance to the pump bowl, or it may have a special seal. For non-sealed impellers, this clearance is critical and must be adjusted when the pump is installed. After an initial run-in (about 100 hours), the clearance should be readjusted according to the manufacturer's specifications. Adjustments to the impellers are made by tightening or loosening the adjustment nut at the top of the head assembly. A maladjusted impeller clearance causes inefficient operation if the impellers are set too high. Mechanical damage results from impellers set so low that they touch the bowls.

Turbine Pump Installation

Laws in many states require that pumps be installed by a licensed well driller or pump installer. Deep-well turbine pumps must have correct alignment of the well casing, pump head, and the power unit. Alignment of the pump in the well casing is most important so that no part of the pump

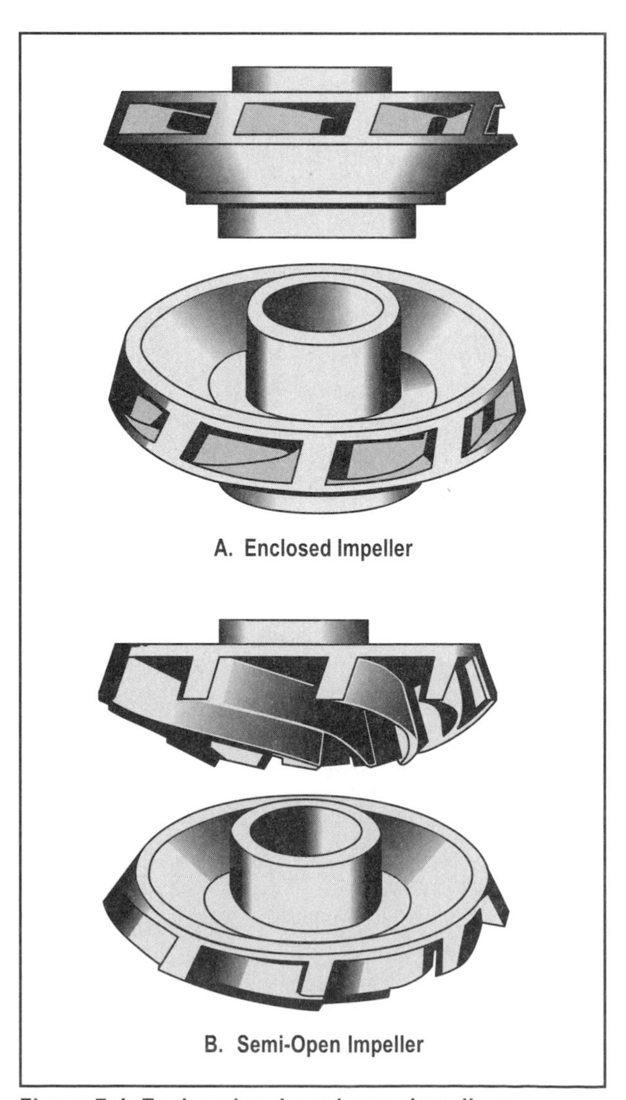

Figure 7-4. Enclosed and semi-open impellers.

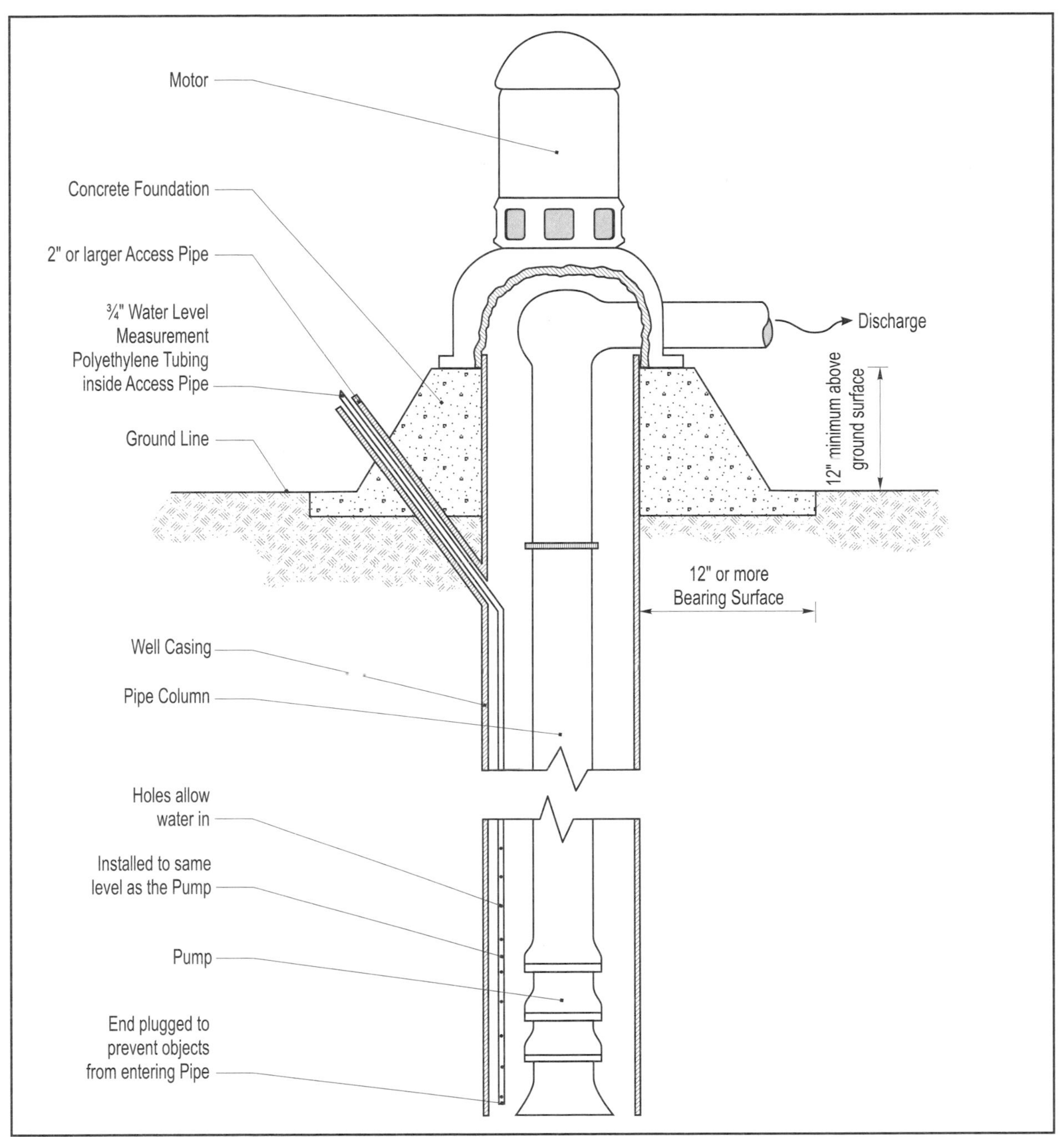

Figure 7-5. Recommended concrete base with access pipe.

assembly touches the well casing; otherwise vibration will wear holes in the well casing where the two touch.

Figures 7-5 and 7-6 shows a pump mounted on a good foundation to maintain alignment between the pump, drive, and well casing. A concrete foundation is the most permanent and trouble-free type of foundation. It must be large enough for the pump and drive assembly to be securely fastened. Extend the foundation at least 12 inches beyond the well casing, or to a distance as required by a state well construction standard.

A well casing access tube, as seen in Figures 7-5 and 7-6, also is very important for maintenance and is required in many states. These tubes are 2 to 4 inches in diameter with a removable cap. An access tube helps the operator or contractor measure water levels and disinfect the well.

Figure 7-6. Using an access pipe for a well drawdown meter.

Submersible Pumps

A submersible pump is a turbine pump with a submersible electric motor attached to the bottom of the bowl inlet (Figure 7-7). Both the pump and motor are suspended in water, which eliminates long line shaft and bearings of deep-well turbines. Operating characteristics are similar to those of deep-well turbine pumps.

The pump has enclosed impellers. A screened water inlet is between the motor and pump. Water must flow past the motor for cooling purposes, and some assemblies contain a shroud to direct intake water past the motor housing.

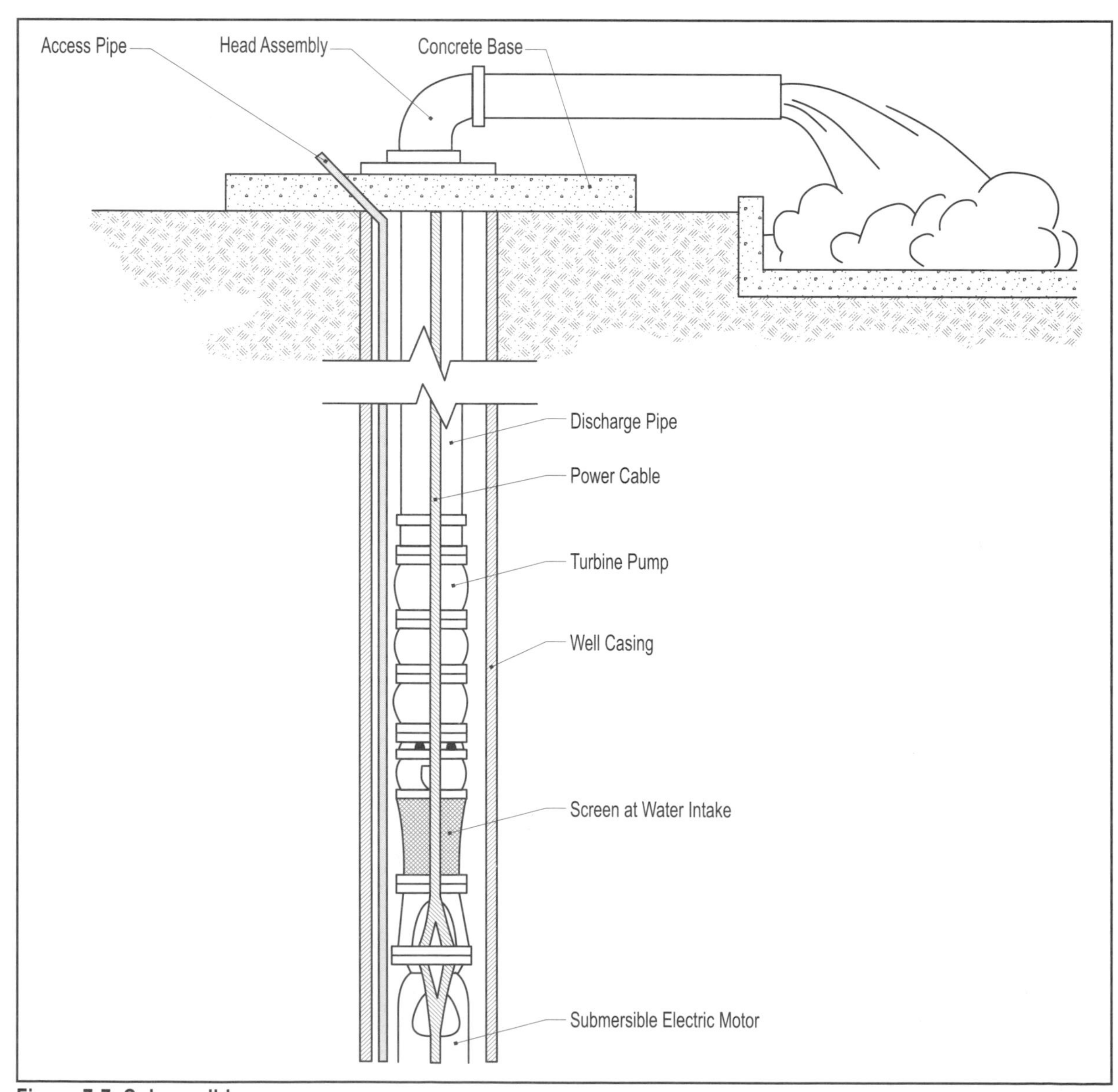

Figure 7-7. Submersible pump.

Submersible motors are small in diameter and are much longer than ordinary motors. Pump motors are generally classified as dry motors or wet motors. Dry motors are sealed to exclude water and are filled with a high dielectric oil. In wet motors, the rotor and bearings operate in the water. Drop electric wiring for the motors should be sized and installed by a licensed electrician.

Submersible pumps come in a range of capacities for 4-inch wells or larger. Submersible pumps

- Are suitable for deep-wells.
- Need no priming.
- Can be adapted to some misaligned wells.
- Are easier to install than turbines.
- Have no long drive shaft.
- Are higher priced in larger sizes than deep-well turbines.
- Must be removed from the well for impeller adjustments.
- Are more susceptible to lightning damage.
- Usually operate at lower efficiency than does a deep-well turbine.

Pump Sizing and Selection

After the type of pump has been selected, the total head (TH) requirement for the chosen sprinkler system, pipeline, and water supply needs to be determined to help select an efficient pump from a manufacturer's catalog. This selection process should be done with an experienced pump supplier or irrigation system designer.

Head, an energy term commonly used with pumps, refers to the height of a vertical column of water that pumps raise. It is expressed as units of feet or pounds of pressure. Pressure and head are interchangeable concepts in irrigation, because a column of water 2.31 feet high is equivalent to 1 pound per square inch (psi) of pressure. The total head produced by a given pump is composed of several types of head that help define the pump's different operating characteristics.

Determining Total Head

Total head (TH) for an operating pump is the sum of total lift, pressure head, friction head and velocity head, (usually negligible). Total head also is commonly called total dynamic head. Figure 7-8 illustrates these components. Knowledge of the total head is necessary to select both the proper pump and power unit and to estimate potential pumping costs.

Total head can be calculated by one of the following equations:

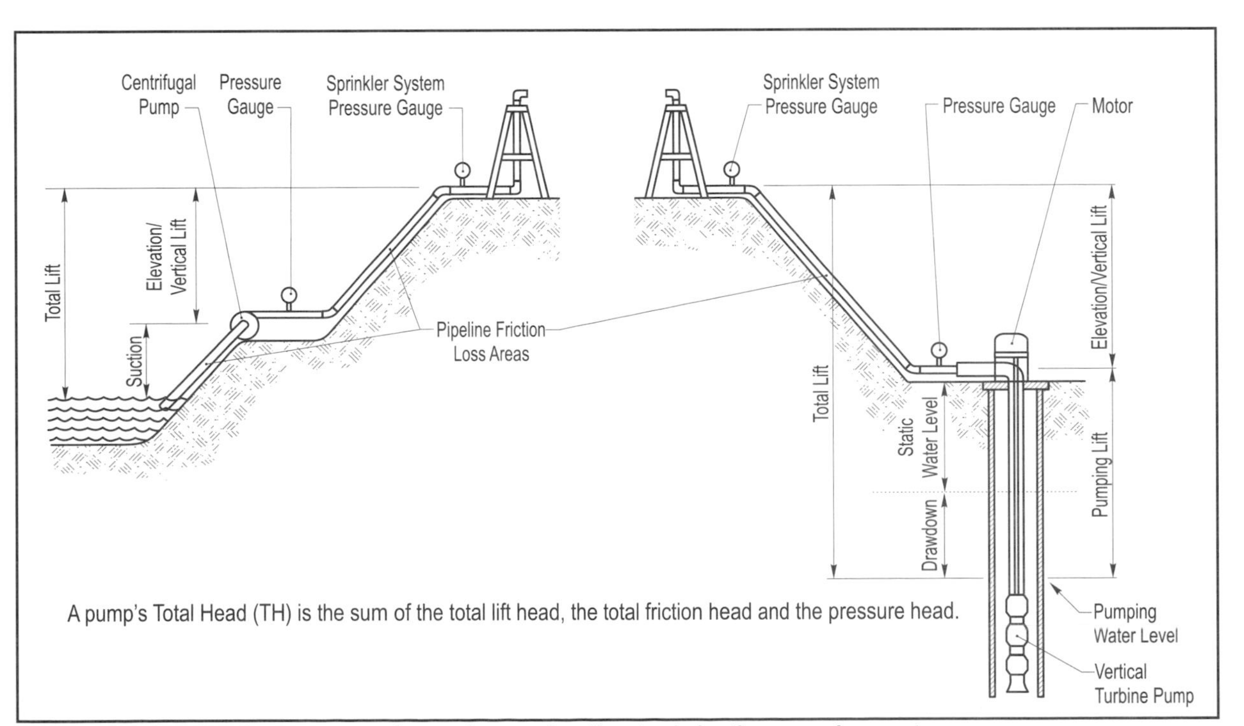

Figure 7-8. Total head components for surface and well water irrigation pumping systems.

Equation 7-2. Total Head.
Total head = total lift + pressure head at system + pipeline friction losses

Equation 7-3. Total Head for Centrifugal Pumps.
Total head = suction lift + elevation lift + sprinkler system head + suction line and pipeline friction losses

Equation 7-4 Total Head for Deep-Well Turbine Pumps.
Total head = pumping lift because of drawdown + elevation lift + sprinkler system head + pipeline friction losses

Total lift head is the total vertical lift distance in feet that the pump must raise the water as shown in Figure 7-8.

This also is called total static head. When pumping from an open water surface, this would be the total vertical suction lift from the water surface, to the pump discharge head (suction head), plus the vertical distance from the pump head to the entrance to the sprinkler system.

When pumping from a well, the total lift head is the distance from the pumping water level in the well to the ground surface, plus the vertical distance that the water is lifted from the ground surface to the sprinkler system discharge point (or highest discharge elevation). The pumping water level lift in a well is a combination of the static water level and the drawdown distance that occurs for a given pumping rate. The drawdown distance usually increases over time for a given well

Example 7-1. Determining the total head for a traveler and surface water supply.
Determine the total head for a traveling gun operating pressure of 100 psi at the hose entrance for a 500 gpm discharge. Water is pumped from the reservoir through 1,000 feet of 6-inch aluminum pipe. The traveler will be located at its highest elevation 50 feet above the reservoir edge. The pump will be located 15 feet above the water surface and uses a 25 foot, 6-inch aluminum suction pipe.

Solution:
Using Equation 7-3, the total head is calculated to be:
Suction Lift = 15 feet
Elevation Lift = 50 feet
Sprinkler system head = 100 psi x 2.31 feet head per psi = 231 feet
Friction loss (6-inch pipe) = 2.4 feet head per 100 feet of pipe (From Table (7-3) = (2.4 feet head per 100 feet of pipe) x (1,000 feet pipe + 25 feet suction pipe) = 25 feet
Total head = 15 feet + 50 feet + 231 feet + 25 feet **= 321 feet of head**

Example 7-2. Determining the total head for a center pivot and water well.
A center-pivot needs an operating pressure of 40 psi at its base for 1,000 gpm discharge. Water will be delivered to the pivot from a deep-well turbine pump with a 10-inch column pipe into 1,000 feet of 8-inch PVC pipe. Static water level in the well is 40 feet and the drawdown is 30 feet for 1,000 gpm after 48 hours.

Solution:
Using Equation 7-4, the total head is calculated to be:
Pumping Lift = 30 feet + 40 feet = 70 feet
Sprinkler system head = 40 psi x 2.31 feet head per psi = 92 feet
Friction loss = 1.4 feet head per 100 feet of pipe (from Table (7-3) = (1.4 feet head per 100 feet of pipe) x 1,000 feet pipe + 1.5 feet head per 100 of column pipe (from manufacture's catalog) x 70 feet = 15 feet
Total head = 70 feet + 92 feet + 15 feet **= 177 feet of head**

so a reasonable maximum estimate must be selected when determining the TH for selecting a pump size.

Pressure head is the pressure in feet of head necessary at the sprinkler system entrance to overcome losses (lateral pipeline friction and elevation) and operate sprinklers at their design pressures. Pressure in pounds per square inch (psi) is easily converted to feet of head with the use of Table 7-2 or the following equation.

Equation 7-5. Head Conversion.

Head (feet of water) = psi x 2.31

or

Head (psi) = feet of water x 0.43

Table 7-2. Head conversion.

Head (ft of water) = (psi) x 2.31

Head (psi) = (ft of water) x 0.43

psi	feet equivalent	feet	psi equivalent
0	0	0	0
5	11.5	10	4.34
10	23.1	20	8.67
15	34.6	30	13.0
20	46.1	40	17.3
25	57.7	50	21.7
30	69.2	60	26.0
35	80.7	70	30.3
40	92.3	80	34.7
45	104	90	39.0
50	115	100	43.4
55	127	110	47.7
60	138	120	52.0
65	150	130	56.4
70	161	140	60.7
75	173	150	65.0
80	185	160	69.4
85	196	170	73.7
90	208	180	78.0
95	219	190	82.4
100	231	200	86.7

Friction head is the energy loss or pressure head reduction due to friction when water flows through straight pipe sections, fittings, valves, water meters, screens, and suction pipes; around corners, and where pipes increase or decrease in size. Estimated values for these losses can be calculated from Tables 7-3 and 7-8, and from Figure 7-12, but during final design, friction losses for the pipeline and fittings should be obtained from the manufacturer's friction loss tables (also see Pipe Sizing section in this chapter for information on friction loss). The friction head for a piping system is the sum of all friction losses.

The velocity of water and inside surface roughness of the pipe or fitting affects friction loss significantly. To reduce water hammer potential and avoid excessive friction losses, generally avoid velocities above 5 feet per second when selecting the best pipe size.

Velocity head is the energy of the water in motion. This is a very small amount of energy and usually is negligible

Table 7-3. Friction loss, feet of head per 100 feet of pipe.

Flow rate	4-in			6-in			8-in			10-in			12-in		
(gpm)	Steel	Alum.	PVC	Steel	Alum.	PVC	Steel	Alum.	PVC	Steel	Alum.	PVC	Steel	Alum.	PVC
100	1.20	0.90	0.60												
150	2.50	1.80	1.20	0.30	0.20	0.20									
200	4.30	3.00	2.10	0.60	0.40	0.30	0.10	0.1	0.1						
250	6.70	4.80	3.20	0.90	0.60	0.40	0.20	0.1	0.1	0.1	0.1				
300	9.50	6.20	4.30	1.30	0.80	0.60	0.30	0.2	0.1	0.1	0.1				
400	16.0	10.6	7.20	2.20	1.50	1.00	0.50	0.3	0.2	0.2	0.1	0.1	0.1		
500	24.1	17.1	11.4	3.40	2.40	1.60	0.80	0.6	0.4	0.3	0.2	0.1	0.1	0.1	0.1
750	51.1	36.3	24.1	7.10	5.00	3.40	1.80	1.3	0.8	0.6	0.4	0.3	0.2	0.1	0.1
1000	87.0	61.8	41.1	12.1	8.60	5.70	3.00	2.1	1.4	1.0	0.7	0.5	0.4	0.3	0.2
1250	131.4	93.3	62.1	18.3	13.0	8.60	4.50	3.2	2.1	1.5	1.1	0.7	0.6	0.4	0.3
1500	184.1	130.7	87.0	25.6	18.2	12.1	6.30	4.5	3.0	2.1	1.5	1.0	0.9	0.6	0.4
1750	244.9	173.9	115.7	34.1	24.2	16.1	8.40	6.0	4.0	2.8	2.0	1.3	1.2	0.9	0.6
2000	313.4	222.5	148.1	43.6	31.0	20.6	10.8	7.7	5.1	3.6	2.6	1.7	1.5	1.1	0.7

Flow rates below the horizontal line for each pipe size exceed the recommended 5 feet per second velocity.

when computing losses in an irrigation system, except possibly in the suction pipeline.

Pump Performance Curves

The best pump for a particular job is one that operates near its highest efficiency at the required flow rate and TH needed. Pump efficiency is the measure of the pump shaft input energy (brake horsepower) required to create the output energy in the discharging water flow (water horsepower). Manufacturers have standard pump designs for many head and capacity ranges. Manufacturers run tests to determine the operating characteristics of their pumps and publish the results in pump performance charts commonly called "pump curves."

Figures 7-9 and 7-10 illustrate typical centrifugal and deep-well turbine pump performance curves. An example showing how to read pump curves appears later in this section. All pump curves are plotted with the flow rate in gpm on the horizontal axis and the head in feet for a single impeller on the vertical axis. A pump operates over a wide range of conditions, but optimum operating conditions are specific and depend on the design criteria for a given system and site.

To select an acceptable pump for the situation, examine the manufacturer's performance characteristic curves (see Examples 7-3 and 7-4) from several pumps to identify a pump that will operate near maximum efficiency under the desired irrigation system discharge capacity and total dynamic head. In studying each of the pump curves, consider the following factors:

- Compare the discharge pressure head from each curve for the same pumping capacity and operating speed to identify which pumps can produce the desired TH.
- For multi-stage turbine pumps, determine the number of impellers needed to produce the desired TH and what impeller diameter will deliver water at the closest TH value.
- Identify each pump's energy efficiency rating and brake horsepower requirements at the point of desired operating conditions.
- For each curve, identify the type of impeller used to create the curve.
- If operating speed can be changed, identify the necessary operating speed for an impeller to produce the desired TH at the needed flow rate.

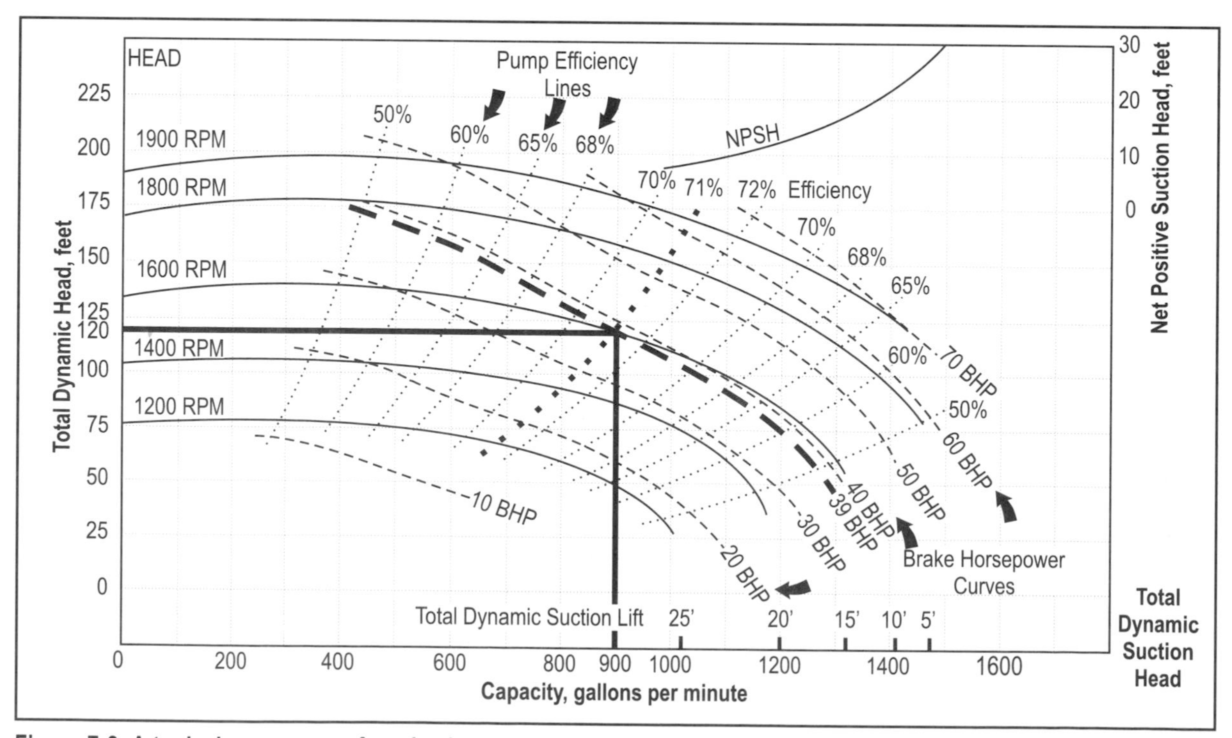

Figure 7-9. A typical pump curve for a horizontal centrifugal pump. NPSH is the net positive suction head required by the pump and the TDSL is the total dynamic suction lift available (both at sea level).

The performance chart in Figure 7-9 is for a centrifugal pump tested at different RPMs. Each of the RPM curves indicates the gpm versus TH relationship at the tested speed. Turbine pump charts, such as those in Figure 7-10, are similar but generally are shown only for a single speed of 1,760 RPM and with three or more impeller diameters.

Overlaid on the head performance curves are the equal pump energy efficiency curves; wherever the efficiency curve crosses the pump head curve is the energy performance efficiency for that operating point.

Efficiency curves show how the pump operates under a particular condition of discharge, speed, and total dynamic head. Most pump impellers can reach peak efficiencies between 75 and 85 percent.

Brake horsepower (BHP) curves are typically shown at the bottom of a pump performance chart or plotted over the head and efficiency curves. The BHP curves are calculated using the real values of TH and efficiency for a given pumping flow rate. At the top of most performance charts is an NPSH curve with its scale usually expressed on the right side of the chart.

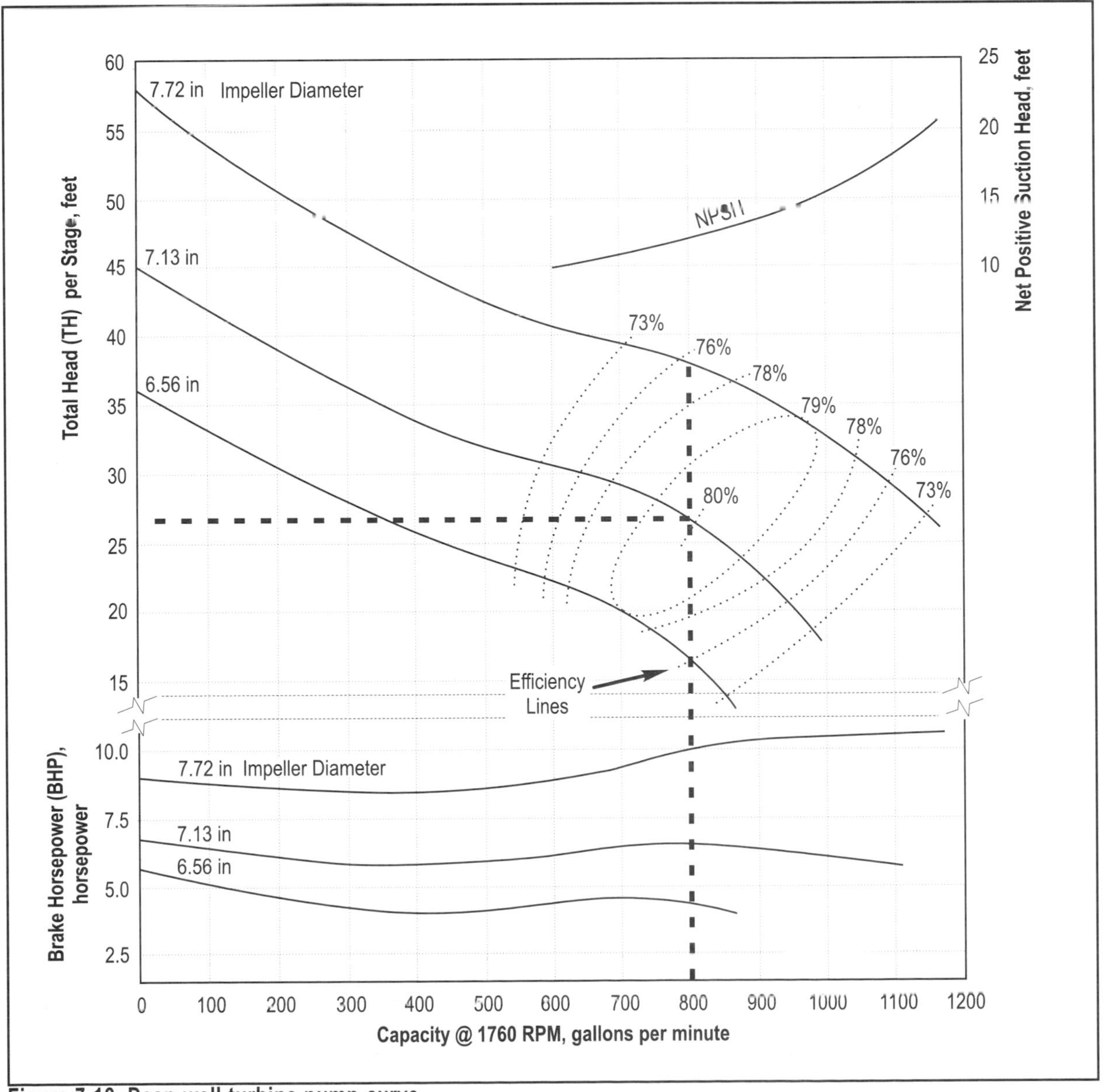

Figure 7-10. Deep-well turbine pump curve.

Speed Change and Pump Performance

Performance of a pump varies with the speed at which the impeller rotates. Theoretically, varying the pump speed results in changes in water discharge flow rate, total head (TH), and brake horsepower (BHP), according to the following formulas. Any change could affect the pump efficiency, therefore, the best information comes from the characteristic curve for a particular pump.

Example 7-3. Determining impeller speed, brake horsepower, and efficiency of a centrifugal pump.

Determine the impeller speed, brake horsepower, and efficiency of the centrifugal pump in Figure 7-9 at 900 gpm discharge and 120 feet of total head.

Solution:

Follow the vertical line up from 900 gpm on the bottom axis to the 120 feet dynamic head horizontal line. For this given impeller diameter, the pump impeller must rotate at 1,600 RPM. At this speed it is operating at 71% efficiency and requires about 39 horsepower on the input shaft.

Example 7-4. Determining the number of stages and impeller diameter for a deep-well turbine pump.

Determine the number of stages and impeller diameter of the deep-well turbine pump in Figure 7-10 at 800 gpm discharge and 133 feet of total head.

Solution:

Start by determining the total head per stage with different impeller diameters. Follow the vertical line up from 800 gpm on the bottom axis and identify the TH for each impeller size:

6.56 in. impeller = 16.5 ft TH
7.13 in. impeller = 27 ft TH
7.72 in. impeller = 38 ft TH

Determine the number of stages required for each impeller size by dividing the TH required by the TH for an impeller :

$$\text{6.56 in. impeller: } \frac{\text{133 ft TH}}{\text{16.5 ft per stage}} = \text{8.1 stages}$$

$$\text{7.13 in. impeller: } \frac{\text{133 ft TH}}{\text{27 ft per stage}} = \text{4.9 stages}$$

$$\text{7.72 in. impeller: } \frac{\text{133 ft TH}}{\text{38 ft per stage}} = \text{3.5 stages}$$

It is best to select an impeller diameter that will operate within its peak efficiency area and produce just slightly more pressure than the total head requirements for the system. The 7.13 in. impeller diameter from this pump curve series offers the best fit in that it will operate in its peak efficiency zone during the desired flow rate and will require 5 stages to produce the desired system pressure. The larger diameter impeller could also be made to work by trimming the impeller diameter so as to require only 4 stages. However this impeller would be operating in a lower efficiency area than the 7.13 impeller.

Equation 7-6. Affinity Law: Flow Rate.

$$GPM_{final} = \left(\frac{RPM_{final}}{RPM_{initial}}\right) \cdot GPM_{initial}$$

Equation 7-7. Affinity Law: Total Head.

$$TH_{final} = \left(\frac{RPM_{final}}{RPM_{initial}}\right)^2 \cdot TH_{initial}$$

Equation 7-8. Affinity Law: Brake Horsepower.

$$BHP_{final} = \left(\frac{RPM_{final}}{RPM_{initial}}\right)^3 \cdot BHP_{initial}$$

Where:

$RPM_{initial}$ = Initial revolutions per minute setting
RPM_{final} = Final revolutions per minute setting
GPM = Gallons per minute (subscripts same as for RPM)
TH = Total head (subscripts same as for RPM)
BHP = Brake horsepower (subscripts same as for RPM)

Diameter Change and Pump Performance

The performance of a pump impeller can be changed by reducing the diameter of a full impeller (trimming impellers). Typically, the impellers are trimmed to achieve a specific flow rate. Theoretically, the trimming of a pump impeller's diameter will result in changes in the water discharge flow rate, the total head (TH), and the brake horsepower (BHP), similar to changing the speed of rotation.

This can be estimated by applying the following formulas. However, the best information is the pump characteristic curve for a particular pump impeller diameter that has been tested by the manufacturer because the efficiency of the modified impeller is not easily determined.

Equation 7-9. Flow Rate: Constant Speed, Change in Impeller Diameter.

$$DIA_{final} = \left(\frac{GPM_{final}}{GPM_{initial}}\right) \cdot DIA_{initial}$$

Equation 7-10. Total Head: Constant Speed, Change in Impeller Diameter.

$$TH_{final} = \left(\frac{DIA_{final}}{DIA_{initial}}\right)^2 \cdot TH_{initial}$$

Example 7-5. Determining the effects of changing pump speed.

Determine the effect of increasing pump speed by 50% on discharge, head, and BHP.

Solution:

Use Equations 7-6, 7-7, and 7-8 to determine the GPM_{final}, TH_{final}, and BHP_{final}

$$GPM_{final} = \left(\frac{1.5}{1}\right) \cdot GPM_{initial} = 1.5 \cdot GPM_{initial}$$

$$TH_{final} = \left(\frac{1.5}{1}\right)^2 \cdot TH_{initial} = 2.25 \cdot TH_{initial}$$

$$BHP_{final} = \left(\frac{1.5}{1}\right)^3 \cdot BHP_{initial} = 3.38 \cdot BHP_{initial}$$

Note how rapidly BHP increases with increases in the RPM. It is therefore very easy to overload an existing power unit by increasing the pump speed. Pump performance relationships shown in Equations 7-6 through 7-8 also are commonly known as the Affinity Laws.

Equation 7-11. Brake Horsepower: Constant Speed, Change in Impeller Diameter.

$$BHP_{final} = \left(\frac{DIA_{final}}{DIA_{initial}}\right)^3 \cdot BHP_{initial}$$

Where:

$DIA_{initial}$ = Initial diameter of curve reported impeller
DIA_{final} = Final diameter
GPM = Gallons per minute (subscripts same as for RPM)
TH = Total head (subscripts same as for RPM)
BHP = Brake horsepower (subscripts same as for RPM)

Pipelines

For most sprinkler irrigation applications, water is transported from the source to the irrigation system through pressurized pipelines. Pipelines are either laid on the surface for portability, or they are permanently buried with risers installed as needed. Buried pipelines reduce labor and are out of the way, but they involve higher initial costs and have little salvage value except in association with the land.

Pipe selection and sizing depends on portability needs, field topography, flow rate, water velocity, operating pressure along the pipeline, and the cost of energy and piping materials. The best pipe size or fitting is not always the lowest cost unit. An assessment of the fixed cost of the pipe and fittings as well as the operating cost should be done to determine the most economical pipeline arrangement.

Buried pipelines require special knowledge and experience to assure proper installation and operation. For example, concrete thrust blocks may be needed at certain locations to transfer the thrust to undisturbed soil in the trench and keep the pipe from moving or breaking (see section on thrust blocks). Buried PVC pipes greater than 4 inches that are drained before freezing usually need at least 30 inches of cover to protect them against machinery loading. Working with an experienced installer and supplier is the best choice when using buried pipelines.

Thrust Blocks for Buried Pipelines

As water is forced to change direction or to stop, it exerts forces that can make the pipeline move. Thrust blocks are masses of concrete used to transfer the thrust to the undisturbed soil in the trench and keep the pipeline from moving or breaking.

Figure 7-11 shows some typical arrangements of concrete thrust blocks. Detailed specifications are outlined in ASAE Standard S376.1, *Design, Installation and Performance of Underground, Thermoplastic Irrigation Pipelines*. Thrust occurs at these locations:

- Changes in flow direction such as at tees, bends, and curves.
- Changes in size, such as at reducers.

Example 7-6. Determining the effects of impeller trimming.

How much does an 8.00 inch impeller have to be trimmed if the desired flow rate is reduced from 820 gpm to 775 gpm? And what effect does this have on the TH and BHP?

Solution:

Use Equations 7-9, 7-10 and 7-11 to determine DIA_{final}, TH_{final} and BHP_{final}.

$$DIA_{final} = \left(\frac{775\ gpm}{820\ gpm}\right) \cdot 8.00\ in\ = 7.56\ in$$

$$TH_{final} = \left(\frac{7.56\ in}{8.00\ in}\right)^2 \cdot TH_{initial}\ = 0.89 \cdot (TH_{initial})$$

$$BHP_{final} = \left(\frac{7.56\ in}{8.00\ in}\right)^3 \cdot BHP_{initial}\ = 0.84 \cdot (BHP_{initial})$$

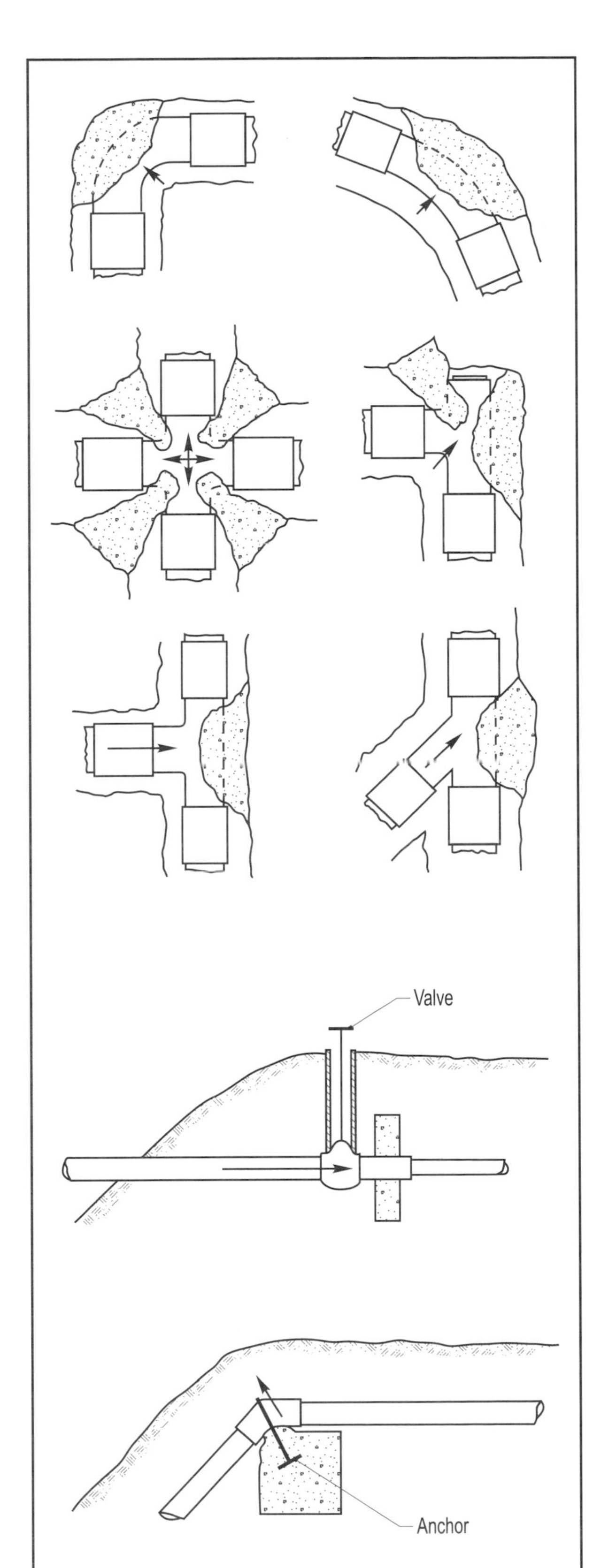

Figure 7-11. Location of thrust blocks.
Source: ASAE S376.1

- Stops, as at a dead end (end cap).
- Valves.

The direction of movement that would occur must be determined, and then the thrust block placed to counteract that thrust.

The thrust block must fit against the pipe and cover an adequate area on the trench wall to counteract the thrust. The size of a thrust block depends on several factors:

- Soil type in the trench.
- Size of the pipe.
- Type of fitting or curve under consideration.
- Working pressure of the pipe.

Water Hammer

Water hammer is the conversion of flowing energy in the water to elastic energy that causes a series of negative and positive pressure waves to move up and down the pipeline until friction damps the energy or until rupture occurs. Water hammer can be caused by closing a control valve too quickly in a flowing pipeline or filling a pipeline too quickly as it becomes fully pressurized. Water hammer surges in irrigation pipelines are similar to the banging that occurs in household pipes except that the result of water hammer is much larger in irrigation lines and can be very damaging. Uncontrolled water hammer can explode pipelines, burst pumps, and blow pivots apart. Water hammer adds to the normal operating pressure of the system. Without proper precautions, water hammer pressures can exceed 400% of the normal operating pressure of a system and exceed the pressure rating of a pipe. Negative water hammer pressure also is possible and can collapse pipelines. To prevent water hammer, many irrigation systems are installed with a manually or automatically controlled flow throttle valve near the pump to minimize flow rate during startups thus reducing water hammer risks.

Water hammer surge potential is the potential of a given installation to create a situation where water hammer can occur. The length of irrigation pipeline is one indicator of water hammer surge potential.

Short pipelines have less water hammer surge potential than longer pipelines.

For example, a producer is pumping water at a rate of 1,000 gpm with an irrigation system that includes a pipeline with a long straight stretch (2,000 feet) between the pump and an elbow (changes direction). If the pump stops operating for whatever reason, water rushes to the elbow, creating a *positive* pressure surge at the elbow. If the elbow, pipeline, or connection does not rupture, then the increasing positive pressure created at the elbow will eventually stop the movement of water rushing towards the elbow and will cause the water to reverse directions towards the pump.

The result of the water rushing towards the pump will create a suction or *negative* water hammer at the elbow. Once water arrives at the pump, the positive pressure will cause the water to stop and reverse direction towards the elbow again. The water oscillating between the elbow and pump will eventually lose energy and dampen to a stop.

Other factors that affect water hammer are pipe size and water flow rate (velocity) changes, and whether air is trapped in the pipeline.

Pipeline water velocity is directly related to flow rate and, thus to water hammer. Recommendations are that water velocity be less than one foot per second when filling pipelines. Maximum velocity once the line is full should be less than five feet per second.

Air pockets in a pipeline can cause water hammer when the air is compressed and moves along the pipeline. Pipelines that have been shut down for even a few minutes often will have air in them due to vacuum relief valves. It is very important to have air release valves, as shown in Figure 7-15, where needed along pipelines to safely discharge trapped air.

Manufacturers usually provide a surge pressure rating for pipes that is determined by using an ASTM standardized test. This test procedure includes subjecting the pipe to high pressures to determine when the pipe will burst. The manufacturer usually publishes the surge pressure rating along with other pipe specifications.

Materials

Irrigation pipes are most commonly made from aluminum, steel, polyethylene (PE) or polyvinyl chloride (PVC) material. Table 7-4 shows some of the available sizes and pressure ratings of each of the materials and their respective wall thickness and dimensions. For the specifics of a given pipe product and size options, consult the respective manufacturer's technical manuals.

Aluminum

Aluminum pipe is light and portable. It is useful for temporary installation of a main line or for hand-move sprinkler lateral systems. Several types of couplers and latch assemblies are used on aluminum pipe. The ring lock coupler is the most popular and offers the greatest pressure rating.

Steel

Steel pipe can withstand very high pressure (150 psi and higher depending on pipe wall thickness). Because of its strength, it is used on systems with a large lift or on those with a high pressure requirement. Generally expensive in small sizes, steel pipe may be the most economical alternative for large pipes. It is available in many diameters and in 20- to 30-foot lengths. Steel pipe sections generally are welded together, but rubber gasket joints are available. Welded joints, especially when buried, should be coated or painted to protect them from rusting. Painting the entire pipe, including fittings, above ground can reduce rust.

PE

Polyethylene (PE) plastic pipe is furnished most often in rolls of black tubing 50 to 200 feet in length and available in a limited number of pressure ratings and diameters under 5 inches. PE is normally used for buried submains or mainlines in small acreage systems.

PVC

Polyvinyl chloride (PVC) plastic pipe is the most commonly used material because it is generally the lowest cost material. PVC is lightweight and has minimal friction loss.

Table 7-4. Pipe characteristics for selected pipe sizes and materials.

Pipe material	Nominal size (in)	Pressure Rating (psi)	Wall thickness (in)	Outside diameter (in)	Inside diameter (in)
Aluminum	2	150	0.050	2.0	1.90
	3	150	0.050	3.0	2.90
	4	150	0.050	4.0	3.90
	6	100	0.051	6.0	5.90
		150	0.058	6.0	5.88
	8	100	0.051	8.0	7.90
		150	0.058	8.0	7.88
		175	0.072	8.0	7.86
	10	100	0.064	10.0	9.87
		150	0.094	10.0	9.81
Steel	6	12 Gauge	0.105	6.0	5.79
		12 Gauge	0.105	6.63	6.42
		180 (Schd. 40)	0.280	6.63	6.07
	8	160 (Schd. 40)	0.322	8.63	7.98
PE tubing	2	100 (SDR9)	0.230	2.53	2.07
	2.5	100 (SDR11.5)	0.215	2.90	2.47
PVC	6	180 (Schd. 40)	0.280	6.63	6.07
	8	160 (Schd. 40)	0.322	8.63	7.98
	10	140 (Schd. 40)	0.365	10.75	10.02
PVC (surface)	3	100	0.092	3.0	2.82
PVC (iron pipe size)	2	160 (SDR26)	0.091	2.38	2.19
	4	160 (SDR26)	0.173	4.50	4.15
	6	100 (SDR41)	0.162	6.63	6.30
		125 (SDR32.5)	0.204	6.63	6.22
		160 (SDR26)	0.255	6.63	6.12
		200 (SDR21)	0.316	6.63	5.99
	8	100 (SDR41)	0.210	8.63	8.21
		125 (SDR32.5)	0.265	8.63	8.10
		160 (SDR26)	0.332	8.63	7.96
		200 (SDR21)	0.316	8.63	7.81
	10	100 (SDR41)	0.262	10.75	10.23
		125 (SDR32.5)	0.331	10.75	10.09
		160 (SDR26)	0.413	10.75	9.92
		200 (SDR21)	0.511	10.75	9.73
PVC (plastic irrigation pipe)	6	80 (SDR51)	0.120	6.14	5.9
		100 (SDR41)	0.150	6.14	5.84
		125 (SDR32.5)	0.189	6.14	5.76
		160 (SDR26)	0.236	6.14	5.67
	8	80 (SDR51)	0.160	8.16	7.84
		100 (SDR41)	0.199	8.16	7.76
		125 (SDR32.5)	0.251	8.16	7.66
		160 (SDR26)	0.314	8.16	7.53
	10	80 (SDR51)	0.200	10.2	9.80
		100 (SDR41)	0.249	10.2	9.70
		125 (SDR32.5)	0.314	10.2	9.57
		160 (SDR26)	0.392	10.2	9.42

Source: NDSU - Irrigation Pipe Selection AE-95, ASAE Standard 376.1 - Design, Installation and Performance of Underground PVC Thermoplastic Irrigation Pipelines; Irrigation Technical Manual-Engineering Data 1975; Hal Werner, SDSU Extension.

Lengths of 20 to 40 feet are most common, with various end configurations for easy installation. PVC pipe is used mainly for underground applications because most PVC pipe loses strength when exposed to sunlight. However, some PVC compounds have been specially formulated to be durable for surface applications and are used as a substitute for aluminum pipe, especially in solid-set laterals.

PVC is available in several diameters and pressure ratings to allow selection of the most economical pipe for an application. The pressure rating assigned to a given pipe is a function of wall thickness, outside diameter, and the type of PVC compound used in manufacturing the pipe.

PVC pipe is available in two size designations: plastic irrigation pipe (PIP) and iron pipe size (IPS). IPS is plastic pipe with the same outside diameter as iron pipe so that iron pipe fittings can be used. The outside diameter for PIP pipes as shown in Table 7-4, is always smaller than the IPS pipe for a similar nominal pipe diameter. Likewise, the inside diameter for a PIP pipe of similar size and pressure rating always is smaller than IPS pipe.

Most PVC pipe products are sized to fit one of nine industry accepted Standard Dimension Ratios (SDR) that describe a pipe's pressure design class rating and its maximum allowable operating pressure for that pipe. Standard dimension ratio is simply the ratio of the outside pipe diameter to the thickness of the wall.

Table 7-5 shows the most common plastic pipe SDR ratings and their respective maximum operating pressure ratings. The operating pressures shown in Table 7-5 are the highest pressure a PVC pipe should be allowed to carry if the potential surge pressure is not known. Surge pressures should not be allowed to exceed 28% of a pipe's design class rating.

Table 7-5. Design and operating pressure ratings (psi).

SDR rating	Design class	Maximum operating[a]
17.0	250	180
21.0	200	144
26.0	160	115
32.5	125	90
41.0	100	72
51.0	80	58
81.0	50	36

[a]Maximum operating pressure is the highest pressure a pipe should be designed to carry if the potential surge pressure is not known.
Source: ASAE Standard 376.1 - Design, Installation and Performance of Underground PVC Thermoplastic Irrigation Pipelines.

Buried Pipe Connections

Gasket connections that allow slight slippage to overcome thermal expansion and contraction are used on most buried PVC pipe. Some PVC designs may warrant using glue-welded joints. The rubber gasket is positioned in a specially flared and shaped groove so that it does not roll or twist when the pipe is inserted. The pipe must be lubricated before assembly to assure a correct fit. Most pipe manufacturers provide marks to show proper insertion distance. When the pipeline becomes fully pressurized with water, the rubber gasket forms a complete seal between the two pipe pieces.

When the pipeline is not under pressure, some water may leak from the surrounding soil-packed trench through the gasket seal. If a pipeline is placed in an area of high water table and left empty during the off season, some outside water may leak into the pipeline. This could cause pipe damage if the water freezes.

Surface Pipe Connections

Aluminum pipe has several types of couplers and latch assemblies in combination with a rubber gasket joint available, depending on where it is to be used and the needed pressure rating. Most common types include band-and-latch or ring-lock couplers. Each type of latch assembly has a maximum pressure rating for safe use. The ring-lock coupler is very popular for high pressure applications along mainlines, but the pipe must be laid carefully so that it cannot snake and cause uneven pressure, which will break the band.

The band-and-latch is most commonly used with lateral pipeline sections with solid-set and hand-move systems. When the couplers are not under pressure, the rubber gaskets leak freely and allow most of the water in the pipeline to be spilled onto the ground around each joint. This may cause very wet spots along some joints, depending on the lay of the pipeline.

Aboveground PVC plastic pipelines used in solid-set sprinkler systems come with a band-and-latch similar to the type used on aluminum pipe laterals.

Steel joints are most commonly welded and should be painted with a non-rusting product or coated after installation. Optional equipment, such as water meters or chemigation valves, is best connected on a steel pipeline with the use of bolted flanges to enable easy removal for maintenance or replacement. For a plastic PVC-to-steel connection, special steel gasket joint couplers can be welded to the steel pipe end to make a water-tight fitting above or below the ground.

Pipe Sizing

Determining the most acceptable pipe diameter for a given irrigation system should consider at least these two criteria:

- The flow velocity in the pipe should be less than 5 feet per second.
- The pressure loss must beless than 1 foot of head per 100 feet of pipe.

To avoid excessive friction losses, extreme surge problems, and possible pipeline failure, the velocity of water in the mainline pipe generally should not exceed 5 feet per second. Piping systems with higher velocities could require special pressure relief devices to help control surge and water hammer pressures. For example, the pipeline of a center pivot and linear-move lateral will most often have its water velocity exceed this value, but the sprinkler nozzles serve as relief valves to reduce water hammer potential. However, if water hammer pressures are too high, some sprinkler heads with pressure regulators can be damaged.

Table 7-6 shows the flow rate for several pipe sizes when water velocity is held at 5 feet per second. Figure 7-12 can be used to determine and compare water velocities of specific pipe diameters and flow rate.

Comparing sizes of pipe made of the same material shows larger sizes cost more initially, but the water flow rate will be less, resulting in lower pressure losses and decreased overall operating costs. Consider the cost of piping materials as well as energy costs to pump at the desired rate.

Pressure Loss

Pressure loss from friction is the pressure drop in a length of pipe caused by friction between the pipe walls and the water. Excessive friction loss raises the pump pressure requirements and increases horsepower needs, thus raising overall pumping costs.

Total pressure loss in a pipeline depends on several factors:

- Flow rate.
- Length of pipe.
- Pipe inside diameter.
- Pipe material and inside roughness.
- Number and types of fittings and bends in the pipeline.
- Number of discharging outlets along submains or laterals.

Information from the manufacturer is the best source to determine friction loss in a specific pipeline. If manufacturer's data are not available, Table 7-3, Figure 7-12, and Equation 7-12 can be used to estimate friction losses in either aluminum, steel, or plastic pipe.

Equation 7-12. Pipe Friction Loss.

$$f = \frac{1044 \cdot \left(\frac{Q}{C}\right)^{1.85}}{D^{4.87}}$$

Table 7-6. Maximum flow rate. Water velocity of 5 feet per second.

Nominal size (in)	2	3	4	5	6	8	10	12	16
Flow rate (gpm)	50	110	200	310	440	780	1225	1760	3140

Source: Pacific Northwest Extension PNW 290-1986. Sizing Irrigation Mainlines and Fittings

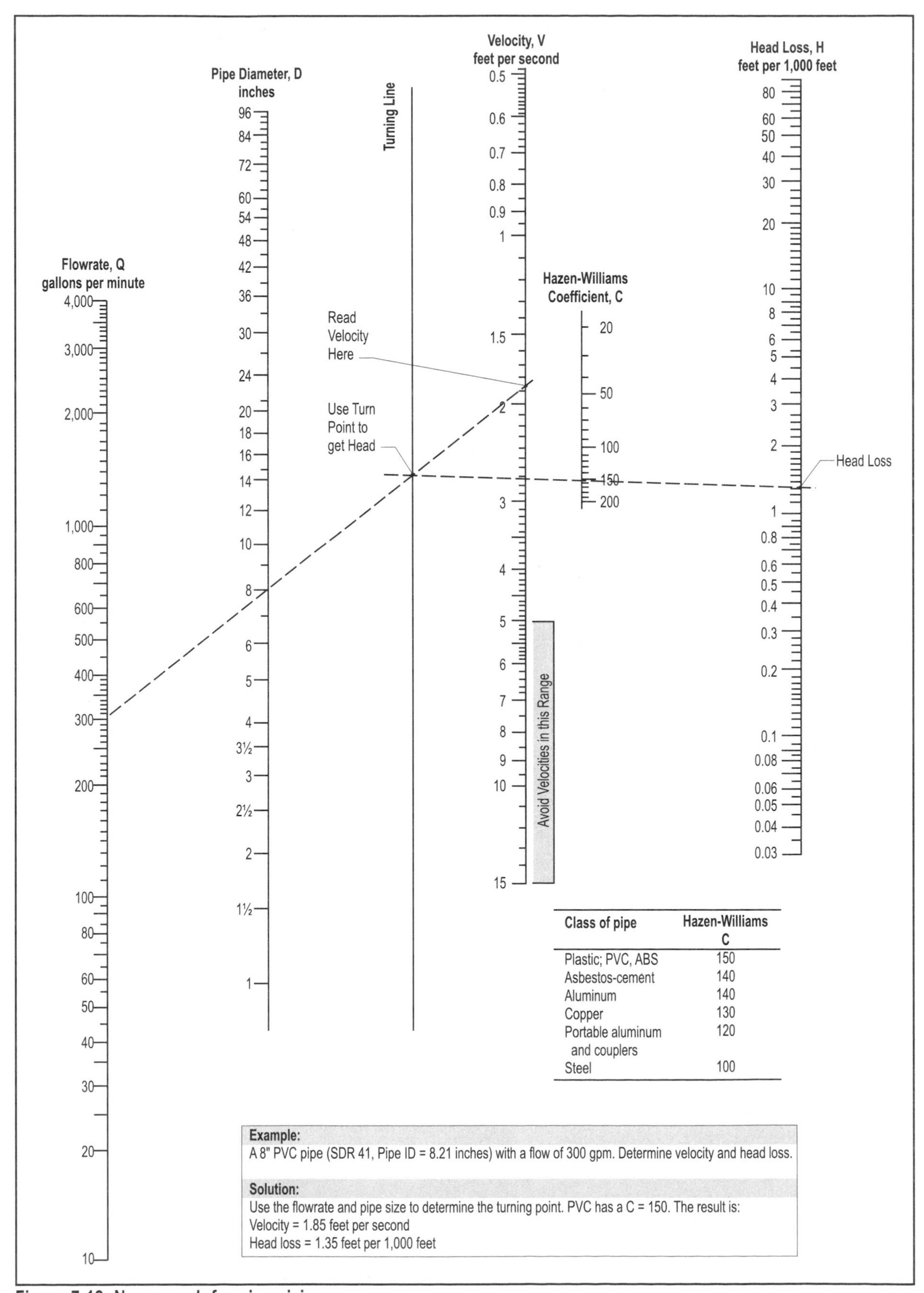

Class of pipe	Hazen-Williams C
Plastic; PVC, ABS	150
Asbestos-cement	140
Aluminum	140
Copper	130
Portable aluminum and couplers	120
Steel	100

Example:
A 8" PVC pipe (SDR 41, Pipe ID = 8.21 inches) with a flow of 300 gpm. Determine velocity and head loss.

Solution:
Use the flowrate and pipe size to determine the turning point. PVC has a C = 150. The result is:
Velocity = 1.85 feet per second
Head loss = 1.35 feet per 1,000 feet

Figure 7-12. Nomograph for pipe sizing.

Where:

f = Friction loss, feet per 100 feet of pipe
Q = Flow rate, gpm
C = Friction coefficient for type of pipe [100 for steel pipe
120 for aluminum pipe with couplers
150 for plastic pipe]
D = Inside diameter of the pipe, inches

Except in a few cases, friction loss should not exceed one foot of head or 0.43 psi per 100 feet of pipe. Economic considerations may make the most economical friction loss less than those values.

Multi-outlet Submain or Lateral Pressure Loss

The pressure loss along a multi-outlet submain pipeline or a solid-set lateral with evenly spaced outlets is equal to only a fractional percentage of the total pressure loss that would have occurred in a pipe of the same length, diameter, and total flow but with only an end outlet. Table 7-7 shows the fractional values for pipelines of different outlet numbers. Multi-outlet pipelines should be sized so the total pressure drop is less than 20% of the operating pressure. Figure 7-13 illustrates how friction losses vary with different sprinkler system laterals.

Table 7-7. Fractional values for multiple outlets.

Number outlets	Fractional percentage	Number outlets	Fractional percentage
1	1.00	8	0.42
2	0.64	9	0.41
3	0.53	10 to 11	0.40
4	0.49	12 to 14	0.39
5	0.46	15 to 20	0.38
6	0.44	21 to 35	0.37
7	0.43	greater than 35	0.36

Source: USDA-NRCS National Engineering Handbook, Irrigation Chapter 11, Sprinkler Irrigation.

Center pivots have a similar pressure loss relationship, but it is much more complex to calculate because most of the water in a center pivot is discharged at the outer sections where most of the land area is covered. Hence, friction loss for a center pivot lateral can be reasonably estimated by taking 54% of an equivalent length of pipeline with all discharge at the end for the given flow.

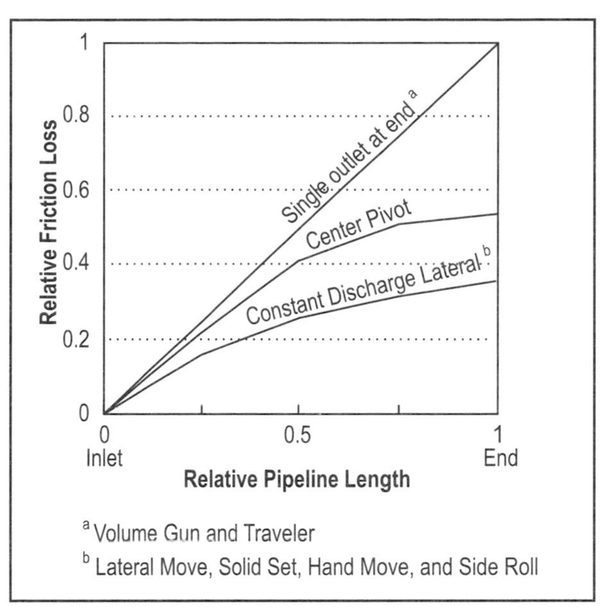

Figure 7-13. Relative friction losses in sprinkler laterals.

Pipeline Fittings and Accessories

Pipelines are the major component of a water delivery system but the pipeline itself must have appropriate fittings, connections, and associated devices to deliver and control the water as intended. These associated devices protect the pipe and the water supply, and allow for branching the pipeline.

Several fittings and connectors placed near the pump make irrigation management more convenient. Figure 7-14 illustrates one arrangement of fittings.

Fittings and accessories produce friction losses like pipelines, but sometimes at a greater rate because of the additional water turbulence their design creates. When determining the total friction losses for a pipeline system, include pressure losses from accessory valves and fittings to ensure the irrigation pump is sized appropriately (Example 7-7, page 132).

Friction loss data for a given accessory or fitting are best obtained from the respective manufacturer's literature. If not available, Table 7-8 can give some general guidelines for fittings and valves. The losses for a valve or fitting may seem small, but over an entire

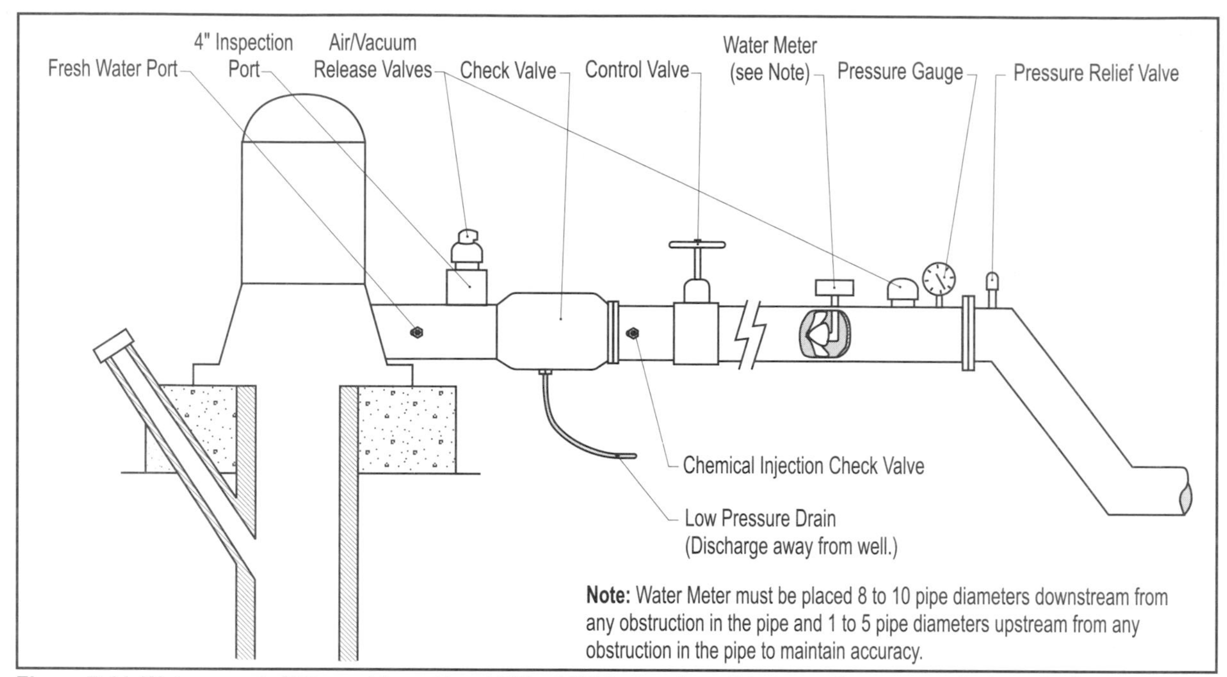

Figure 7-14. Water supply fittings. Allow at least 20 feet (30 feet preferred) between the pump and where the line goes underground or connects to distribution lines. Allow for control valves, chemigation equipment, water meters, and other fittings.

Table 7-8. Effect of friction associated with pipe fittings and valves.

Fitting type	Pressure loss (psi)	Equivalent pipe length (ft)
90° long sweep elbow	0.05	14
90° standard elbow	0.08	20
45° elbow	0.03	8
Sudden entrance	0.06	19
Swing check, wide open	0.16	50
Gate valve, wide open	0.02	5
Gate valve, half open	0.41	130
Foot valve	0.25	76
Flow meter	0.14	32
Chemigation valve	0.20	60

system, the total loss may be substantial. A general recommendation is to select fittings the same size as the main pipeline, but some valves might be more economical if a smaller diameter is selected.

Most valves and other pipeline accessories are best installed in the pipelines with flanged couplers so they are more easily removed from the line for regular maintenance or replacement. Several pipeline accessories, for example water flow meters, are available to help an operator better manage the irrigation system.

Some irrigation systems may be used to apply a portion of a crop's fertilizer or plant protection needs, insecticides for example, through the irrigation watering system. This requires a chemical injection port downstream. Figure 7-14 shows an example of a possible state-approved arrangement. Figure 8-9 in Chapter 8 shows a more detailed cross-section of a anti-backflow chemigation check valve. A garden hose valve installed upstream from this chemigation valve also might be useful for supplying water to the chemigation supply tank, or as a safety wash if chemicals come into contact with the equipment operator. This is shown as the freshwater port in Figure 7-14.

The transition from above-ground to underground pipe is most commonly done with a Z-pipe (dogleg) assembly shown in Figures 7-14 and 7-16, or a vertical, large-diameter stilling steel tube. The Z-pipe to or from underground pipe should have appropriate bends (commonly 45^0) and adequate thrust blocking to prevent movement when starting or stopping the pump. Access ports should be installed at all low points in a pipeline to allow the pipe to be completely drained of water or pumped out for winter.

Hydrants

Hydrants deliver water to an intermediate point along the pipeline. They must be sized and installed to deliver the flow volume at the correct pressure. Numerous styles are available. Hydrants should deliver water at the proper location and height. Hydrants on a permanent line must be marked and protected from damage during field operations, from livestock, and from erosion.

Valves

Valves are installed in a pipeline to either control the amount of flow or to protect the pipeline. Manual and automatic flow control valves typically are of the same nominal size as the pipeline so that pressure drop through the valve is minimal. Valves may be one size smaller than the pipeline to reduce costs.

A flow control valve should be installed in the pipeline near the pump for each water source. In some cases this valve may be used for continuous control, but more importantly, it can be used to reduce the rate of flow into a pipeline during startup until all air is purged from the line. Trapped air can cause water hammer and surges that produce high pressures and damage the pipeline. For the most efficient operation, the control valve should be fully open once the pipeline is up to pressure; pumping against a partially closed valve causes a pressure drop, which wastes energy and increases operating costs.

Automatic flow and pressure control valves are available to maintain a constant flow rate or pressure in the line. This helps ensure safe, unattended operation of the irrigation system. Most simple piping systems do not need automatic valves, but they are useful on systems where flow volumes vary depending on how many sprinklers are operating, or for some sprinkler systems equipped with an automatic startup feature.

Check valves prevent the reverse flow of water. Placed between the pump and the pipeline, they prevent the backwards flow of water after shutdown that may cause damage to a pump. Check valves also help prevent sand formation around a well screen. Some valves have a counterweight or spring to close the valve before flow reversal, which also reduces water hammer stress on the pipeline and fittings, (see Figure 8-9 in Chapter 8). If chemigation is to be used, state law usually will require a higher quality check valve.

Chemigation Check Valves

Chemigation is the practice of using an irrigation system to apply an agrichemical (fertilizer or pesticide) to the soil or plant foliage. The chemical is injected into the irrigation water. The EPA requires any irrigation system that applies an approved pesticide compound to contain an anti-backflow chemigation check valve in the main pipeline between the water supply and the point of chemical injection (see Figure 8-9). Most states also require a similar backflow check valve be installed when applying a fertilizer compound through irrigation. Figure 7-11 shows a typical chemigation valve along with other irrigation accessories.

Some state regulations allow only certain anti-backflow chemigation valve models for irrigation water supplies and require other types, such as a reduced pressure zone valve (RPZ), when connected to a public water supply system. Many of these states also have a permit application and inspection procedure. Contact a local Cooperative Extension office or an irrigation equipment supplier to receive more detailed information on chemigation regulations in a given state.

Air and Pressure Relief Valves

Trapped air reduces the capacity in a water line and can result in water hammer or contribute to high pressure surges; thus, as a pipeline is filled, air must be purged from the line. In a similar manner, air must be allowed to enter a pipeline as it drains to prevent pipe collapses due to a vacuum. To meet these needs, provide an air/vacuum relief valve between the pump discharge and the check valve on well installations to allow water to freely flow back into the well when the pump is shut off. Figures 7-14

Figure 7-15. Air/vacuum release valve.

and 7-15 illustrate this practice.

In addition, provide air/vacuum relief at all high points and at the end of the pipe, Figure 7-16. Some irrigation pipelines may require the addition of a continuous air release valve at the high points to remove trapped air during operation. For pipeline diameters of 8 inches or less, install air/vacuum relief valves with at least a 1-inch outlet diameter. For larger pipelines, provide at least a 2-inch outlet diameter.

Install pressure relief valves at the end of the pipeline and on the downstream side of a check valve to relieve surge pressures. The smallest pressure release valve should be at least one-fourth the size of the smallest pipe. It should be set to open at a pressure no greater than 5 psi above the pipe's pressure rating.

Pressure Gauges

A working pressure gauge is one of the most important indicators of proper irrigation system operation; therefore, a pressure gauge should be installed near the pump and at the entrance of the sprinkler irrigation system. Pressure gauges can provide very important information when an irrigation pumping system is having problems. Liquid-filled pressure gauges are the best option and should be able to indicate an accurate reading for several years before needing to be replaced. Pressure gauges that are not liquid filled are lower in cost but usually do not last more than one year before losing their accuracy.

Installing a pressure gauge with a small shut off valve at the pressure tap can greatly extend the life and accuracy of the gauge. To read the pressure, just open the valve then shutoff the valve when done reading. This will remove the pressure fluctuations that cause damage to the gauge over long-term use. When operating more than one irrigation system or pumping station, install small shutoff valves at all pressure taps, and share one or two accurate pressure gauges between all systems.

Waterflow Meters

Increasing energy costs and decreasing water supplies point out the need for better water management and to know flow rates, application depths, and volumes. Water measurement is one of the first steps that must to be taken in a total water management program. Water measurement will pay off in the long term with water savings, optimum yields, and lower energy costs.

Good irrigation water management starts by knowing how much water is applied. Waterflow measurement is important in conserving water and energy, as well as:

- Improving irrigation efficiency and water management for growing the greatest amount of crop per unit of water applied.
- Monitoring pumping plant efficiency to minimize fuel used in pumping water.
- Detecting well and pump problems.
- Satisfying legal reporting requirements for water use.

Several instruments are available to measure waterflow in pipelines. These include propeller meters, venturi shunts, paddle wheel meters, and ultrasonic meters. The most common types for water management on farms are the propeller meters and venturi shunts. Ultrasonic meters are expensive and more commonly used by consultants and farms that need a portable pipeflow measuring device.

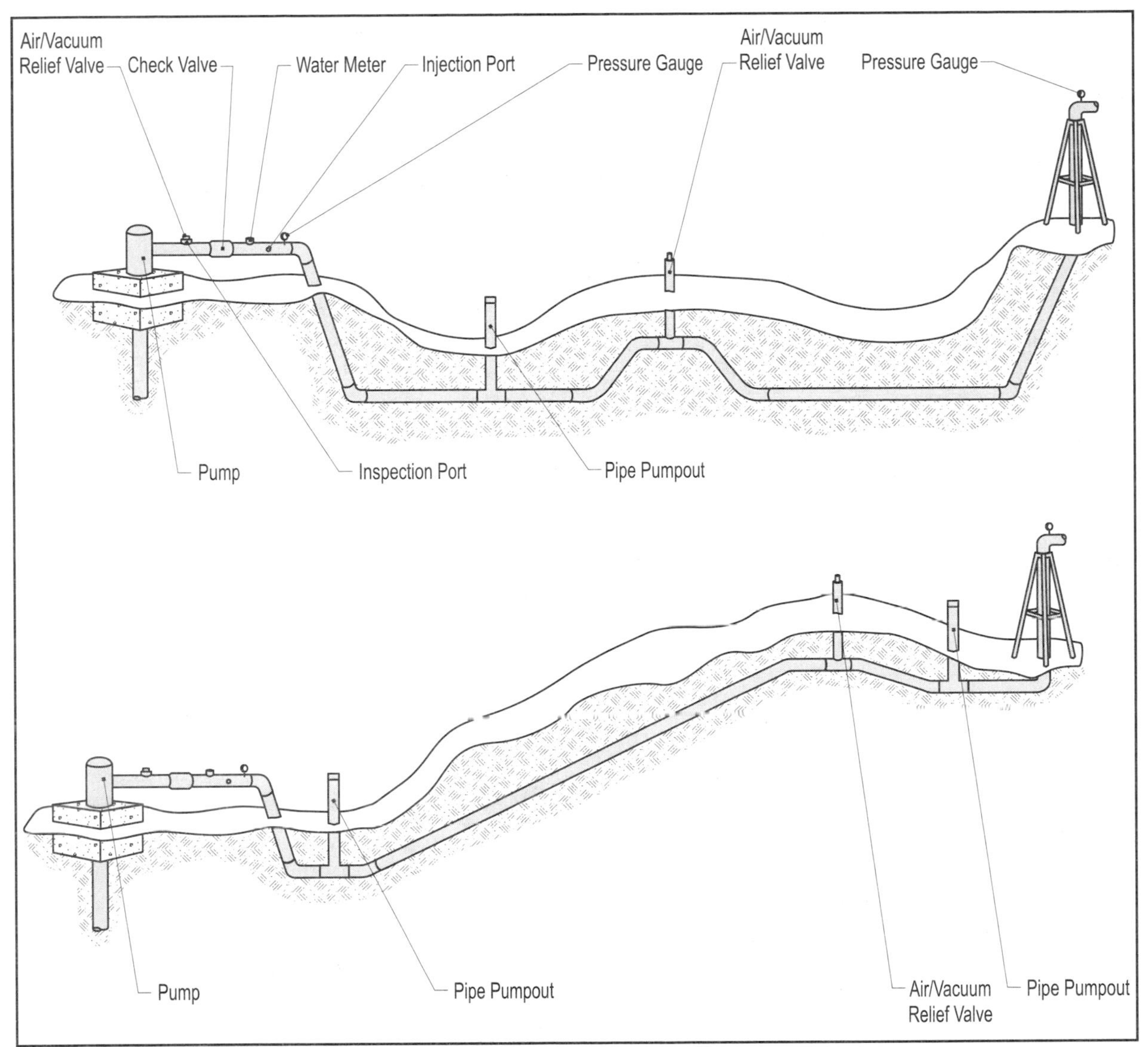

Figure 7-16. Placement of air vent/vacuum-relief valves.

Figure 7-17. Typical pressure gauge.

Propeller Flow Meters

Propeller-flow meters use multi-blade propellers whose speed of rotation is determined by the speed of the water flow, Figure 7-18. Thus, the propeller measures water velocity in the pipeline, and the indicator converts this measurement to flow rate based on the inside diameter of the pipeline.

Propeller flow meters require the following:

1. The proper selection of the meter based on pipe size, range of flow, and head loss.
2. Installation of the meter so the pipe is always flowing full, turbulence in the flow is prevented, and the meter is

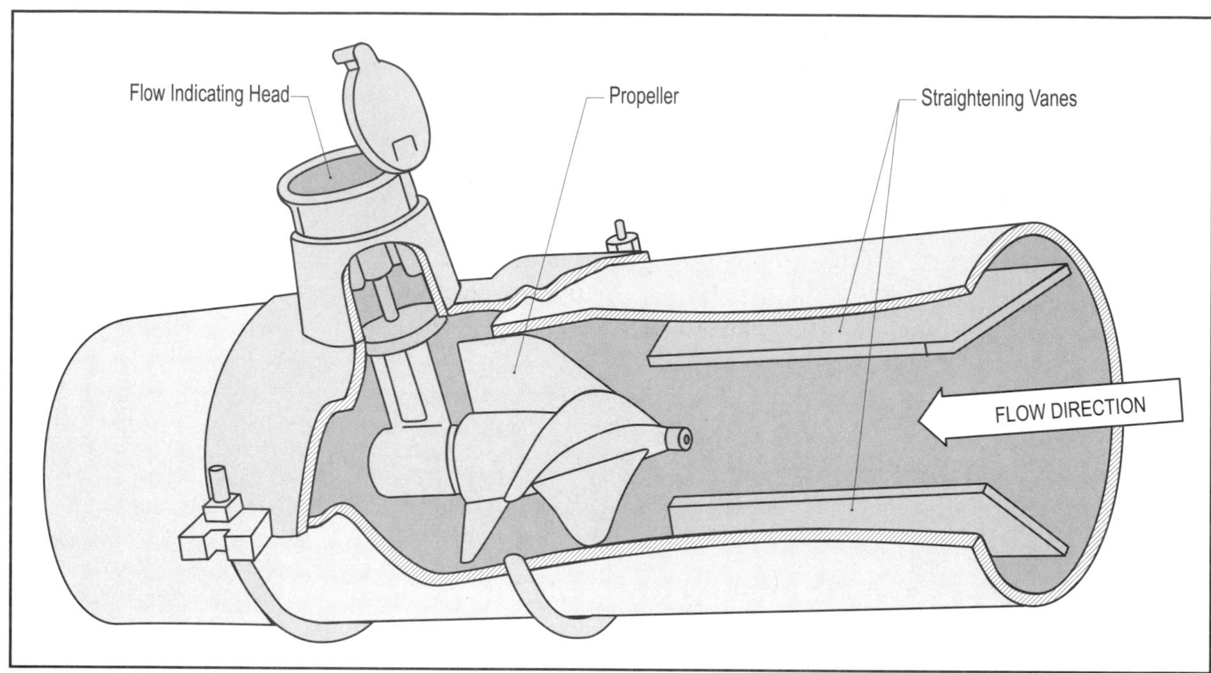

Figure 7-18. Propeller flow meter.

positioned correctly. The meter must be at least 10 pipe diameters downstream from an elbow or valve or any other device that causes turbulence.

3. Proper maintenance to keep the meter accurate and give it a long service life.

A number of commercial companies market these meters in several sizes and styles. Most are built to register both flow rate and total volume of water passing through the meter.

After they are purchased, meters must be calibrated for a particular application. It also is important to have parts and maintenance service available from the manufacturer or manufacturer's representative.

Water meters may vary in size, quality, and design. The size of pipe and range of flow to be measured determine the size of the flow meter needed. The meter must be able to accommodate the expected range of flow so that it can measure the lowest anticipated flow at or near 100% accuracy. Also select a meter that will perform near the mid-range of its capacity.

Most meters now use a magnetic drive between the propeller and the indicator head. This is desirable because it eliminates problems with sealing direct-drive bearings, which can bind from sand or corrosion.

Table 7-9 lists common flow ranges suggested by manufacturers for various flow meters. The size of the propeller usually ranges from 50 to 80% of the pipe's total inside diameter. In pipelines with high flow, small propellers are used when there is little variation in the flow or flow rates. However, larger propellers are more accurate with a wide fluctuation in flow because more flow passes through the propeller.

Table 7-9. Propeller meter flow ranges.

Meter size (in)	Minimum flow		Maximum propeller (gpm)
	Small propeller (gpm)	Regular propeller (gpm)	
4	—	60	400
6	300	100	900
8	500	120	1,200
10	700	160	1,600
12	900	200	2,000

When choosing a meter, make sure the meter gear ratio is selected for the exact inside diameter of the pipe in which the meter will be installed. Otherwise, measurement errors will occur.

For example, if a meter is geared for 6-inch aluminum pipes with a 0.051-inch wall (an inside diameter, or ID, of 5.898 inches) and is installed in a standard 6-inch steel pipe (6.065-inch ID) the meter will register 5% low. Some companies use a correcting factor to account for various IDs rather than changing the meter's gears. Some meters come already installed in a pipe section complete with straightening vanes and should be calibrated by the manufacturer.

Meter Indicator Options

All propeller meters have a volume totalizer on the indicating head. Most companies have the option to add a sweep hand that can be clocked for flow rate calculations and/or a flow rate indicator. Figure 7-19 illustrates two options. The meter can be calibrated in whatever volume units the buyer desires, commonly acre-inches, acre-feet, gallons, or cubic feet. Common flow rate units are gallons per minute and cubic feet per second. To check accuracy, use the volume indicator (totalizer) rather than the sweep hand.

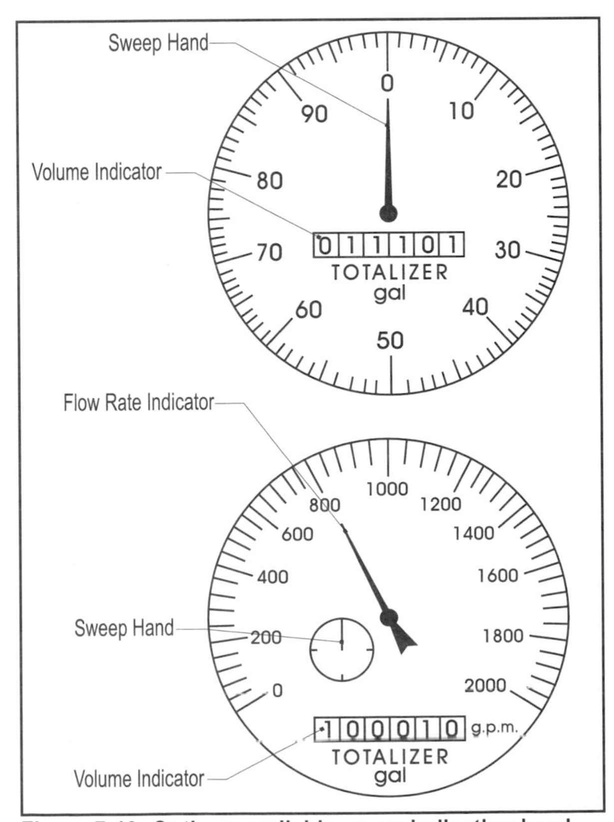

Figure 7-19. Options available on an indicating head.

Installation

Most meter manufacturers offer a wide range of fittings for their meters so they can be installed in essentially any irrigation application. Examples of the types of sections and fittings include flanges, aluminum couplings, weld-on saddle meters, clamp-on saddle meters, or tubes that can be welded in or installed with dresser couplers. Meters can be installed in a buried pipeline with the indicating head extended above ground.

For accurate measurement, the propeller must be positioned in the center of the pipe with the propeller's center line parallel to the pipe's center line.

Propeller meters may be installed in any position convenient on the pipe: vertical, horizontal, or at an angle. Regardless of position, the pipe must always be flowing full for accurate measurement.

Proper installation of flow meters is one of the most important criteria for accurate flow measurement. Figure 7-20a illustrates a correct meter installation, while Figure 7-20b shows an incorrect installation. Spiraling and turbulent flow in the water section caused by valves and pumps, by changes in pipe size, and by tees and elbows will reduce the accuracy of the reading. Because of this, most manufacturers recommend a *minimum of 8 to 10 straight* pipe diameters without obstructions upstream from the propeller, and at least one straight pipe diameter without obstructions downstream from the propeller.

For better results, at least 10 diameters or more are preferred upstream. For an 8-inch meter, this would suggest at least 80 inches of straight pipe be designed between the propeller and an upstream fitting. If space does not allow for 10 diameters, straightening vanes in the pipe section ahead of the meter are recommended. Again, purchasing a meter already installed in a pipe section with straightening vanes is recommended.

Proper installation of meters is usually easiest during the initial installation of the irrigation system rather than as an afterthought. When metering is planned during

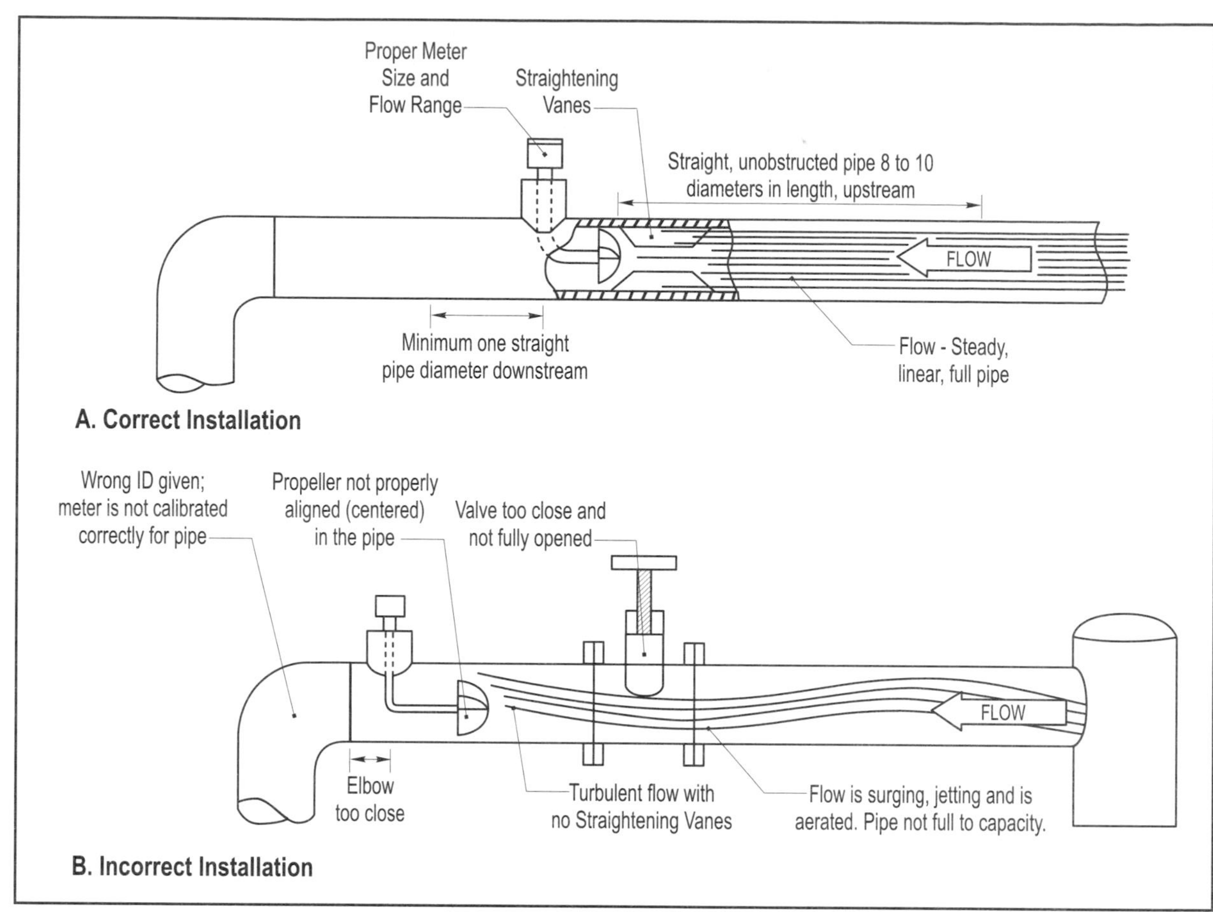

Figure 7-20. Water meter installation.

installation of the irrigation system, enough space can be left for the required length of straight pipe.

Maintenance

Propeller meters are like any other piece of mechanical machinery; they require maintenance and care. Follow the manufacturer's maintenance recommendations. Some meters require periodic lubrication. Anything that causes the propeller to drag will cause inaccurate measurement. Therefore, the meter should be checked periodically to make sure the propeller spins easily. If it does not, check for obstructions that may cause the propeller to bind, or for a worn shaft, bearing, or gear.

A good way to store meters during the off-season is to remove them and place them in a dry place. If contained in their pipe tube, set them on end with a board over the top to keep rodents away from the propeller. Store them in a place where they will not freeze.

Venturi Shunt/Proportional Meters

Venturi shunt/proportional meters divert and measure a small, known percentage of the total flow, usually with a smaller propeller-type sensor, Figure 7-21. These meters require similar installation specifications as do propeller meters.

Paddle Wheel Meters

Paddle wheel meters are mechanical meters that position a small paddle wheel usually just inside the top part of the pipe's cross-section to monitor the water velocity at this point, (Figure 7-22). Paddle meters usually come with fittings for a specific size pipe of known material. This flow meter seems to work best on smaller pipe diameters of 4 inches or less.

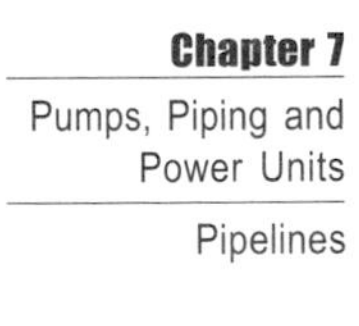

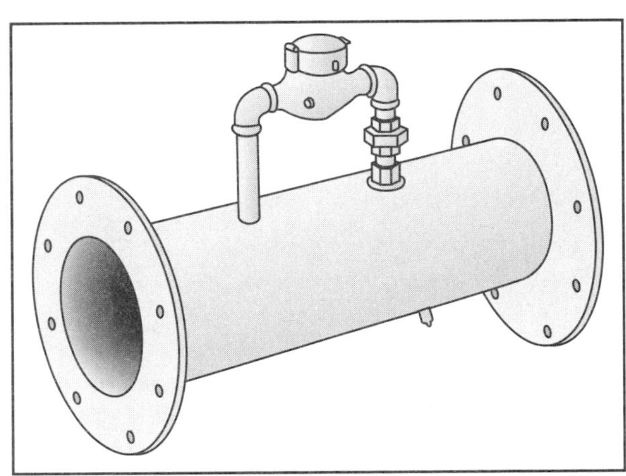

Figure 7-21. Venturi shunt/proportional meter.

The paddle wheel has a magnetic marker that allows the signal reader to track the marker to measure the number of revolutions. The signal reader converts the paddle wheel movement to flow rate by the velocity/cross-sectional area relationship. Like the propeller meters, the paddle wheel should be installed with adequate piping ahead of the meter so that the velocity profile can be established before the water reaches the meter.

Ultrasonic Flow Meters

Ultrasonic flow meters clamp onto the pipe and measure water velocity inside a pipeline without disassembling the pipe line, Figure 7-23. This meter is set according to the type of pipe material, wall thickness, and pipe outside diameter and has a similar placement as propeller meters. These units are portable and can be used on pipelines of aluminum, steel, PVC and other materials. The units are costly and generally are used only by consultants, government agencies, and water equipment suppliers.

With this method, ultrasonic waves are transmitted through the pipe wall and into the flowing water. One burst of energy is transmitted upstream while another burst is sent downstream. The travel times of the two waves are measured and compared. The difference in wave velocity is directly related to the velocity of the water. This type of sensor cannot be exposed to air bubbles in the water flow.

Figure 7-22. Typical paddle wheel meters.

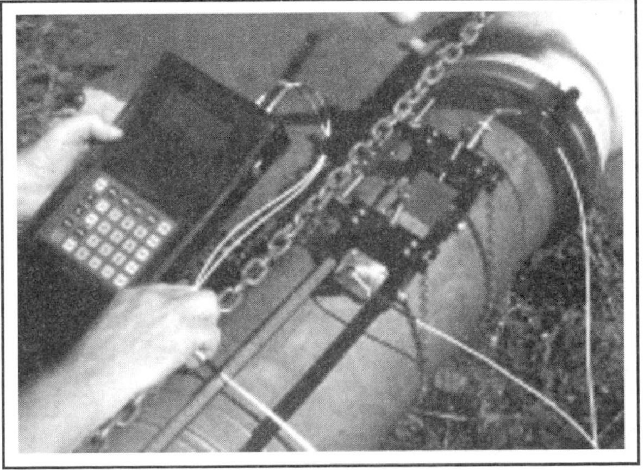

Figure 7-23. Typical ultrasonic flow meter.

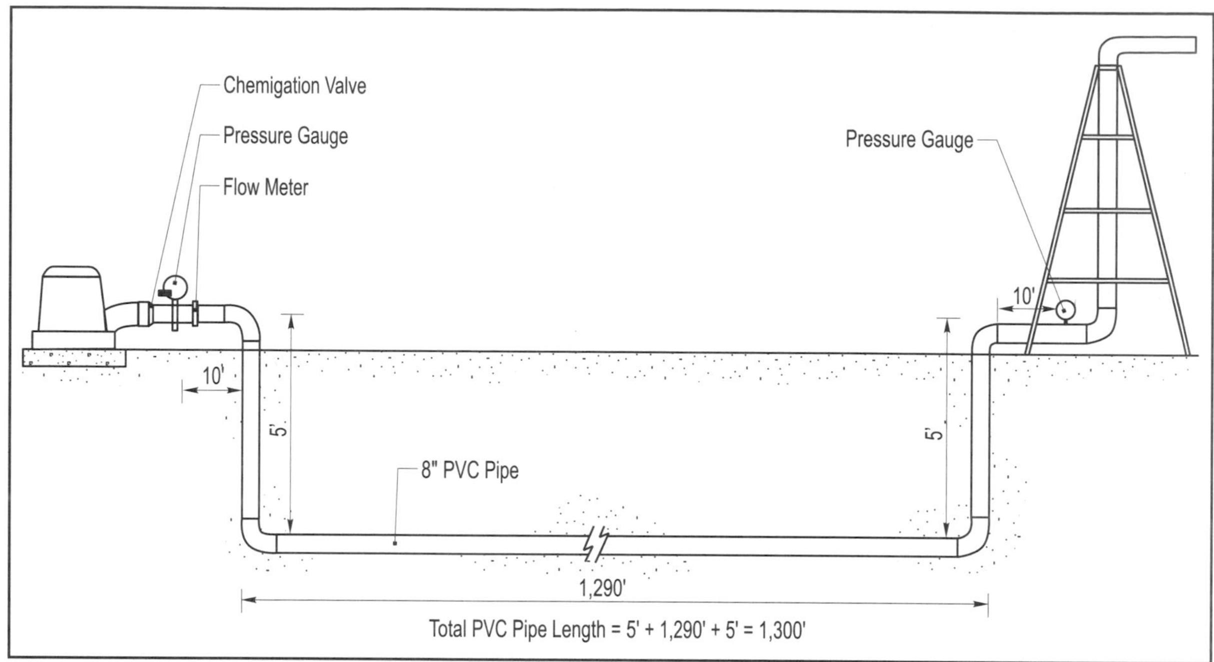

Figure 7-24. Pipe and fitting friction loss example.

Example 7-7. Determining the effects of pipe and fitting friction loss.
A pumping plant is located a quarter mile from a center pivot system used for a 100 acre field. The system has been designed for a water flow rate of 750 gallons per minute. Figure 7-24 shows the layout for the irrigation system. Water is delivered from the pumping plant through 20 feet of 8-inch steel pipe and 1,300 feet of underground PIP class PVC pipe. The system has an 8-inch steel check valve and four 90° steel elbows. What is the friction loss and the resulting difference in pressure between the gauges at the pivot base and the pump discharge head assuming no elevation differences?

Solution:
First, determine the equivalent pipe length for the fittings by using Table 7-8:

Fitting/Valve	Equivalent Pipe Length per Fitting	Number of Fittings	Equivalent Length
Chemigation Valve (steel)	60 ft	1	60 ft
90° standard elbow (steel)	20 ft	4	80 ft
Flow meter (steel)	32 ft	1	32 ft
		Total Equivalent Steel Pipe Length:	172 ft

Next determine the friction loss through the pipe by using Table 7-3:

Pipe	Pipe diameter (in)	Total Pipe Length (ft)	Friction Loss (ft per 100 ft of length)	Total Friction Loss (ft)
Steel	8	20 + 172 = 192	1.80	3.5
Plastic	8	1,300	0.80	10.4
			Total Friction Loss:	13.9

Finally, use Equation 7-5 to convert the total friction loss into the total pressure loss between the gauges at the pivot base and the pump discharge head:

$$(13.9\ \text{ft}) \cdot \left(\frac{0.43\ \text{psi}}{\text{ft}}\right) = 6.0\ \text{psi}$$

If the pump discharge operates at 65 psi, the pressure at the pivot base would be about 59 psi (65 psi - 6 psi = 59 psi).

Power Unit Selection

Common irrigation power units include electric motors, stationary internal combustion engines, and farm tractors. The power unit must fit the irrigation pump requirements to achieve an efficient operation. An engine or motor that is too small cannot deliver the water at the rate and/or pressure required by the irrigation system. An engine or motor that is too large may be inefficient and have excessive initial and operating costs because of the unused capacity.

Typically, industrial internal combustion engines or electric motors are best suited for irrigation operations. A properly designed power unit should meet the power requirements of the irrigation system without supporting any idle capacity. An irrigation load is relatively constant, so the power unit must be suitable for continuous duty at the design load. Electric motors are designed and rated for continuous loads, while internal combustion engines must either be rated for continuous duty at the design load or be derated from some other horsepower rating to a continuous horsepower rating.

Farm tractors should be avoided for permanent installations because they are not designed for the all-day continuous loading typically required for irrigation. When using a tractor temporarily, protect it with safety controls to prevent engine damage, and operate at less than maximum horsepower.

When selecting a power unit, consider the following factors:

- The amount of brake horsepower (BHP) for pumping. This includes the pumping rate, elevation difference between source and sprinkler system point, system operating pressure, friction losses, power unit efficiency, and pump efficiency.
- Hours of operation per season.
- Availability of electricity or fuel source. For electricity, availability of three-phase power may influence selection.
- Cost of fuel or electric energy including requirements to participate in off-peak pumping program.
- Availability of parts and service.
- Power needs for the sprinkler irrigation system, an electric-drive center pivot, for example.
- Pump speed and drive unit.
- The need for portability.
- Labor requirements and convenience of operations.
- Initial cost, including possible electric line construction costs.
- Ease in installing an automatic restart system.

It is important to match the engine horsepower to the requirements of the pump plus any other accessories, such as a generator for electric center pivot tower motors or a hydraulic oil pump. An electrically-driven center pivot many require 15 to 20 additional BHP to run a three-phase generator to provide power to the tower motors and end gun booster pump motor.

Previously used power units should be carefully checked and evaluated as to condition, available horsepower, and speed. Sometimes, an old power unit can cost more to operate than a more expensive unit properly matched to the design load.

Power Unit Requirements

To calculate the actual brake horsepower (BHP) requirements for a power unit requires knowledge of the following for the desired pumping unit:

- Total head (Figure 7-8 and Equations 7-2, 7-3, or 7-4).
- Pumping rate.
- Pump operating efficiency (if for early planning purposes only, use 75%).
- Drive efficiency (Table 7-10).
- Type of power unit.
- Accessories such as a generator for center pivot electric tower motors, a hydraulic pump, or an end-gun booster pump.

The first step to determine BHP requirements is to calculate the water horsepower (WHP) or useful work done by a pump. WHP represents the power required to operate a pump if the pump and drive unit are operating at 100% efficiency. The WHP for a pump is calculated as follows:

Equation 7-13. Water Horsepower.

$$WHP = \frac{(Q) \cdot (TH)}{3960}$$

Where:

WHP = Water horsepower
Q = Discharge, gpm
TH = Total head, feet
3960 = Conversion constant

Brake horsepower (BHP) is the actual horsepower requirement at the drive unit connection and takes inefficiencies of the pump and the drive unit into consideration. The continuous BHP rating of the power unit must provide at least this value at the desired pump speed. The BHP is calculated as follows:

Equation 7-14. Brake Horsepower.

$$BHP = \frac{WHP}{(E_{pump}) \cdot (E_{drive})}$$

Where:

BHP = Brake horsepower required
E_{pump} = Pump efficiency at the operating conditions
E_{drive} = Drive efficiency between the pump and the power unit (Table 7-10)

The pump efficiency must be obtained specifically for an individual pump from the manufacturer's performance curve at the desired operating conditions to determine actual BHP requirements for the pumping unit. If only a preliminary BHP estimation for early planning considerations is desired, use a pump efficiency of 75% (0.75).

Internal Combustion Engines

Internal combustion engines (Figure 7-25) use gasoline, diesel, propane, or natural gas for fuel. Internal combustion engines must be used if an electric power source is not available. Internal combustion engines

- Are portable.
- Provide a variable-speed power source.
- Require on-site fuel storage except with natural gas.
- Have higher service requirements, compared to electric motors.
- Need an electric generator for an electric-drive center-pivot system.
- Have a higher initial cost relative to an electric motor.

Manufacturer's performance curves show horsepower ratings at various speeds for each engine. At laboratory conditions of 60° F air temperatures and sea level elevation, the curves are developed with a bare engine to produce the most horsepower per unit of engine weight. Curves should be corrected for power loss from accessories, elevation differences, and air temperature; internal combustion engines also are corrected to compensate for continuous loading of irrigation pumping (Figure 7-26). Actual reductions of bare engine horsepower are in Table 7-11.

Some manufacturers publish both dynamometer (maximum) and the continuous brake horsepower curves. When only one curve is shown, it is usually the dynamometer horsepower and must be reduced for continuous horsepower.

The most efficient operating load for an internal combustion engine is at or slightly under the continuous brake horsepower curve. Running an engine under much lighter loads (50% or less) usually results in a poorer fuel economy because too much horsepower is need to overcome the engine friction and throttling losses. However, the

Table 7-10. Drive efficiencies of typical units.

Drive	Efficiency
Direct	1.00
V-belt	0.95
Right-angle gear	0.95

Table 7-11. Corrections to bare engine horsepower.
Developed under laboratory conditions.

Condition	Correction
For each 1,000' above sea level deduct	3 %
For each 10 F above 60 F deduct	1 %
For accessories (generator, etc.) deduct	5 %
For radiator and fan deduct	5 %
For continuous operation deduct	20%

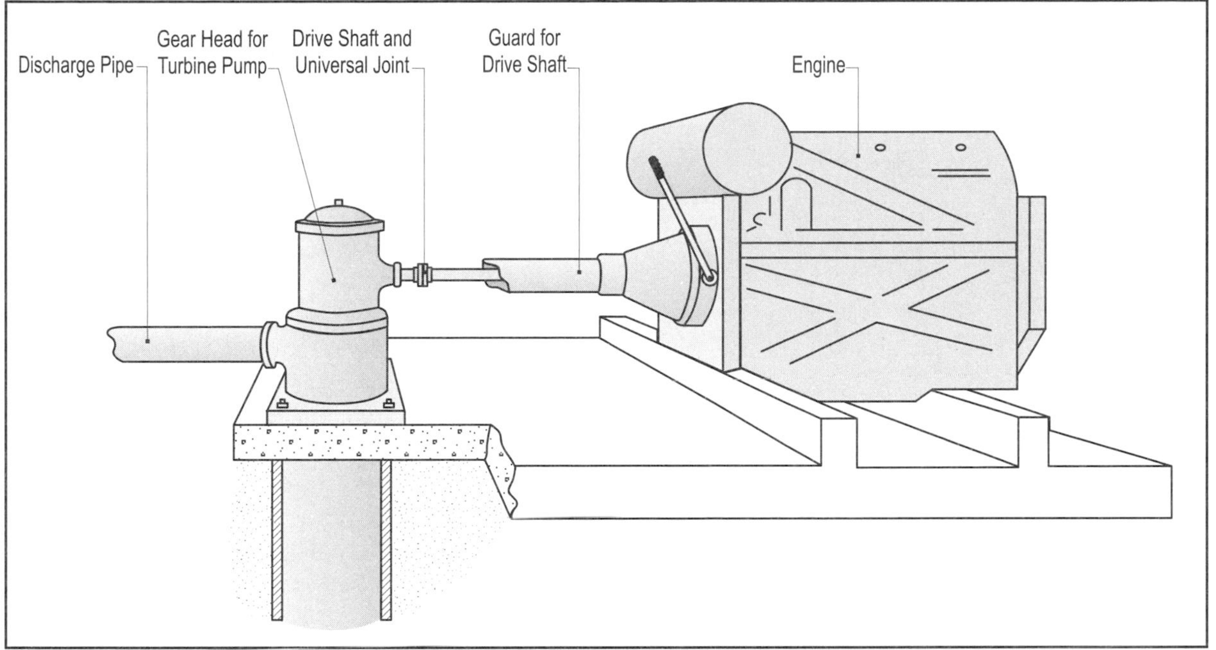

Figure 7-25. Internal combustion engine on a deep-well turbine pump.

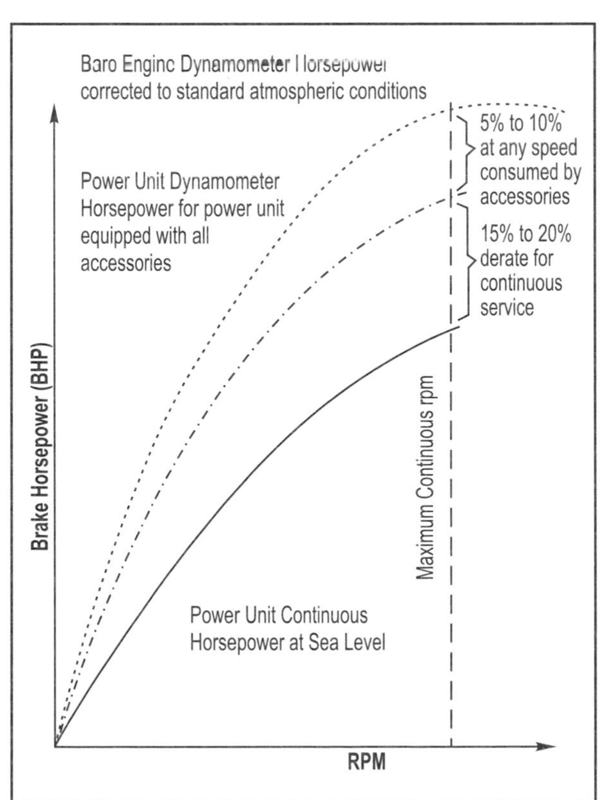

Figure 7-26. Example horsepower vs. RPM curve for internal combustion engines.

engine life expectancy may be higher. Running over the engine's maximum horsepower shortens the life and invites engine trouble and excess fuel consumption.

Engines often are connected to horizontal centrifugal pumps by a shaft, but also may be connected by belts. On deep-well turbines, a right-angle gear head is common for transferring horizontal power to the vertical shaft.

Table 7-10 shows some commonly used drive efficiencies for typical pump drive units.

An engine also might be used to operate a generator to provide electricity for one or more small submersible pumps for repriming, chemigation, or other uses, or to run a center pivot's tower motors and end-gun booster pump. **As shown in Figure 7-27, safety shields must be installed around all rotating shafts and belt-drives.**

Fuel storage tanks should be large enough to provide power for at least one to two irrigation water applications. Some state regulations may have size limitations, water supply separation distances, and secondary containment requirements. State regulation information should be available from a local irrigation equipment or fuel supplier.

Figure 7-27. Typical safety shield.

Electric Motors

Electric motors provide many years of service if properly installed and protected. An electric motor delivers full power throughout its life and operates from no load to full load without damage. Electric motors

- Have relatively long life.
- Have low maintenance costs.
- Are susceptible to lightning.
- Are dependable and easy to operate.
- Work easily with automatic and computer-operated controllers.
- Generally operate at a constant pump speed.
- Require electric power supply at each pumping station.
- Usually require a yearly minimum power cost.
- Have a low initial cost compared to internal combustion engines.
- May have downtime due to electric load management.

Motor Types

Most large electric motors for irrigation are of the squirrel cage-induction type, three-phase, 460/480V motors. Pumps may be connected to the motors by direct couplings, right-angle drives, or belts. The best and most commonly used connection is a direct coupling.

Right-angle drives and the belt drives are less than 100% efficient and thus require more energy to operate (Table 7-10).

Most electrical motors connected to centrifugal pumps have a horizontal shaft, shown in Figure 7-1. On deep-well turbine pumps, a vertical hollow shaft electric motor, as depicted in Figure 7-28 is most common. A horizontal electric motor with a hollow-shaft right-angle drive or a twisted v-belt drive also may be used. The hollow-shaft unit is necessary so the pump impellers can be adjusted.

Single-Phase Electric Power Supply Motors

Single-phase power usually supplies electric motors under 10 horsepower. Special large single-phase motors that start under reduced current are available up to 100 horsepower and may be used on some single-phase lines without having adverse effects on other consumers. Three-phase electric motors sometimes may be best served from the single-phase power using some type of a phase converter. Consult the local electric power supplier when considering the single-phase power to know what options are available in the area.

Phase converters change single-phase electric power to three-phase power and are either static or rotary. Static-phase converters are applicable to single motor loads where the load on the pump is nearly constant. These are "balanced out" (the phase amperages to the motor made equal) for a particular motor load.

Rotary-phase converters serve either single or multiple motor loads. They are desirable when motor loads vary, such as when a permanently installed pump draws from a reservoir that varies in elevation.

Converters can serve several motors. For example, the tower drive motors on an electric drive center pivot system are constantly starting and stopping. Rotary converters need a minimum load to operate; therefore, if one large rotary converter is used for both the irrigation pump motor and the pivot drive motors, the pivot drive motors could not run unless the large pump motor also runs. This would prevent movement of the irrigation system. Dry movement may be necessary for better water management. A separate static converter for the pump motor and a small rotary con-

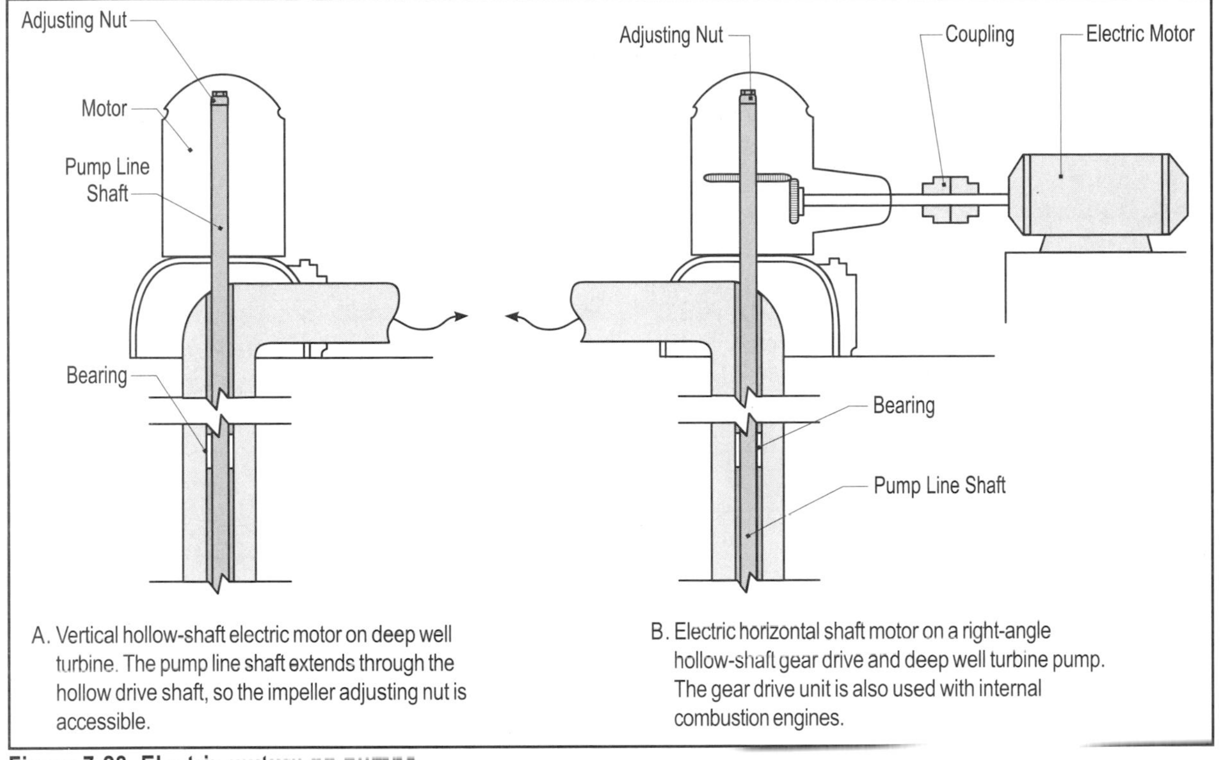

A. Vertical hollow-shaft electric motor on deep well turbine. The pump line shaft extends through the hollow drive shaft, so the impeller adjusting nut is accessible.

B. Electric horizontal shaft motor on a right-angle hollow-shaft gear drive and deep well turbine pump. The gear drive unit is also used with internal combustion engines.

Figure 7-20. Electric motors on pumps.

verter for the center pivot towers would make a workable arrangement.

Proper application of phase converters is critical and should be done only in consultation with the manufacturer, pump supplier, and electric power supplier. Always provide correctly sized overcurrent protection for all three leads to the three-phase motor to help protect the motor from any current imbalance that could result.

Other newer circuitry technologies such as frequency modulation or Written-Pole enables the starting current to be much lower than for regular motors, which increases operating efficiencies and causes power factors to increase. One arrangement can operate any three-phase motor up to 100 horsepower with single-phase power by using only a patented control box containing starting capacitors and running capacitors. Another converter device allows full control of the operating speed of the three-phase motor from start up to full load.

Contact the local electric power supplier or an irrigation motor supplier for information on the latest products available.

Special Provisions for Electrical Installations

All installations should comply with the *National Electric Code* and meet local electrical codes and regulations. For more information, see the standards presented by the American Society of Agricultural Engineers, S362.2, *Wiring and Equipment for Electrically Driven or Controlled Irrigation Machines*, and S397.1, *Electrical Service and Equipment for Irrigation*. Motors, electrical enclosures, and other electrical equipment must be effectively grounded by a separate grounding conductor suitably connected to the power supply grounding system.

Motor Starting

Electric motors draw more amperage to start than to run. Depending on the location, this momentary high amperage may be a nuisance to others on the same electric distribution line. Among other problems, the high draw causes momentary low voltage, light dimming, and television picture distortion.

To minimize starting problems, many electric suppliers require a reduced-voltage starting panel for large electric motors.

These reduced-voltage starters limit the amperage during startup. Four types of reduced-voltage starters are available. The part-winding and wye-delta types require specially wound motors. The primary-resistor and auto-transformer types may be used with any induction three-phase motor.

Protective Measures

The motor power supply should have a fused service disconnect, a motor controller (magnetic or manual), and overcurrent protection. Install single-phase motors with the single-heater type over-current units in the ungrounded conductor. For the three-phase motors, three-heater type overcurrent protection in each of the leads should be provided to the motor windings.

Limit the rating of the overcurrent protection devices to 115% of the motor nameplate amperage. Ambient temperature compensated overload protectors can help offset the effect of sunshine on the motor controls.

Inherent overload protection devices with temperature sensors in the motor windings are highly recommended. They should not be used, however, in place of overcurrent protection unless the motor manufacturer fully guarantees the motor against locked-rotor and overload burnout with only the temperature-sensing element protection.

Electric Power Monitors

Electronic power monitors track the incoming electric power supply and are available at a nominal cost. These monitors are wired into the safety circuit of the electric motor and commonly record loss of any phase, reversed phase sequence, and low and/or high voltage on any phase of a three-phase system.

If a fault or problem is detected in the incoming power supply, the motor will be shut off. Because a three-phase electric motor will attempt to run even when it loses a phase, monitors provide protection beyond in-line overcurrent protection and will increase the life of the motor.

Motor Enclosures

Motors should be either drip-proof or enclosed. Use a drip-proof enclosure if the motor is operated with the shaft vertical unless it is protected from the weather by other suitable enclosures. Rodent screens should be installed either at the factory or when the motor is installed.

Secondary Lightning Arrestors

Secondary lightning arrestors should be installed from ground to each ungrounded conductor in the supply, on the secondary side of the transformer, but ahead of the service disconnect. These devices are especially important for submersible pumps.

Safety Controls

Install safety devices with the irrigation system, and include a list of procedures for protecting the pump and power equipment from damage due to natural causes, sudden loss of water or power, or operator error.

Equip pumping power plants with safety devices (for example, a flow switch or low-pressure shutoff) that stop the motor or engine when there is a break in the suction on centrifugal pumps, or a loss of pressure in the main pipeline. Pumps with water-lubricated bearings can be severely damaged by prolonged operation without water.

Equip engines on sprinkler system pumps with safety devices that stop the engine before damage occurs from

- Overloading.
- Runaway motor from loss of load (for example, loss of prime).
- Loss of oil pressure.
- Overheating.

Couple turbine pumps to power units with a ratchet to prevent rotation of the power plants on high pressure sprinkler systems if a shutdown occurs during operation. Install automatic valves that permit starting or stopping pumps without water damage to high head mainline pipes due to water hammer or surge.

If the irrigation system will be used for chemigation, other safety equipment is required. One part of that equipment is a safety interlock; if the irrigation pump or sprinkler system stops operating, the chemical injection pump also must stop. See Figure

8-2 in Chapter 8 for a typical chemigation system installation and more information on this and other safety arrangements.

Automated Control Systems

Several automatic control systems are available for irrigation pumping plants and sprinkler systems. Some are designed to perform a single operation while others provide for implementation of many tasks. Cost can vary from a few hundred to several thousand dollars. Each has special advantages and limitations that need to be evaluated to determine the best system for each situation.

Restarts

Mechanisms are available that automatically restart an irrigation pump and the system when electric power is restored after an interruption. Loss of power may result from any number of causes including off-peak load interruption, lightning, or other types of power surges. Automatic restart mechanisms eliminate the need to manually push the start switch to restart the system and can save on driving time when systems are located many miles from the farmstead.

Automatic restart mechanisms may be factory kits or assembled installations. The actual installation will vary with the type of equipment and/or type of control used with off-peak power. Installation will use one, two, or three timers with two being the most common.

The first timer is activated when power is restored. After a 3-to-5 minute delay, the irrigation pump is started, and power is restored to the irrigation system. At this point, the second timer is activated. The second timer then runs for 3 to 5 minutes and then activates the low pressure switch, pivot alignment safety, and the rest of the safety circuit.

If the pump is normally valved back during startup, automatic restart is not recommended unless special pressure control valves and sensors are installed. This may be the situation where long pipelines must be filled with water.

Priming

Automatic priming of an electric-powered centrifugal pump is necessary for a sprinkler system with automatic restart. Several automatic priming solutions are available. Commercial packages can be installed but will cost several hundred dollars more than repriming systems using off-the-shelf components.

Two basic types of systems are common: the vacuum-priming method and the pump-fill method. The vacuum method can be used with or without a foot valve, while the pump-fill method can be used only on suction lines with a foot valve. Figure 7-29 shows typical arrangements of priming equipment.

Vacuum method. This requires an airtight check valve on the pump discharge. The vacuum pump evacuates air from the suction line and pump and fills them with water. A water sensor on top of the irrigation pump switches power from the vacuum pump to the irrigation pump when the suction hose and pump have been filled with water.

Pump-fill method. A small submersible or sump pump is used to fill the suction line and pump. An air release valve lets air escape, and a pressure or water sensor switches power from the sump pump to the irrigation pump. The check valve on the pump discharge must have sufficient back pressure to activate the sensor.

For more information on automatic priming systems, obtain South Dakota State University Extension fact sheet #898, *Automatic Priming of Centrifugal Irrigation Pumps.*

Shutdown

Automatic pumping plant shutdown systems provide both safety and management advantages. Shutdown systems are available for all types of sprinkler systems and pumping plants. Typically, a system may be shut down for the following reasons:

- At the end of a run for a big-gun traveler or a center pivot circle.
- At the end of a set application time or water volume pumped for a solid set.
- When a center pivot becomes misaligned.

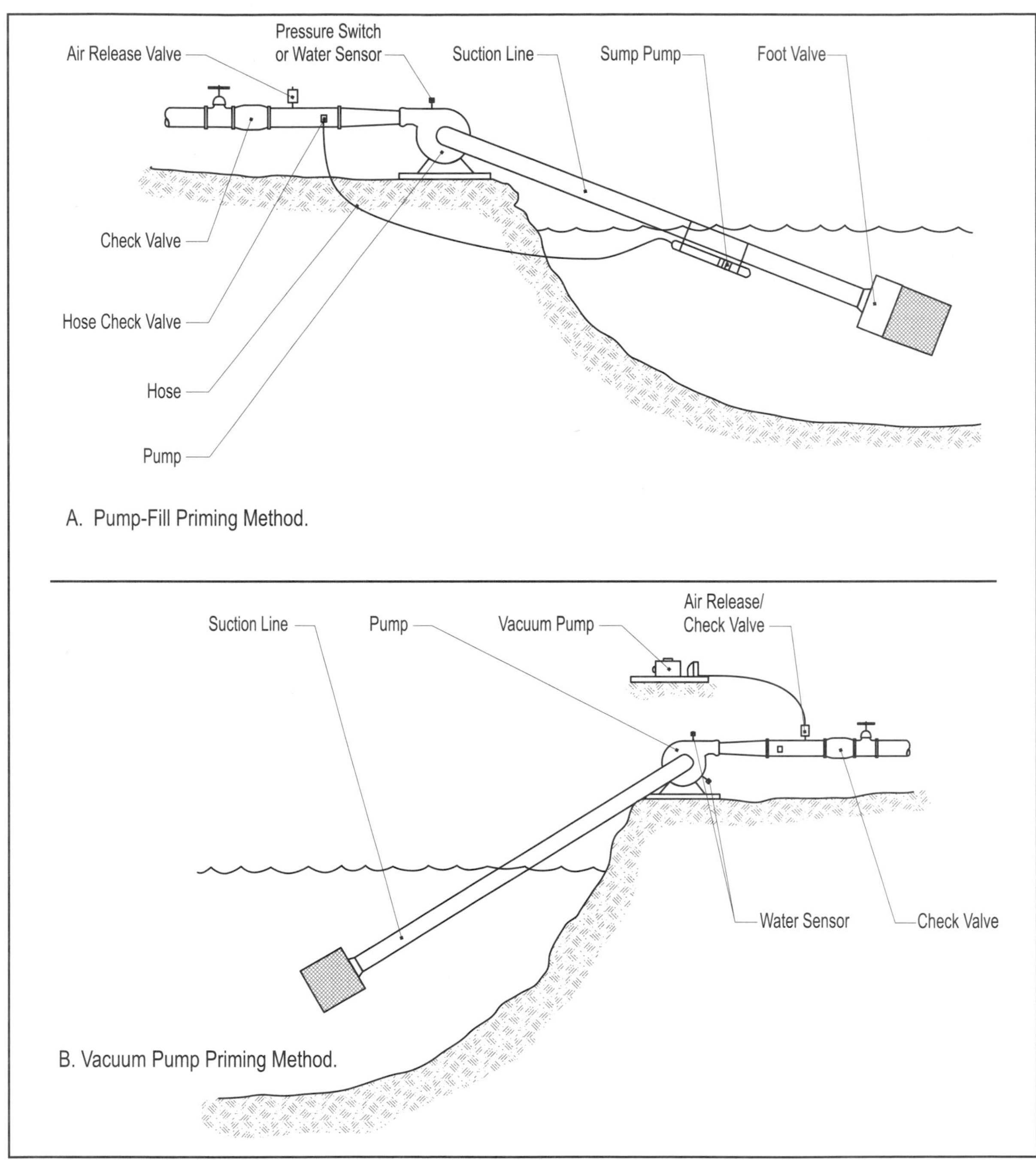

Figure 7-29. Priming methods.

- Due to rainfall.
- Due to wind speed.
- Because of temperature overload in an engine.
- Loss of oil pressure.
- Due to a break in the pipeline.

Shutdown systems can be something as simple as a timer interfaced with the power unit, to something as complex as a combination of an automatic in-line shutoff valve and a high pressure water switch.

Energy Consumption and Pumping Plant Performance Criteria

The cost of operating an irrigation pump depends on energy costs, pump and drive efficiency, power unit energy performance, water horsepower (WHP) required, and hours of operation.

Some accepted energy performance criteria for pumping plants are given in Table 7-12. Energy consumption rates in Table 7-12 are based on accepted power unit energy performance efficiencies shown in Table 7-13, and a pump efficiency of 75%. Some power unit and pump combinations may produce higher operating efficiencies and will, therefore, consume less energy than the guidelines show in Table 7-12.

An estimation of a given irrigation pumping plant's hourly and total seasonal energy consumption can be derived by applying Equations 7-15, 7-16, and 7-17 with Table 7-12, as shown in Example 7-8.

Table 7-12. Accepted energy performance criteria for irrigation pumping plants. Performance: At 75% pump efficiency

Energy source, units	Direct	Right angle or V-belt
	(WHP-hr. per unit of energy)	
Diesel (gal)	12.50*	11.90
Gasoline (gal)	8.66	7.60
Propane (gal)	6.89	6.50
Electricity (kWh)	0.88	0.84
Natural gas (1,000 cu ft)	61.70	58.60

* This value may be higher for some selected diesel engines.
Source: 1982 Irrigation Pumping Plant Performance Handbook.

Table 7-13. Average performance criteria of power units.

Energy Source	Performance BHP-hour per unit of energy
Diesel (gal)	16.66
Gasoline (gal)	11.50
Propane (gal)	9.20
Electricity (gal)	1.18
Natural gas (1,000 cu ft)	82.20

Source: 1982 Irrigation Pumping Plant Performance Handbook.

Management and Maintenance

The irrigation pumping plant and the pipeline together form the main system component for moving water from the water supply to the sprinkler system. If not maintained and operated effectively, the different components can cause reduced profits through unnecessary energy usage, reduced crop yields, and unplanned downtimes with the irrigation system.

Each component has its own maintenance requirements and must be inspected and serviced regularly to maintain the operation of the original installation. Obtain a copy of the manufacturer's operation and the maintenance guide book for each of the irrigation system components. Operator guide books will generally outline the manufacturer's recommended startup, operation, and end-of-season shutdown servicing steps.

Shutdown season maintenance can greatly affect the overall life expectancy of the components and may well make the difference between being able to get the system in operation at the beginning of the next season or having to call for service.

Listed below are several maintenance and management tips that should be considered on a regular basis.

Electric Motors and Control Panels

- Always turn the main power disconnect off and check for stray electric current before servicing the electric motor and sprinkler irrigation system.
- Inspect motor housing and all control panels for rodent invasion and damage. Cover openings with rodent screens and other sealing materials. Rodents also can enter the control panel from underground.

Internal Combustion Engines and Gear Heads

- Regularly monitor oil levels and change as per manufacturer's recommendations.
- Inspect safety control switches and replace immediately when not working.
- Keep PTO and belt safety shields on at all times.
- Remove fuel from the supply tank at the end of the season.

Centrifugal and Turbine Pumps

- Maintain a working water pressure gauge, and regularly monitor operating pressure.
- Regularly inspect drive shaft packing gland tension and adjust when needed.

Example 7-8. Determining seasonal energy consumption.

Sixteen gross inches of water will be applied during the season by a center pivot covering 125 acre. The pumping rate is 850 gpm and the total head is 200. Estimate seasonal energy consumption for: 1) a diesel with a right-angle drive, and 2) an electric direct drive.

Solution:

1. Find the water horsepower requirement using Equation 7-13:

$$WHP = \frac{(850\text{ gpm}) \cdot (200\text{ ft})}{3960} = 43\text{ WHP}$$

2. Find the hours of pumping per season using this equation:

Equation 7-15. Season Pumping Time, hours.

$$PH = \frac{453 \cdot (A) \cdot (D)}{Q}$$

Where:

- PH = Seasonal pumping time, hour
- A = Area irrigated, acres
- D = Gross irrigation depth applied, inches
- Q = Discharge, gpm

Using Equation 7-15:

$$PH = \frac{453 \cdot (125\text{ ac}) \cdot (16\text{ in})}{850\text{ gpm}} = 1065\text{ hr}$$

3. Find the estimated hourly energy consumption using this equation:

Equation 7-16. Estimated Hourly Energy Consumption, units are a function of fuel.

$$HEC = \frac{WHP}{E_{perf}}$$

Where:

- HEC = Estimated hourly energy consumption, units are a function of fuel.
- E_{perf} = Accepted energy performance criteria Table 7-12 for specific fuel.

Using Equation 7-16 and Table 7-12, solve for estimate hourly energy consumption using a diesel engine with right-angle drive:

$$HEC = \frac{43\text{ WHP}}{11.90\text{ WHP-hr per gal}} = 3.6\text{ gal per hr}$$

Then solve using an electric motor and direct drive connection:

$$HEC = \frac{43\text{ WHP}}{0.88\text{ WHP-hr per kWh}} = 48.9\text{ kWh per hr}$$

Example 7-8. (continued)

4. Determine the seasonal energy consumption using Equation 7-17:

Equation 7-17. Seasonal Energy Consumption.

$$SEC = (HEC) \cdot (PH)$$

Where:

SEC = Seasonal energy consumption, units are a function of fuel

$$SEC = (3.6 \text{ gal per hr}) \cdot (1{,}065 \text{ hr}) = 3{,}834 \text{ gal}$$

Then solve for electric motor:

$$SEC = (48.9 \text{ kWh per hr}) \cdot (1{,}065 \text{ hr}) = 52{,}079 \text{ kWh}$$

Chapter 8

Chemigation

Chapter 8 Contents

Chapter 8 Contents Continued

To reduce input costs, many sprinkler systems can be used to apply agricultural chemicals mixed with irrigation water. This method of chemical application reduces costs by eliminating additional trips over the field to apply fertilizer or pesticide. Proper chemigation equipment and procedures are important for the safe and efficient application of chemicals through the irrigation system. Protection of the environment, whether it be in the area where chemicals are being applied or in the water source, is of utmost importance.

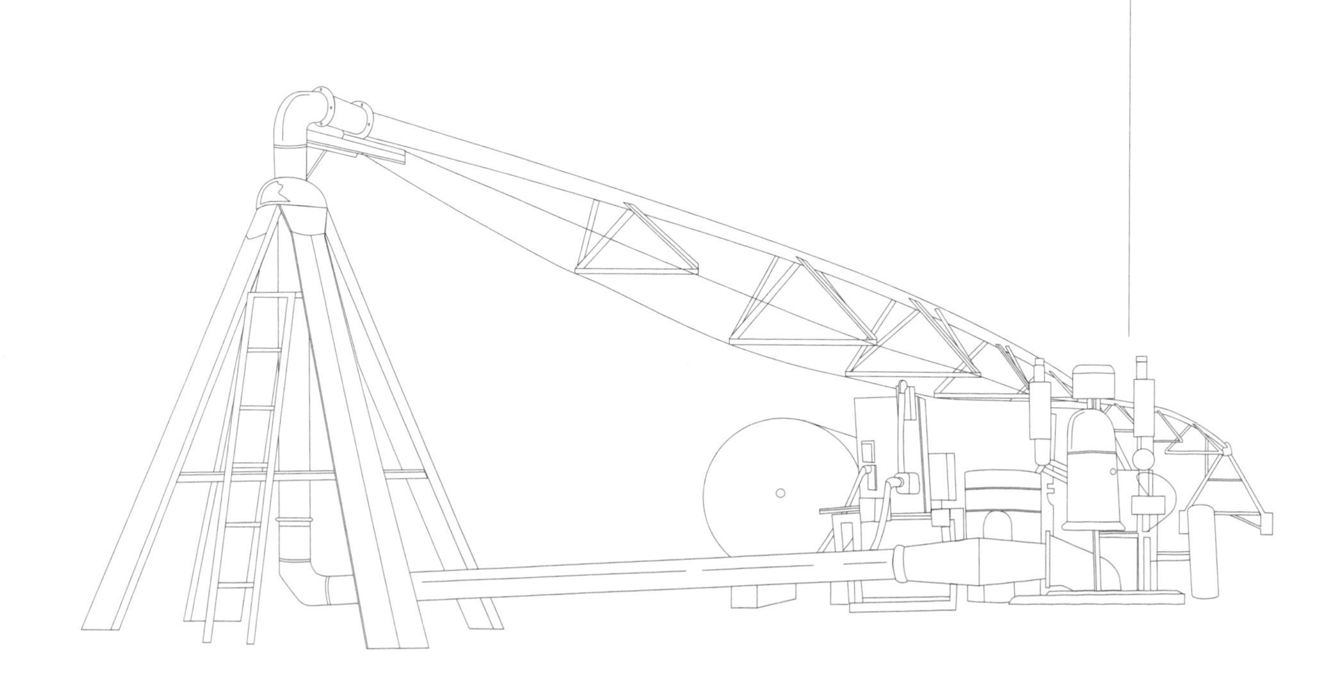

Overview of Chemigation

Chemigation is the process of applying an agricultural chemical (fertilizer or pesticide) to the soil or plant surface with an irrigation distribution system by injecting the chemical into the irrigation water. Depending on the type of agricultural chemical being applied, chemigation may be referred to as fertigation, herbigation, insectigation, or fungigation.

A common example of chemigation is applying nitrogen fertilizer through a sprinkler irrigation system. This technique has been used for many years and is recognized as a best management practice for some crops grown on sandy soils.

An irrigator must consider the information in this chapter when deciding if a given field and sprinkler irrigation system can use chemigation and, if so, what practices need to be followed to chemigate safely and effectively. This chapter describes the general chemigation safety measures and management practices to minimize the potential risk of accidental flow of an injected chemical back into the water source or discharge onto the land, which can create a public health problem. Detailed information related to a specific chemigation system must be obtained from the respective manufacturers.

This chapter does not discuss the specific protection requirements for chemigation required in each state or what is needed to connect to a potable well or public water supply system. Specific details on the safety devices and measures for public water systems should be obtained from the appropriate local agency before installing any chemigation equipment.

Like other chemical application methods, chemigation has advantages, limitations, and risks that a farm manager must consider when deciding which method of application is the best choice for the desired fertilizer or pesticide treatment. The greatest risk with chemigation is the potential for accidental backflow of all or part of the chemical into the irrigation water source if the chemigation system is not properly set up, operated, and maintained.

Advantages of Chemigation

Chemigation has these advantages:

- Equal or better uniformity of chemical application than other methods when the irrigation system's nozzling package is properly selected and maintained.
- Flexibility in timing the chemical application when the field is too wet for a tractor or an aircraft is unavailable.
- Increased activity and effectiveness of some compounds.
- Lower environmental risks associated with having a large portion of a crop's fertilizer N supply available for potential leaching into ground water after a major rainfall.
- Lower application costs in some situations.
- Less mechanical damage to plants than that caused by ground sprayer wheels.
- Less soil compaction than that caused by ground application methods.
- A possible reduction in chemical applications.

Disadvantages of Chemigation

Chemigation has these disadvantages:

- Chemical application depends on water application uniformity.
- A longer application time than other chemical application methods.
- Inability of some pesticides to be applied by sprinkler irrigation.
- Potential risk of chemical back flow into the irrigation water source (ground or surface) if two or more of the required safety devices malfunction while chemigating.
- Additional investment in safety equipment.
- Possible adherence to a state chemigation user permit.

In summary, chemigation can be an effective way of applying certain agricultural chemicals to some irrigated crops. However, the irrigation system must be able to apply the chemical/water mixture uniformly and at the proper amount. In addition, all of the anti-pollution safety devices to protect the water source must be utilized.

Chemigation Requirements

Most states have mandated chemigation regulations and a permitting program for pesticide and fertilizer application. These laws and rules generally include provisions governing the following:

- The type and quality of chemigation antipollution and safeguard devices.
- Installation specifications.
- Power interlocking controls.
- Chemical injection hose and backflow prevention at the injection port, chemical storage tank location, secondary containment units, and systems connected to a potable and public water system.
- Inspections.
- Reporting.
- User training procedures
- Structured management practices.

At a minimum, the decision about whether to use to chemigation should include an evaluation of the following factors:

- Site location.
- Pesticides.
- Regulations.
- Fertilizers.
- Land and soil characteristics.
- Type of irrigation system.

Site Location

Chemigation is an application option only for some fields and will not be right for all situations. Chemigation should not be used if the irrigation system will cause off-target spray or drift to adjacent homes or occupied buildings, surface water sources, wetlands, neighboring crops, or roadways.

Pesticides

Some, but not all, pesticides can be applied through an irrigation system. The choice of which pesticide product is best to use to protect the crop from a potential problem always should be made first. In April 1988, the Federal Environmental Protection Agency (EPA) required all pesticide labels to state if they are allowed for application through an irrigation system. Check the pesticide label for information to see if it can be applied through chemigation. The label also states minimum safety devices for the irrigation system for the states that do not have their own regulations. Estimates are that less than 1% of the sprinkler-irrigated land is equipped with the proper equipment to apply pesticides (usually insecticides or fungicides),

The effectiveness of any pesticide via chemigation depends on the ability of the system to apply the recommended amount of water and chemical solution uniformly throughout the field. For example, a premergence herbicide may work best with 0.4 to 0.75 inches of water.

Regulations governing pesticide applications change often, so before chemigating check the local Cooperative Extension office, state department of agriculture, or chemical company representative to verify current regulations and recommendations. Some states may require posting of the treated field at all times during the chemigation application and a specified re-entry time.

Fertilizers

Several sources of commercially prepared fertilizer are available for supplementing a crop's nutrient needs. However, not every fertilizer source can be injected into irrigation water safely or mixes well with irrigation water. Some fertilizer sources, ammonium sulfate for example, have limitations on application rate to prevent plant damage.

Other fertilizers, like some formulations of potassium or phosphate, may not mix well and can plug sprinkler nozzles. Contact the local Cooperative Extension office or a fertilizer supplier for advice on which fertilizer formulations are mixable with local irrigation water.

More than two-thirds of the sprinkler irrigation systems in most states are equipped to apply 28% liquid urea, ammonium nitrate, which is the most common source of N injected into irrigation water. It maintains a constant concentration without agitation and is easy to transport and store. This N source supplies three pounds of N fertilizer for each gallon applied.

Granular fertilizers, ammonium nitrate or ammonia sulfate, can be used in batch-loading situations with solid-set sprinklers or trickle systems.

Anhydrous ammonia or any other N fertilizer that has free ammonia should not be applied through sprinkler systems. Ammonia can react chemically with the dissolved salts in the water and form precipitates that coat the inside of pipelines and possibly plug sprinklers. Some of the ammonia also can be lost to the atmosphere due to volatilization while water is sprayed into the air.

Nitrogen (N) amount and timing of a chemigation application depend on several factors: crop type, crop stage of growth, yield goal, soil texture, previous N applications, and in-field N credits from manure and previous crop. The total amount of N fertilizer applied by chemigation may vary between 20 to 60 pounds per acre depending on the N requirements for the crop. With larger amounts, the application should be split into two or three chemigation events.

For corn, chemigation is the best way to apply the last one-sixth to one-third of the crop's N needs. Nitrogen application by chemigation should be started at or just before the plant's peak N uptake time. Peak N use time for corn starts between the 12th and 16th leaf stage of development and slows down substantially by the time the silk turns brown.

Late N chemigation for potatoes generally should be applied based on petiole nitrate status. Nitrogen demand by potatoes is generally greatest between initial tuber growth and tuber enlargement (three to eight weeks after emergence).

The total N applications or another nutrient for a given crop should be based on recommendations from the Cooperative Extension in your state.

Land and Soil Characteristics

Certain soils and topographies are not suitable for chemigation. For example, if the land is very hilly with a lot of elevation variations, the chemical-water mixture may not be distributed uniformly to the plant or soil surface. Hilly land also may cause the chemical water mixture to run off into low-lying areas where it could cause injury to the crop. Ponding also may cause some chemical to leach into the groundwater in the ponded areas or into surface water due to runoff.

Type of Irrigation System

Chemigation should be done only with systems that can provide a uniform water application over the entire field and at an application rate that does not exceed the infiltration rate of the soil. An irrigation system that causes water to flow down plant rows and exceeds the intake rate of the soil will not provide an adequate distribution of applied pesticides and fertilizers.

An irrigation system also should have the flexibility to apply at a wide variety of application depths. Some pesticides only work when applied with a very shallow application depth of water, for example, 0.15 to 0.25 inches.

If an irrigation system is three or more years old, the water distribution pattern should be evaluated with an in-field catch test, as seen in Figure 8-1, before using with chemigation. Specific characteristics of different types of irrigation systems are discussed in a later section of this chapter.

Injection and Anti-Pollution Equipment

Special equipment is needed to prevent accidental backflow of chemicals into the water supply. Most state regulations require the owner and/or operator of any irrigation system used for chemigation to install several antipollution and safety devices, comply with local or state separation

Figure 8-1. In-field catch test.

distances, and implement several management measures. Specific antipollution and safety measures are described beginning on page 154 of this chapter.

The owner/operator of the injection equipment and safety devices must take the time to regularly inspect, maintain, and repair each component to ensure the chemical is applied correctly and safely.

Calibration

The chemigation operator must have the knowledge and skills to calibrate the irrigation and chemigation system to achieve an accurate application of chemical. The injection pump must be easy to calibrate and adjust during application. An in-line calibration tube should be used to assist in calibration. Details on how to calibrate equipment begin on page 160 of this chapter.

Weather

Winds can cause irrigation water droplets to drift. Moderate to strong winds may cause uneven application of water and chemicals. Do not chemigate if winds are strong enough to cause drift onto non-target areas.

Suitable Irrigation Systems

Sprinkler systems, such as the electric- or hydraulic-drive center pivot and the linear-move, can distribute water and chemicals uniformly if the sprinkler package is properly selected and maintained. Water-driven center pivots, however, should not be used because the application around each drive tower usually is higher than between the towers. Center pivots and linears can be equipped with several types of sprinkler packages (10 to 60 psi), and each can provide adequate water distribution. Spray packages that direct the flow of water downward to the plant or soil surface provide the least risk from wind drift.

The end gun on a center pivot should be operated during an chemigation event only if it can provide a uniform application of water and can be controlled to spray water only within the field boundary.

Traveling guns and set-move sprinkler systems (wheel-roll or hand-move lateral) do not offer a water distribution over the field that is as uniform as center pivot because of the overlaps between moves. These systems should not be used to apply any type of pesticide but may be used to apply fertilizer when the wind is very low and no other method is available.

Chemigation Equipment

To chemigate safely and effectively, the irrigation system must be equipped with the appropriate chemical injection system, approved antipollution devices, and safety measures. Figure 8-2 shows a common safety equipment arrangement.

Chemigation can pollute the water source in three ways if it is not protected:

- The chemical in the supply tank and in the irrigation pipeline could flow or be siphoned back into the water source when the irrigation system shuts down (Figures 8-3 and 8-4).
- The chemigation pump could continue to inject chemical into the irrigation pipe line when the irrigation pump shuts down, causing the chemical solution to flow back into the water source or spill onto the ground (Figure 8-5).
- The chemigation pump could shut off while the irrigation pump continues to operate and force water back into the chemical supply tank, causing it to overflow and spill onto the ground (Figure 8-6).

The chemigation operator and farm manager/owner are responsible to minimize these pollution risks and others with the proper injection equipment, safety devices, and management measures. Before injecting any pesticide, always do a trial run with water to check the performance of the irrigation system, injection equipment, and safety devices.

Injection Equipment

A chemical injection system consists of an injection meter or pump, chemical supply hose, supply tank, calibration equipment, antipollution devices, and safety devices. Any equipment that comes in contact with chemicals, including hoses, seals, and gaskets, must be resistant to all formulations

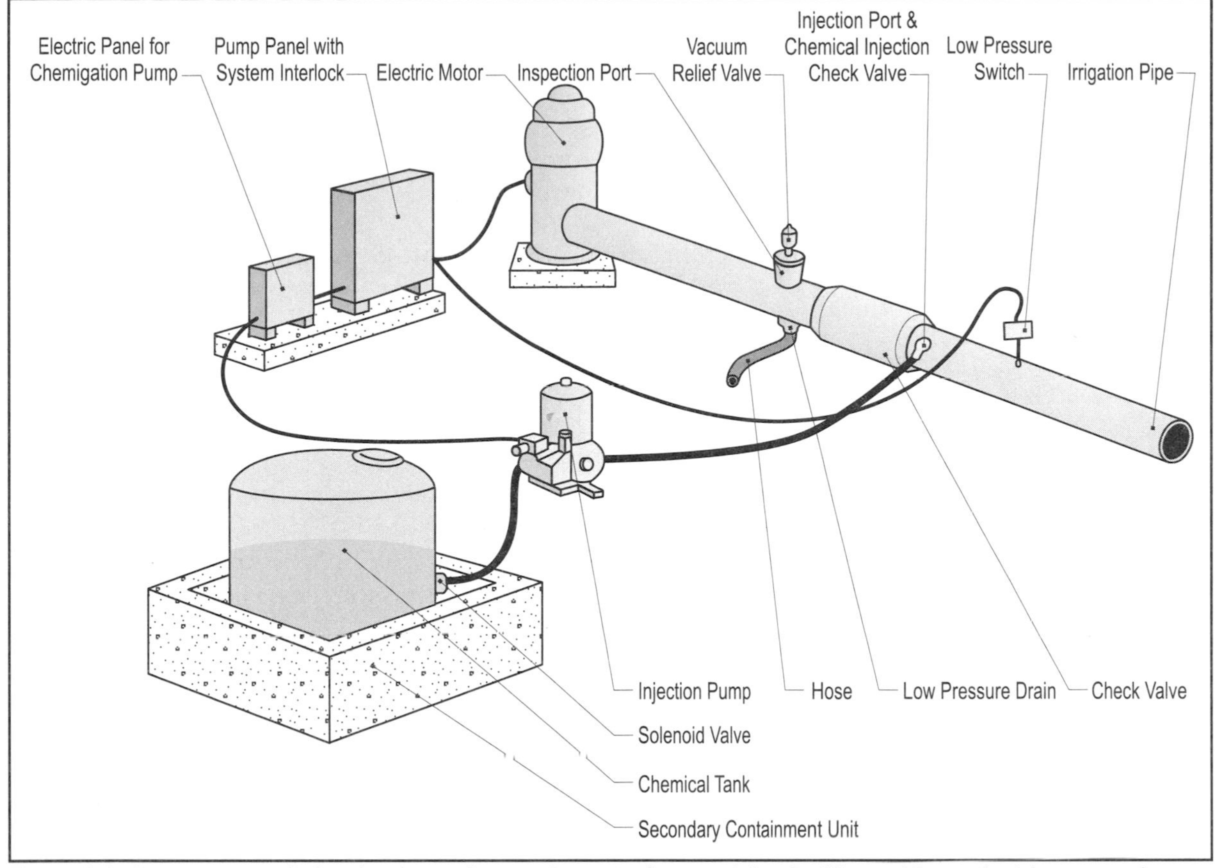

Figure 8-2. Arrangement of chemigation safety equipment when applying fertilizer with an electric-powered irrigation system connected to an irrigation well.

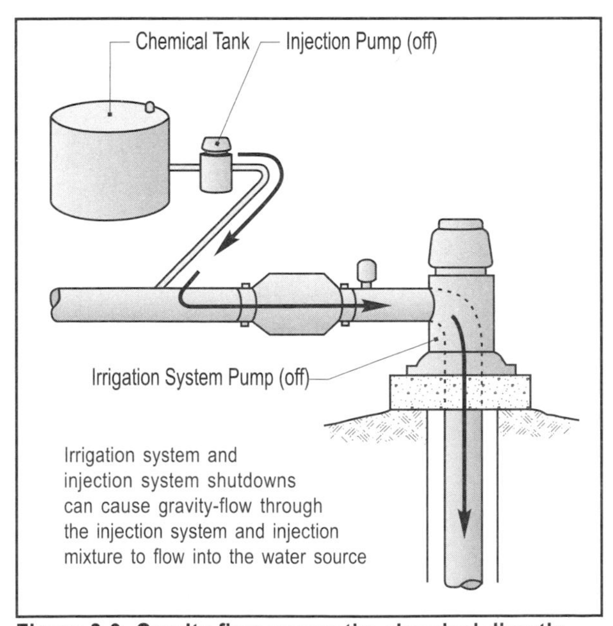

Figure 8-3. Gravity flow moves the chemical directly into the water source after irrigation and injection systems are shut down.

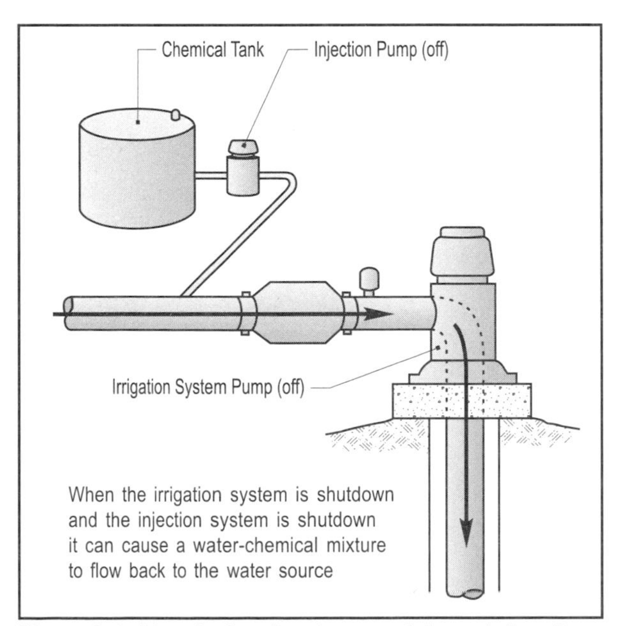

Figure 8-4. Chemical and water are siphoned directly into the water source after irrigation and injection systems are shut down.

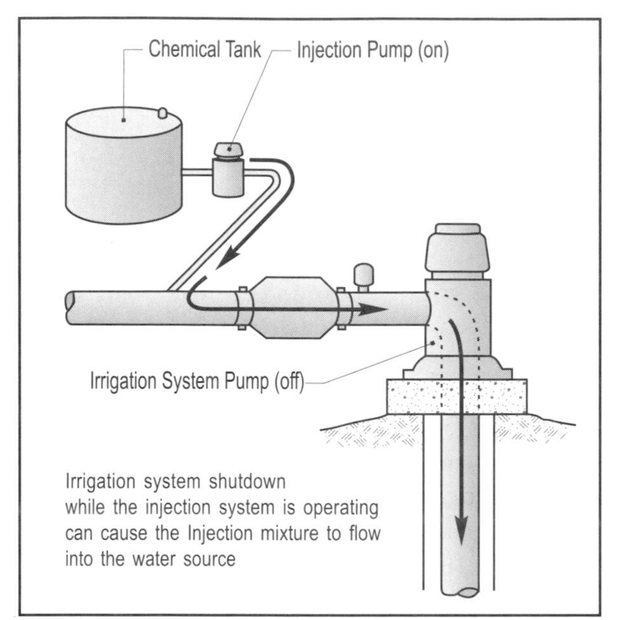

Figure 8-5. Injection system pumps chemical directly into the water source after the irrigation system is shut down.

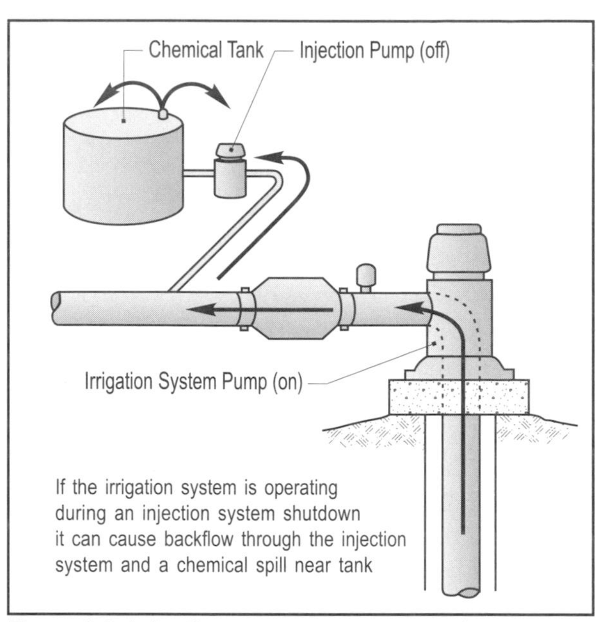

Figure 8-6. Irrigation system pumps water into chemical supply tank after the injection system is shut down. Can result in chemical tank overflowing.

being applied. This includes emulsifiers, solvents, and other carriers as well as the active ingredient.

Injection Pump

A chemical injection pump or metering device should be easy to adjust for different injection rates. It should be sized to meet the injection rates of the specific system and chemical. No single pump can do all jobs, since application rates may range from pints to several gallons per hour. Some systems may require a pair of injectors to provide the necessary chemical injection rate. Pumps should not be operated at their maximum or minimum settings as this may result in inaccurate injection rates. A filtering device always should be located on the inlet side of the suction line to prevent the pump and injection hose check valves from clogging.

The main types of injection pumps and metering devices are diaphragm pumps, piston pumps, venturi injectors, and differential mixing tanks. **Diaphragm pumps**, Figure 8-7, are the best all-around metering device for chemigation even though they are more expensive than piston or venturi units. They have fewer moving parts, are less subject to corrosion and

Figure 8-7. Typical diaphragm pump.

leaks, and are easily adjusted during the chemigation process. **Piston pumps** can not be easily recalibrated during chemigation, and the piston parts are more likely to wear where they are in contact with the chemical. They must be stopped to make a calibration adjustment. **Venturi injection units** usually are lower in cost, but maintaining an accurate or consistent injection rate with these types of pumps may be difficult.

Piston and diaphragm pumps have positive displacement injection. Several manufacturers market both types, and they are available in at least two or more injection rate ranges.

Three-phase electric motors are the most common source of power for piston and diaphragm pumps. Some injection pump models can be driven by belt power or a water motor.

Venturi-shunt injectors come in several sizes and can be operated under different pressures. Most venturi-shunt systems are set up in a shunt pipeline parallel to the main irrigation pipeline because they require at least a 20% differential pressure to work properly. Venturi-shunt injectors do not require external power to operate, but some chemigation units use a small booster pump in the shunt pipeline to produce differential pressure. Venturi-shunt systems can be adjusted easily during operation to change the rate of chemical injection.

Irrigation systems like the center pivot require a metered pump that can provide a constant injection rate. The positive displacement pumps and the venturi-shunt meter are the most common injection equipment when applying nitrogen with a system like the center pivot. The metered pump for a given system should be able to apply the desired chemical rate per acre (for example, 10 to 30 pounds of N) at a reasonable water depth (1/3 to 1 inch). The diaphragm pump provides the most effective and controlled system when using a pesticide with irrigation water.

Solid-set sprinklers can be used to apply a chemical at a constant rate, but they are more commonly operated with a batch-loading method. The venturi-shunt injector is suited for either type of injection. Batch-loading also can be done with a pressure-differential mixing tank for systems that cover small acreages.

Pressurized differential mixing tanks are available in only a few sizes, with the largest tank suited for only a few acres at a time. Pressurized mixing tanks require diversion of some water from the main irrigation line into the tank, then returning the mixed solution back into the main line at a point of lower pressure. The rate of backflow into the main line is controlled by a regulating valve, but chemical concentration will be slowly reduced over the injection time period. Some mixing tanks are equipped with a collapsible bag that separates the chemical from the water. This modification allows the chemical to be injected at a more constant rate. These systems may require repeated fillings to complete an application.

Supply Tank

The chemigation supply tank should be made of noncorroding materials such as stainless steel, fiberglass, nylon, or polyethylene (Figure 8-8). Avoid materials like iron, steel, copper, aluminum, or brass, which can corrode. Depending on the chemical formulation used, the tank may need mechanical or hydraulic agitation to keep the different chemicals properly mixed. The outlet of the tank should contain a manual shutoff valve. Some states may require tanks to be located a specific distance from a well head or require tanks to be placed in a secondary containment unit.

Injection Line Strainer

A chemical-resistant filter should be located on the chemical suction line or hose to remove foreign materials that could plug or damage the injection meter or pump or chemical injection line check valve.

Hoses, Clamps, and Fittings

All components in contact with the chemical mixtures should be constructed of materials that are resistant to chemicals and sunlight degradation. The pressure rating of all components should be adequate to

Figure 8-8. Typical supply tank.

withstand all operating pressures. Hoses and fittings should be protected from mechanical damage.

Calibration Equipment

A calibration tube or in-line flow meter installed on the chemigation injection hose line provides an easy way to measure the rate of flow of the chemical being injected into the irrigation system. A calibration tube should be placed on the suction side of the injection device with the necessary valves and fittings so the injection rate can be checked during a chemigation. Typically, a calibration tube is a clear tube with markings in milliliters or fluid ounces. It is used with a stopwatch to measure the flow rate.

An in-line flow meter can be used in either the suction or discharge side of the injection device. Its markings typically are expressed in flow units of volume over time.

Anti-Pollution Devices and Measures

Some states require the owner/operator of any irrigation system used for chemigation (pesticide or fertilizer) to obtain a user permit, install several antipollution and safety devices, comply with local or state well separation distance rules, and implement several management measures. The safety devices are necessary to prevent pollution of the water supply as described in Figures 8-3, 8-4, 8-5, and 8-6.

Figure 8-2 shows a typical arrangement of basic safety devices. Some requirements will vary depending on type of water supply system and the location of the water source.

Basic safety devices and measures suggested to protect the irrigation water source from pollution pathways are as follows:

- Install a check valve in the main irrigation pipeline.
- Use a power interlock.
- Install a check valve in the chemical injection line.
- Have a low pressure shutdown switch.
- Use a chemical supply tank.
- Post chemigated fields.

For specific requirements, check with the appropriate local or state agency.

Irrigation Main Pipeline Check Valve

In most states, a single chemigation check valve assembly (Figure 8-9) must be used in the main pipeline when injecting a pesticide or a fertilizer. For pesticides, some states also require a second chemigation valve assembly or a reduced pressure zone (RPZ) backflow preventer (Figure 8-10) in the main irrigation water supply pipeline of any system directly connected to an irrigation water well or a surface water source.

The check valve or RPZ assembly must be located between the point of chemical injection and the irrigation water supply pump. The main purpose is to keep the water and chemical mixture from flowing back or siphoning back into the water source. Check valves should be installed with flanged fittings that allow for easy removal for maintenance or repair, or replacement. The check valve assembly may be installed as a portable unit and moved to other irrigation systems during chemigation.

Each check valve assembly must contain an air vacuum relief valve and an automatic low pressure drain immediately upstream of the check valve flapper. The check valve assembly also must have an inspection port that can be opened easily to inspect the check valve flapper and the low pressure drain when the irrigation system is shut down.

The air vacuum and relief valve allows air to enter the pipeline when the water stops flowing. This prevents the creation of a partial vacuum that could cause siphoning of the water and chemical mixture back into the water supply. The valve also allows air to leave at start-up so it doesn't cause surging.

The low pressure drain must be located on the bottom of the pipeline on the supply side of the check valve and have a fully functioning drain opening of at least a 3/4-inch diameter. It must open automatically whenever the irrigation water flow stops. This provides a secondary safety backup to prevent any chemical and water mixture from entering the water source if the check valve should leak. Any drainage must be directed away from the water source during shutdown. A hose, pipe, or open conduit can be used to direct the drain discharge.

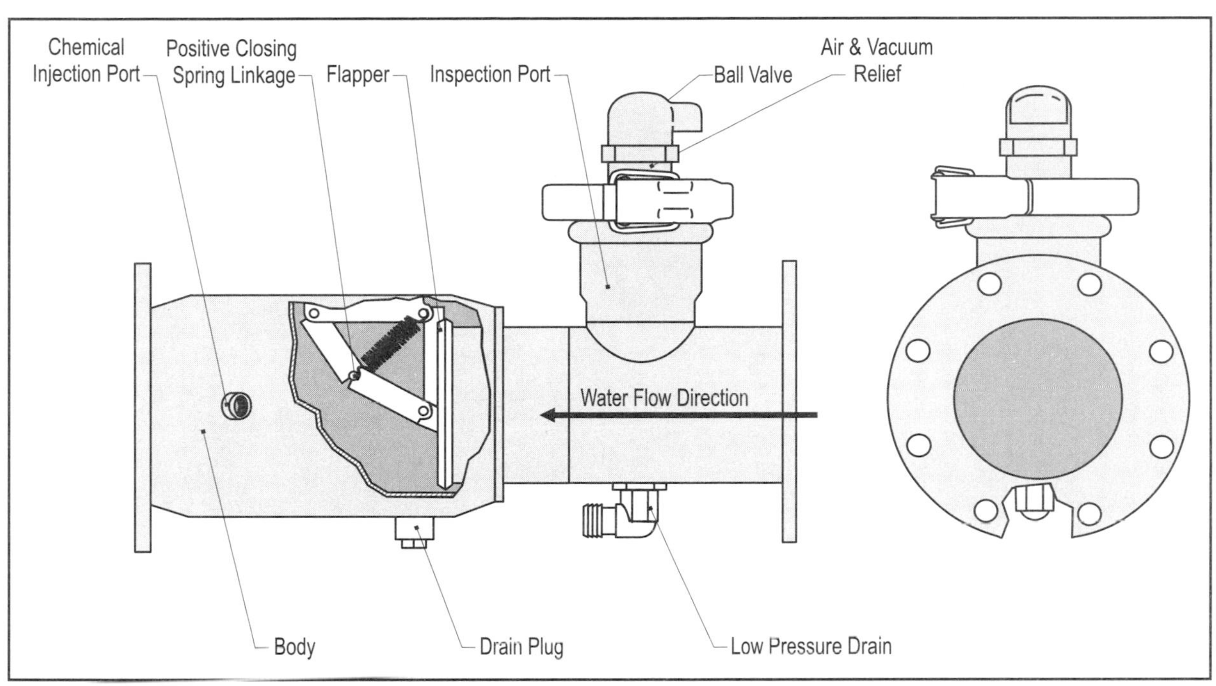

Figure 8-9. Spring-loaded check valve.

Figure 8-10. Reduced pressure zone (RPZ) backflow preventer.

Some states may require using only approved check valve assemblies that meet certain design and operating standards and are certified by an independent testing laboratory. Check valve assemblies must be quick-closing by spring action, provide a water-tight seal, and be constructed of material resistant to corrosion or protected to resist corrosion. Check valves also must be easy to maintain and repair.

Power Interlock

The chemigation injection system must be interlocked with the irrigation system's power or water supply so it will shut down any time the irrigation system or pumping plant stops operating, or the water flow is disrupted. In all cases, this measure must prevent the injection of chemicals into the main irrigation pipeline after the water supply stops flowing.

If electric motors are used for both the irrigation and chemigation systems, the control panels for the two systems must be electrically interlocked. This interlock must be set up so the injection pump motor stops whenever the irrigation or pump stops.

Irrigation pumps driven by an internal combustion engine can be interlocked with an injection pump by being belted to the drive shaft or an accessory pulley on the engine. If the injection pump is electrically powered, it should be connected to the engine's generator or electrical control system.

Some chemigation systems use the flowing water or water pressure to power the injection meter or pump. In most cases these systems will stop injecting a chemical when the irrigation water supply stops flowing.

If chemical flow from the supply tank could possibly continue after shutdown, a normally closed solenoid valve that is closed in its normal position should be provided in the chemical injection line, preferably on the suction side of the injection pump. The solenoid valve must be interlocked and powered by the irrigation system control panel, water supply pressure, or the injector power supply.

Chemical Injection Line Check Valve

The chemical injector's discharge line or hose must contain a positive-closing check valve that will not allow flow either way when the injection system is not operating. The chemical injection line check valve, Figures 8-2 and 8-11, must be between the injection meter and the point of the chemical injection into the irrigation pipeline.

This valve must perform two functions:

1. Stop the flow of water from the irrigation system into the chemical supply tank if the injection system stops.
2. Prevent gravity-flow from the chemical tank into the irrigation pipeline following an unexpected shutdown.

To provide two-way protection, the valve must have a watertight sealing check valve with a minimum opening (cracking) pressure of 10 pounds per square inch. It also should be constructed of a material resistant to chemical corrosion.

If irrigation water is allowed to flow back into the chemical supply tank, it could overflow the tank and cause the chemical to spill onto the ground. Likewise, if chemical in the supply tank is allowed to flow into the irrigation pipeline by gravity or be siphoned when the irrigation system is not operating, a potential danger to the crop could be created. Chemicals also could leak unto the ground, possibly getting into a surface or groundwater source.

Low Pressure Shutdown Switch

The irrigation system must have a low pressure device on the main pipeline that will shut down the irrigation system and the chemigation system if the operating pressure drops to an unsatisfactory level.

Chemical Supply Tank

The chemigation supply tank is best located where it is easily accessible for refilling and servicing and as far away from the water source as possible. Some states require that a tank must be at least 150 feet from a well head during preparation or filling of a chemical unless appropriately safeguarded.

The supply tank should be placed away from the water source in such a way that if a spill occurs, the chemical will not move directly to the source.

The chemical supply tank must be constructed from materials such as fiberglass, polyethylene, or stainless steel. The tank also must be resistant to the chemical being stored and resistant to degradation by sunlight. If not contained in a secondary unit, the tank should be located and landscaped to direct any leakage away from the water source. The tank also should be protected from damage from farm machinery, livestock, and soil erosion.

Some states require that a chemigation supply tank must be housed in a secondary containment unit if the tank storage meets at least two of the following conditions:

1. The supply tank has a rated capacity of more than 1,500 gallons,
2. The tank is located within 100 feet of a water supply.
3. The supply tank storage is located at the site for more than 30 consecutive days.

The minimum required capacity for a secondary containment unit is 125% of the tank capacity (110% if under a roof). The walls and base of the containment unit may be made of ferrous metal, reinforced concrete, solid reinforced masonry, synthetic lined earth or prefabricated metal, or synthetic materials. Synthetic liners must have a minimum thickness of 30 mils.

The unit must be leakproof and built to withstand the hydrostatic pressure from the release of a full tank. The walls or base must not contain a drain. Design specifications for some types of units are described in *Designing Facilities for Pesticide and Fertilizer Containment*, MWPS-37, from MidWest Plan Service. This bulletin is available at many Cooperative Extension offices.

Posting of Field

Fields being treated with a pesticide through the irrigation water generally require posting with signs during the complete chemigation treatment. Some states require the signs to contain the signal word from the pesticide label, name of the pesticide, date of treatment, and reentry date as described by the pesticide label. An example of a sign can be obtained from a local chemical supplier or Cooperative Extension office.

Signs should be posted at usual points of entry and at property corners immediately adjacent to public transportation routes or other public or private property. Some states require the signs be placed no greater than 100 feet apart for a field that is located adjacent to a public area such as a park, school, or residential area. If more restrictive instructions for posting are described on the label, those restrictions must be followed.

Additional Protection Measures

In addition to the safety equipment and measures previously described, several other devices and measures can ease the management of the chemigation operation and reduce the potential risks to the environment.

- Portable chemigation system and chemical supply tank.
- Injection meter or pump.
- Chemigation system location.
- Bleed valve.
- Injection port location.
- Injection line flow sensor.
- Two-way interlock.
- Solenoid valve.

Portable Chemigation System and Chemical Supply Tank

Install the chemigation injection meter or pump and chemical supply tank on a portable trailer or truck. Construct a secondary containment unit of appropriate size on the bed of the trailer or truck.

Injection Meter or Pump

Locate the injection meter or pump within the chemical supply tank containment unit when available and possible.

Chemigation System Location

When developing a new irrigation system, always try to locate the chemigation system, chemical supply tank, injection port, power interlock controls, etc., at least 150 feet from the irrigation water supply.

Bleed Valve

Locate a bleed valve (Figure 8-11), upstream and next to the injection line check valve. The valve will help relieve "locked-in" pressure in the chemical injection line whenever the injection line is disconnected. This will prevent the operator from being sprayed with chemical in the line during line removal.

Injection Port Location

When possible, locate the port for chemical injection higher than the chemical supply tank but lower than the lowest sprinkler outlet to prevent siphoning from the tank. In all cases, the injection port must be located downstream from the main pipeline check valve.

Figure 8-11. Typical injection port and check valve.

Injection Line Flow Sensor

An injection line flow sensor installed just upstream from the chemical injection line check valve and interlocked with the injection device can be used to shut down the injection system if flow in the injection line ceases. This safety measure will prevent continuous operation if the injection device loses prime or fails, the supply tank is emptied or the injection port becomes plugged, or a line or hose ruptures or disconnects. The flow sensor also could be interlocked with the irrigation system to shut down the entire system if the injection line flow stops.

Two-Way Interlock

A two-way interlock arrangement between the irrigation system and the injection system will stop either system if the other system stops. This interlock will eliminate untreated areas in the field by stopping the irrigation pump and sprinkler system if the injection system stops or malfunctions. The interlock can be done electrically with the use of a flow sensor on the discharge side of the chemical injection device. When there is no flow in the injection line, the irrigation system and pumping plants will shut down.

Solenoid Valve

A solenoid valve that is closed in its normal position installed on the suction side of the injection device can provide a good backup. It also functions as an automatic shutoff valve on the injection line when the injection pump is not in use. The solenoid valve must be interlocked with the injection device power supply to open or close properly.

Management Practices

A chemigation system requires regular maintenance and supervision to apply fertilizer or chemical safely and effectively. **The owner or operator must see that all equipment components function properly and the pesticide is applied according to the label. The operator must operate the system so that the chemical is applied only on the target area and no drift is allowed off the area. The operator must follow all of the safety requirements shown on the EPA label, such as wearing the appropriate protection clothing and having ready access to clean water for eyewashing and other cleanup.** Following are several management tips that should be considered each time a chemigation system is used to apply a pesticide.

1. Review Operation of Irrigation System.

Periodically observe the irrigation system's water distribution pattern, and conduct a water distribution test of the spray pattern. Remember that the uniformity of the chemical distribution will be no better than the distribution of the water.

Adjust the irrigation system, such as the end-gun, to prevent spray going beyond the boundaries of the target field. Shut down the irrigation system if wind will carry chemical drift off target. Manage the irrigation system so runoff or deep percolation of the water-chemical mixture does not occur.

Do not chemigate in areas containing wetlands and other surface water bodies. Do not apply any pesticide that is not labeled for use in an irrigation system. Such applications are illegal and may adversely affect wildlife, non-target plants, and water quality.

2. Inspect Safety and Antipollution Equipment.

Inspect all components of the chemigation and irrigation systems before each use. Before chemigating, repair or replace any components not working at the time of the inspection. Routine inspections should minimize the potential for failure of any component during chemigation.

Follow the procedures listed below to inspect the irrigation pipeline check valve, low pressure drain, and injection line check valve for leaks, and evaluate the low pressure switch's sensitivity before chemigation occurs. RPZ backflow preventers and other types of check valves will require a different approach for inspection.

- Connect the chemigation system power unit to the irrigation system, but leave the chemical injection line or hose disconnected from the injection port check valve. Add fresh water only to the supply tank for this test procedure.
- Start the irrigation pump, and pressurize the irrigation system to its normal operating pressure.
- Observe the injection line check valve to see if any water is leaking back out the inlet side of the check valve. There should be no leakage observed when

Figure 8-12. Inspect all chemigation and irrigation components.

the irrigation system is operating or when shut down.

- Connect the chemical injection hose to the injection check valve and start the chemigation system. Operate the chemigation system only with clean water at this point.
- Turn down the main pipeline control valve (reducing the operating pressure) until the low pressure switch shuts down the irrigation system. The pressure switch should be set to cause the irrigation pump and system to shut down when the normal operating pressure has been reduced by 15 to 25%. If no flow control valve is present, shut off power to the pump and/or irrigation system.
- Immediately after shutdown, observe if any water is flowing from the low pressure drain(s). Some drainage for a short period of time after shutdown is normal, then drainage should stop.
- Check to see if the chemigation injection device has stopped operating. This device should stop when the irrigation system and pump shuts down. If the chemigation system has an agitation system, this unit does not have to shut down when the injection device stops.
- Open the inspection port at the main pipeline check valve assembly after the low pressure drain has stopped flowing. Inspect for any leakage from the check valve flapper. Also check for the proper functioning of the flapper valve assembly.

3. Fill Supply Tank and Mix Agricultural Chemicals Properly.

Closely inspect the supply tank and plumbing fixtures each time before filling the tank. Fill the supply tank to no more than 95% of capacity. Monitor the supply tank during chemigation for development of any leaking.

Some states may have coded separation distances for chemigation supply tanks. For example, they should be located at least 150 feet from any water well during filling unless housed in the appropriate safeguard unit. To determine if any specific requirements exist contact the appropriate local or state agency.

Triple rinse pesticide containers at the time of use, and add the rinse water into the supply tank. Rinse containers above the opening of the supply tank to minimize the risk of spilling pesticide on the ground.

4. Keep Chemigation Site Uncontaminated.

To facilitate safe monitoring of the chemigation operation, do not allow the irrigation system to spray water and chemical into the chemigation equipment area. This may mean plugging nozzles on the irrigation system near the chemigation equipment site.

5. Use Accurate Calibration.

Accurate calibration of the chemical injection device is essential for proper application. Minor differences in the injection rate over an extended period can cause too much or too little of a chemical to be applied. This may produce unsatisfactory results or cause potential pollution or crop damage when large applications are made. Periodically recheck the calibration setting of the injection device. Follow the calibration procedures described by the chemical label, chemigation equipment manufacturer, or Cooperative Extension office in your state.

6. Empty the Chemigation Supply Tank.

Remove leftover pesticide mixtures from the supply tanks. Store the leftovers in an appropriate place for later use, or immediately apply them to another crop or site listed on the label. Rinse the empty tank, and apply the rinse water to the irrigated crop or to another labeled site.

7. Flush Injection Equipment.

Flush the chemigation injection device, hoses, and check valve with clean water or the appropriate solvent after each use. Put all rinse water in the irrigation system while it is in operation to be applied to the field. Clean the strainer after use.

8. Flush Irrigation System.

After chemigation is finished and the injection system is cleaned and flushed, operate the irrigation pump as long as necessary to flush the system free of chemical. This may take 10 to 15 minutes for most systems.

9. Report Accidental Spills.

If an accident occurs, regardless of size, avoid personal contamination. Take action to keep spills to a minimum, and report the incident to the appropriate authorities.

Calibration

Proper calibration of the irrigation system and the chemigation injection pump is essential for an effective, safe, and economical application. Minor differences in the calibration and application rate over a period of time can cause the chemical application rate to be too high and potentially damaging to the crop and/or the environment. An application rate that is too low, may make the chemical treatment ineffective.

It is important to have in-field measurements of the field size, travel time at the desired water depth, and rate of chemical application per acre to calibrate all of the equipment. To finalize the calibration of the injection pump, the whole system must be operated under pressure to set the pump dial for the desired injection rate and to check its real flow.

The chemical injection rate worksheet (page 164 and 165) shows the eight steps in calculating the proper chemigation injection rate for a center pivot. The worksheet has calculations for both the fertilizer and chemical. Example 8.1 illustrates information on the worksheet. The following steps further explain the calibration calculations.

Calibration Step #1

If the acreage of the field is unknown, calculate the area using the formulas in Appendix 2. For example, a center pivot that runs a full circle covers an area determined by the equation shown for a full-circle without end-gun in Appendix 2.

In the case where the area covered is only part of a circle, multiply the full circle area by the percentage of the coverage. If an end-gun or corner system is used, calculate the wetted area for both on and off operations, and estimate the percentage of time that each is applying water.

If assistance is needed to calculate the area, contact an equipment dealer, or the local Cooperative Extension office.

Calibration Step #2

This step sets up the volume of chemical or fertilizer that must be applied per acre based on the recommended application rate.

Calibration Step #3

This step determines the total volume of chemical or fertilizer needed for the entire application area.

Calibration Step #4

Some chemical products, like a wettable powder pesticide, may need to be placed in a water solution for mixing purpose to enable injection of the product. Some liquid pesticides may need to be mixed with water to increase the total volume of applied solution so the injector pump can be operated within its best working range.

Calibration Step #5

This step sets up the water application depth based on the chemical or fertilizer being applied.

Calibration Step #6

Knowing the correct travel time to cover the field is very important in making an accurate calibration. Two methods can be used to time the application of a center pivot. The first method involves operating the system wet at the same travel speed (percent timer setting) that is planned for chemigation, and measuring the time to make one revolution, or to cover the desired part of the field. Some dealers or parts manufacturers provide a chart for the percent timer versus the travel time.

With the second method, measure the distance from the pivot point to the outer tower wheel track. Next, operate the system wet at the desired travel speed (percent timer setting) planned for chemigation. Measure the time it takes the outer tower to travel a preset distance, such as 100 feet. To calculate the time to cover a circle, use the following formula:

Equation 8-1. Wheel Track Circumference.

Wheel track circumference (ft)

$$= 6.28 \cdot \left(\text{Distance from pivot to outer wheel track}\right)$$

Equation 8-2. Rotation Time.

Rotation time

$$= \left(\text{Wheel track circumference}\right) \cdot \left(\frac{\text{Time between markers}}{\text{Travel distance}}\right)$$

Calibration Step #7

This step determines the injection rate. If the chemical needs dilution water, then divide the volume from Step 4 by the time from Step 6. If dilution water is not needed then divide Step 3 by Step 6.

Calibration Step #8

This step selects the injection pump setting. Setting selection is based on manufacturer's curves for the injection rate calculated in Step 7. Once the system is operating, check the delivery rate of the injector with a calibration tube to make certain it is injecting at the proper rate. Adjust the setting as needed to obtain the calculated injection rate.

Example 8-1.
Determining the fertilizer and chemical injection rate for a chemigation system.

A producer has two fields each with a center pivot. The producer wants to apply fertilizer with Center Pivot #1 and chemical with Center Pivot #2. The wetted radius of each center pivot is 1,320 feet, and each system will have the center pivot run a full circle.

With Center Pivot #1 the producer wants to apply 30 pounds of N per acre using a 28% liquid nitrogen (urea - ammonium nitrate). The liquid urea contains 3 pounds of N per 1 gallon of liquid urea. The fertilizer supplier recommends a water application depth of 0.75 inches for the product. To apply water at a rate of 0.75 inches, the producer determined that it took 13.33 minutes for the outer track to travel 25 feet. A piston pump will be used.

With Center Pivot #2 the producer wants to apply 2 pints of pesticide plus 2 pints of oil per acre for a total of 2 quarts of pesticide product per acre. The manufacturer recommends using 7 parts of water to 4 parts of pesticide product. The chemical supplier recommends a water application depth of 0.2 inches for the product. To apply water at a rate of 0.2 inches, the producer determined that it took 14.69 minutes for the outer track to travel 100 feet. A diaphragm pump will be used.

(Continued on the next page)

Example 8-1 (continued)

The producer wants to know the injection pump dial setting to use for the chemigation injection system. Determine the fertilizer and chemical injection rate for a producer to use with the manufacturer's injection pump curves.

Solution

Step 1: The wetted area for both center pivots is:

$$\text{Wetted area} = \frac{3.14 \cdot (1{,}320\text{ ft})^2}{43{,}560} = 125.6\text{ ac}$$

or approximately 126 acres.

Step 2: The volume of chemical or fertilizer per acre is:

Center Pivot #1

$$\frac{30\text{ lbs of N per acre}}{3\text{ lbs N per gal of liquid nitrogen}} = 10\text{ gal of liquid nitrogen per acre}$$

Center Pivot #2
2 quarts of product

Step 3: The total volume of chemical or fertilizer per acre is:

Center Pivot #1

$$(10\text{ gal of liquid nitrogen per acre}) \cdot (126\text{ ac}) = 1{,}260\text{ gal of liquid nitrogen}$$

Center Pivot #2

$$(2\text{ qt of pesticide product}) \cdot (126\text{ ac}) = 252\text{ qt of pesticide product}$$

Step 4: The volume of solution water needed is:

Center Pivot #1
Dilution water not needed

Center Pivot #2
Based on the manufacturer's information of using 7 parts of water to 4 parts of pesticide product, the total amount of solution is:

$$\text{Total amount of solution} = (252\text{ qt}) \cdot \left(1 + \frac{7\text{ parts water}}{4\text{ parts product}}\right)$$

$$= 693\text{ qts of solution, or } = 173.25\text{ gal of solution}$$

Step 5: The water application depth recommended by the supplier for the product:

Center Pivot #1
0.75 inches

Center Pivot #2
0.2 inches

Example 8-1 (continued)

Step 6: To determine the travel time of the irrigation system, the producer must first determine the travel distance for one rotation of the outer track. Because both systems have the same wetted radius, both systems will have the same travel distance of the outer track (Equation 8-1):

$$\text{Wheel track circumference (ft)} = 6.28 \cdot (1{,}320 \text{ ft}) = 8{,}105 \text{ ft}$$

The total travel time for each pivot is:

Center Pivot #1

$$\text{Rotation time} = (8{,}105 \text{ ft}) \cdot \left(\frac{13.33 \text{ min}}{25 \text{ ft}}\right) = 4{,}322 \text{ min, or } = 72 \text{ hrs}$$

Center Pivot #2

$$\text{Rotation time} = (8{,}105 \text{ ft}) \cdot \left(\frac{14.69 \text{ min}}{100 \text{ ft}}\right) = 1{,}191 \text{ min, or } = 19.8 \text{ hrs}$$

Step 7: To determine the injection rate, divide the travel time (Step 6) by the volume of solution (Step 4) per rotation. If no solution is being used, divide travel time by the total volume of chemical or fertilizer for the system (Step 3).

Center Pivot #1

$$\text{Injection rate} = \frac{1{,}260 \text{ gal of liquid urea}}{72 \text{ hr}}$$

$$= 17.5 \text{ gal/hr, or } 1{,}100 \text{ ml/min, or } 5.4 \text{ sec/100ml}$$

Center Pivot #2

$$\text{Injection rate} = \frac{173.25 \text{ gal of solution}}{19.8 \text{ hr}} = 8.75 \text{ gal/hr,}$$

$$\text{or } 552 \text{ ml/min, or } 10.9 \text{ seconds per } 100 \text{ ml}$$

Step 8: Based on injection rates calculated in Step 7 and using the manufacturer's supplied information for chemigation injection pumps, the producer was able to determine the following settings:

Center Pivot #1

Setting for a piston pump of 7.5

Center Pivot #2

Setting for a diaphragm pump of 90%

Chemigation Injection Rate Worksheet

Steps	Example	
	Nitrogen	Pesticide
1. Number of acres under center pivot or linear system (wetted area)	126 ac	126 ac
2. Volume of chemical or fertilizer per acre	10 gal (30 lb)	2 qt
3. Total volume of chemical or fertilizer for system Multiply Step 1 by Step 2	1,260 gal	252 qt
4. If chemical needs some dilution with water, use total amount of solution	---	173 gal
5. Selected water application depth per product recommendation	0.75 in	0.2 in
6. Determine travel time of the irrigation system for one pass to apply water amount in Step 5	4,322 min (72 hr)	1,191 min (19.8 hr)
7. Determine injection rate by dividing Step 3 or Step 4 by Step 6	17.5 gal/hr, or 1,100 ml/min, or 5.4 sec/100 ml	8.75 gal/hr, or 552 ml/min, or 10.9 sec/100 ml
8. Select injection pump dial settings based on Step 7	7.5(piston pump)	90% (diaphragm pump)

Additional Notes:

	Pivot 1	Pivot 2
Distance of Traveler:	25 ft	100 ft
Travel Time:	13.33 min	14.69 min
Speed	1.88 ft/min	6.81 ft/min
Timer Setting	28%	100%

Chemigation Injection Rate Worksheet

Steps		
1. Number of acres under center pivot or linear system (wetted area)		
2. Volume of chemical or fertilizer per acre		
3. Total volume of chemical or fertilizer for system Multiply Step 1 by Step 2		
4. If chemical needs some dilution with water, use total amount of solution		
5. Selected water application depth per product recommendation		
6. Determine travel time of the irrigation system for one pass to apply water amount in Step 5		
7. Determine injection rate by dividing Step 3 or Step 4 by Step 6		
8. Select injection pump dial settings based on Step 7		

Additional Notes:

Chapter 9

Sprinkler Application of Effluent

Chapter 9 Contents

Irrigation is an efficient means of moving large volumes of effluent to application sites. A properly designed and calibrated sprinkler irrigation system can uniformly apply effluent with minimum labor and usually minimizes the capital and operating costs for land application. Effluent irrigation can increase cash crop yields by providing needed water and nutrients. The increased value of the crop can help pay for the application system.

There are many similarities between designing an effluent irrigation system and a conventional water irrigation system. The majority of the chapter will focus on the equipment needed for the proper design of an effluent irrigation system. The chapter begins with a basic overview of soil-plant-effluent relationship. A simple procedure is outlined to estimate basic application rates and crop nutrient uptake. Some knowledge of effluent-soil-plant interaction is needed beforehand to most effectively use this chapter. A more detailed discussion of soil-plant-effluent relationship can be found in the Livestock Waste Handbook, MWPS-18, and the NRCS Agricultural Waste Management Field Handbook, Part 651.

Irrigating with Effluent

Effluent is the liquid discharge from a effluent treatment process, such as from a livestock lagoon, municipal treatment plant, or food processing plant. **Effluent irrigation** is the application of effluent to the soil without detrimental effects to surface water, ground water, soils, or crops.

A **soil-plant filter** is the land where effluent is applied and crops are grown. Designing and managing the soil-plant filter to effectively remove nutrients, metals, and salts that are present in the effluent is key to these systems. Crops that are grown can be harvested for livestock feed, bedding, or industrial uses.

Because of the nutrient content, effluent irrigation may require a larger application area than water irrigation. Agricultural effluent is often applied using a portable irrigation system. The producer usually owns all the equipment, and either owns or rents the land.

Commercial or municipal effluent typically has a much lower concentration of nutrients per volume than agricultural effluent. Because of the lower nutrient content, commercial or municipal effluent usually can be applied at higher rates for a longer period of time during the year than can agricultural effluent. Commercial or municipal effluent irrigation systems usually can use more permanent application equipment. Municipalities or processors usually own or manage the land and equipment.

Advantages of Effluent Irrigation

As stated in the chapter introduction, effluent irrigation can be a safe and efficient means of handling and applying large volumes of agricultural, municipal, and industrial effluent to an application site. With careful operation and monitoring, an effluent irrigation system prevents runoff and erosion.

The land application of effluent can be an effective means of breaking the disease cycle. Several diseases that might infect both animals and humans can be transmitted in waterborne livestock and human effluent. Land application can successfully interrupt infection cycles.

Effluent irrigation can also save the soil from damage. Field equipment such as heavy manure wagons can compact wet soil, alter soil structure, and reduce water movement thus reducing yields of cash crops. Effluent irrigation keeps heavy equipment off the field during application.

Challenges of Effluent Irrigation

As with any effluent management system, there are potential problems:

- Application of some effluents to crops may adversely affect plant growth.
- More management is required to prevent over-application.
- Fine-textured soils may not have enough permeability to receive effluent rapidly.
- The application of effluent may conflict with crop management options such as cutting alfalfa, harvesting corn or beans due to a late fall, or tillage requirements.
- Many states regulate the total amount of nitrogen, phosphorus, and other nutrients and the time of year that nitrogen can be applied, thus, the depth of annual application depends upon the nutrient analysis of the effluent and the annual crop uptake potential.

The challenges most commonly associated with the content of the effluent are:

- Nutrient build-up.
- Odors.
- Heavy metals.
- Salts.
- Bacteria.

Nutrient Build-up

Applying excess effluent can be harmful to crops, contaminate the soil, cause surface and groundwater pollution, and cause a loss of nutrients. While most soils have a tremendous capacity to absorb phosphorus, very high soil phosphorus levels can interfere with plant nutrition by inhibiting plant uptake of metallic trace elements such as iron, zinc, and copper. When plant residue or effluent is added to soil, there is an immediate and marked drop in oxygen and an increase in carbon dioxide in the soil air, which can inhibit plant growth.

Nitrogen in excess of crop requirements can leach through the soil. Excess nutrients in surface water can cause algae blooms, fish kills, and odors. Application should be done when the soil is not saturated and at a rate such that runoff does not occur and deep percolation does not contaminate groundwater sources. During summer months nutrient runoff is less likely to occur where effluent has been applied and incorporated. Nutrient losses from runoff are more likely to occur in the spring where effluent has been applied on frozen ground. Late fall and winter applications may result in a 25 to 50 percent total nitrogen loss (from leaching and denitrification). Fall applications allow soil microorganisms time to more fully decompose effluent and release nutrients for the following cropping season. Drainage tile may be needed to control a seasonal high water table at some locations.

Odors

Odors can be severe, depending on the type of effluent and management. Nitrogen loss as ammonia from land is greater during dry, warm, windy days than during humid or cold days. Ammonia loss is usually greater during the spring and summer months. Avoid irrigating on days with high humidity, no wind, or when the wind blows toward areas of concern. It is preferable to irrigate when temperatures are warming (8 a.m. to 3 p.m.) to encourage odor dissipation. Avoid irrigation during evening and night hours when odors dissipate slowly. Check local and state regulations regarding buffer distances from wells, drain wells, water supplies, residences, property boundaries, etc.

Heavy Metals

Heavy metals may be a problem in some municipal, industrial and agricultural product-processing applications. Effluent from industrial sites may be very high in heavy metals. Heavy metals in an effluent treatment process are usually retained in the sludge but some heavy metals will stay suspended in the effluent. Even though most irrigated effluent will have a very low concentration of heavy metals, unless the plants are matched to remove the heavy metals from the application site, over time heavy metals can build up in the soil.

Because the handling of effluent high in heavy metals is beyond the scope of this chapter, a qualified engineer, crop specialist or soil scientist should be consulted on these systems.

Salts

Soils high in inorganic salts can reduce seed germination and yields. Heavy effluent applications can increase soil salinity, especially in arid areas where little or no leaching occurs and in areas with a restrictive layer at the base of the root zone. This is not generally a problem in the humid Midwest where salts are leached from the soil by rainfall. In the drier Western states where rainfall is low, this may be an important factor in designing an effluent irrigation system.

Consult a qualified engineer, crop specialist or soil scientist to determine if loading rates will result in a salinity problem.

Bacteria

Typically the soil-plant filter is a good method of controlling bacteria, but problems can arise if large quantities of bacteria are present in the effluent. All effluent from human sources should be disinfected before being applied by irrigation. If chlorine is used as the disinfectant, a chlorine demand test should be conducted at least once a month. Ozonization and ultraviolet disinfection are also methods that can help eliminate bacteria. If bacteria is a concern then a qualified engineer should be consulted.

Soil-Plant-Effluent Relationship

Effluent can help build and maintain soil fertility. Effluent can also improve tilth, increase water-holding capacity, lessen wind and water erosion, improve aeration, and promote beneficial organisms. In livestock operations where the effluent includes runoff or dilution water, effluent can supply needed water as well as nutrients to crops.

Chapter 2 discusses in detail the relationship of soil-plant-water relationship. The soil-plant-effluent relationships is very similar because effluent is largely comprised of water. Unlike water, effluent has solid materials that can have an effect on plants, soil, and equipment. The approach to designing the effluent application system is slightly different. Instead of determining the water needs of the plants and then determining the amount of water to apply to the plants for a water irrigation system, the effluent irrigation system should be approached from the stand point of first determining the nutrient content of the effluent then determining the amount that can be applied to the land. Knowing the nutrient content of the effluent and soil, and knowing the nutrient uptake of the crop is essential to having a successful effluent irrigation system.

Effluent Characteristics

Testing is essential to account for all effluent nutrients, and to prevent over-application of nutrients or harmful elements or compounds that could decrease the potential productivity of the land resource. Contact the laboratory testing effluent for the proper sampling procedures.

The test items listed and discussed on page 18 in Chapter 3 under Irrigation Water Quality also should be included in an effluent analysis, especially when the effluent contains mostly water, common with operations that use a lot of flush or wash water. In most cases, especially livestock effluent, the most important components needed for testing are:

- Nitrogen (total Kjeldahl nitrogen and ammonia nitrogen).
- Phosphorus (P, or P_2O_5).
- Potassium (K, or K_2O).
- Total suspended solids (TSS).
- Volatile Suspended Solids (VSS).

An analysis of nitrogen, phosphorus, and potassium is essential in determining the proper application rate of effluent based on soil and crop conditions. Nitrogen should be measured as total Kjeldahl nitrogen (TKN) and ammonia nitrogen (NH_3-N). TKN is the total organic nitrogen and ammonia nitrogen in the effluent. Organic nitrogen is slow-releasing nitrogen. Ammonia nitrogen is the gaseous form of nitrogen. Ammonium nitrogen is equivalent to commercial fertilizer and, except for that lost to the air, can be used in the same year it is applied.

Phosphorus (P) is often expressed in terms of P_2O_5 and potassium (K) is often expressed in terms of K_2O so comparisons can be made to commercial fertilizer formulations. Nearly all the phosphorus and potassium in animal effluent is available for plant use the same year it is applied.

An analysis of TSS and VSS can indicate the possibility of soil clogging. Soil clogging can be a problem if effluent is high in solids. Clogged soil pores can lead to ponding of effluent on the surface and slow the biological break down of the effluent. An analysis that has high TSS and low VSS is an indication that soil clogging could be a problem. If VSS approaches TSS, then little soil-clogging should be expected, provided the proper biological breakdown of the effluent occurs.

Biochemical oxygen demand (BOD) is another analysis that can be done. BOD is the indirect measure of the concentration of biodegradable substances present in the effluent. Essentially, BOD is a measure of waste strength and/or the degree or efficiency of waste treatment. Testing for BOD can be expensive and requires at least 5 days for the lab to run the test. BOD has very little to do with application rates and should not to be tested unless needed for lagoon design.

Effluent from Livestock Operations

Livestock effluent contains considerable plant nutrients. Livestock effluent must be applied at a rate consistent with the annual nitrogen utilization of the crop in the land application area. In some areas, phosphorus buildup may be of concern and limit application rates.

Most crops will not utilize all the P and K contained in effluent if the application rate is designed to supply the nitrogen needs of the crop. This will result in a gradual buildup of P in the soil, which may eventually cause problems. An alternative is to apply the effluent at a rate adequate for the element requiring the lowest application rate to supply the projected crop needs and apply commercial fertilizer to supply the balance of the other elements.

Tables 9-1 and 9-2 show typical nutrient values for livestock effluent. See the *Livestock Waste Handbook*, MWPS-18, for a more detailed discussion on livestock nutrient values and for determining land application rates.

Table 9-1. Nutrients in liquid manure (pounds per 1,000 gallons of effluent).
Approximate fertilizer value of manure from liquid handling systems

Species	Handling Method [a]	Dry Matter (%)	Ammonium N	Total N	P_2O_5	K_2O
Swine	Liquid pit	4	33	45	32	25
	Lagoon	0.5	4	5	3	4
Dairy	Liquid pit	8	12	30	15	25
	Lagoon	0.5	3	5	3	5
Beef	Liquid pit	11	24	40	25	35
	Lagoon	1	2	4	3	4
Poultry, layer	Liquid pit	13	42	62	59	37
	Lagoon	0.5	6	7	2	10
Veal Calf	Liquid pit	3	25	30	15	25

[a] Lagoon values include lot runoff water.

Table 9-2. Nutrients in liquid manure (pounds per acre-inch of effluent).
Approximate fertilizer value of manure from liquid handling systems

Species	Handling Method [a]	Dry Matter (%)	Ammonium N	Total N	P_2O_5	K_2O
Swine	Liquid pit	4	900	1,200	870	680
	Lagoon	0.5	110	135	80	110
Dairy	Liquid pit	8	325	815	410	680
	Lagoon	0.5	80	135	80	135
Beef	Liquid pit	11	655	1,090	680	950
	Lagoon	1	55	110	80	110
Poultry, layer	Liquid pit	13	1,140	1,685	1,600	1,005
	Lagoon	0.5	165	190	55	270
Veal Calf	Liquid pit	3	680	815	410	680

[a] Lagoon values include lot runoff water.

Effluent from Municipalities

Some industrial or municipal effluents may have little or no plant nutrient value but contain heavy metals or salts that can be harmful to the land. Effluent with low nutrient value and low or no harmful effects may be applied at the maximum annual rate that the soil can use (see **Soil Infiltration Rate** on page 81). Other effluents, such as those containing high amounts of sodium, may need to be applied at an annual rate that will not significantly degrade the soil. Typical data for municipal effluent are provided in Table 9-3.

Effluent from Food Processing

Effluent from food processing plants may contain considerable plant nutrients. Dairy food processing effluent usually contains higher amounts of plant nutrients than municipal effluent. Tables 9-4 to 9-7 list effluent from food processing.

Table 9-3. Characterization of municipal effluent (percent wet basis). Based on NRCS Agricultural Waste Management Field Handbook Part 651.

Component	Raw Effluent	Secondary Effluent
Moisture	99.95	99.95
Total solids	0.05 [a]	0.05 [b]
Volatile solids	0.035	—
N	0.003	0.002
P_2O_5	0.002	0.002
K_2O	0.001	0.001

[a] Suspended solids 0.03%; dissolved solids 0.02%.
[b] Suspended solids 0.0025%; dissolved solids 0.0475%.

Crop Utilization

Effluent should be applied at a rate that does not exceed crop nutrient needs. Tables 9-8, 9-9, and 9-10 show common crop utilization values. Nutrients, especially nitrogen, are used more efficiently by grasses and cereals than by legumes. Legumes get most of their nitrogen from the air, so additional nitrogen is not usually needed. For the greatest return, apply effluent first to corn and small cereal grains, then to sorghum and forages, and finally to pasture.

Before heavy effluent applications, have your soil tested for fertilizer nutrient needs. Adjust effluent application rates based on soil conditions from nitrogen, phosphorus, and potassium tests. Be aware that not all the nitrogen applied will be available to the plant the first year (see Table 9-11). Over application of effluent may lead to nutrient carryover for years to follow. A manure

Table 9-4. Characterization of dairy food processing effluent (percent wet basis).
Based on NRCS Agricultural Waste Management Field Handbook Part 651.

Component	Industry-wide	Sweet cheese whey	Acid cheese whey	Cheese effluent sludge
Moisture	98	93	93	98
Total solids	2.4	6.9	6.6	2.5
Volatile solids	1.5	6.4	6.0	—
N	0.077	7.5	—	0.18
P_2O_5	0.11	—	—	0.27
K_2O	0.081	—	—	0.061

Table 9-5. Characterization of meat processing effluent (percent wet basis unless noted).
Based on NRCS Agricultural Waste Management Field Handbook Part 651.

Component	Red meat slaughter	Red meat packing	Red meat processing	Poultry [a]	Broiler [a]
Volume	696 gal [b]	1,046 gal [b]	1,265 gal [b]	2,500 gal [b]	—
Moisture	—	—	—	—	95
Total solids	4.7 lb [b]	8.7 lb [b]	2.7 lb [b]	6.0 lb [b]	5.0
Volatile solids	—	—	—	—	4.3
N	—	—	—	—	0.30
P_2O_5	—	—	—	—	0.18
K_2O	—	—	—	—	0.014

NOTE: Slaughter is killing and preparing the carcass for processing; Packing is killing, preparing the carcass for processing, and processing; Processing is butchering, grinding, and packaging.
[a] Quantities per 1,000 lb product
[b] Per 1,000 lb live weight killed

Table 9-6. Characterization of meat processing effluent sludge (percent wet basis).
Based on NRCS Agricultural Waste Management Field Handbook Part 651.

Component	Dissolved air flotation sludge			Effluent sludge
	Poultry	Swine	Cattle	
Moisture	94	93	95	96
Total solids	5.8	7.5	5.5	4.0
Volatile solids	4.8	5.9	4.4	3.4
N	0.41	0.53	0.40	0.20
P_2O_5	0.27	—	—	0.091

Table 9-7. Characterization of vegetable processing effluent (pounds per 1,000 pounds of effluent).
Based on NRCS Agricultural Waste Management Field Handbook Part 651.

Component	Cut bean	French style bean	Pea	Potato	Tomato
Total solids	15	43	39	53[a]	134
Volatile solids	9	29	20	50[a]	—

[a] Total suspended solids.

Table 9-8. Plant nutrient uptake used by the grain portion of grain crops (pounds per bushel).

Crop	Dry Weight, (lbs/bu)	Typical Yield (per acre)	Total N (lbs/bu)	P_2O_5[a] (lbs/bu)	K_2O[b] (lbs/bu)
Corn	56	120 bu	0.902	0.356	0.269
Soybeans	60	35 bu	3.750	0.872	0.137
Wheat	60	40 bu	1.248	0.844	0.374
Sorghum	56	60bu	0.935	0.458	0.282
Oats	32	80 bu	0.624	0.247	0.188
Barley	48	50 bu	0.874	0.370	0.248
Rye	56	30 bu	1.165	0.331	0.329
Sunflower	25	1,100 lb	0.893	0.970	0.333
Rice	45	5,500 lb	0.626	0.245	0.124
Rapeseed	50	35 bu	1.800	0.897	0.456

[a] Equivalent P_2O_5 amount based on the average concentration of P.
[b] Equivalent K_2O amount based on the average concentration of K.

Table 9-9. Plant nutrient uptake used by the plant portion of grain crops (pounds per ton of harvested field-dried material).

Crop	Typical yield (per arce)	Total N (lbs/ton)	P_2O_5[a] (lbs/ton)	K_2O[b] (lbs/ton)
Corn	4.5 tons stover	22.2	9.08	32.2
Soybeans	2.0 tons stover	45.0	9.99	25.0
Wheat	1.5 tons straw	13.4	3.18	23.3
Sorghum	3.0 tons stover	21.6	6.81	31.4
Oats	2.0 tons straw	12.6	7.26	39.8
Barley	1.0 tons straw	15.0	4.99	30.0
Rye	1.5 tons straw	10.0	5.45	16.6
Sunflower	4.0 tons stover	30.0	8.17	70.1
Rice	2.5 tons straw	12.0	4.09	27.8
Rapeseed	3.0 tons straw	89.6	19.5	80.9

[a] Equivalent P_2O_5 amount based on the average concentration of P.
[b] Equivalent K_2O amount based on the average concentration of K.

Table 9-10. Plant nutrient uptake by forage crop (pounds per ton of harvested field-dried material).

Forage crop	Typical yield (tons per acre)	N	P_2O_5	K_20
Corn silage, 35% dry matter	20	22.0	11.4	26.2
Forage sorghum silage, 30% dry matter	20	28.8	8.62	24.5
Sorghum-sudan silage, 50% dry matter	10	27.2	7.26	34.8
Alfalfa hay	4.0	45.0	9.98	44.8
Big bluestem	3.0	19.8	38.6	42.0
Birdsfoot trefoil	3.0	49.8	9.98	43.6
Bluegrass	2.0	58.2	19.6	46.8
Bromegrass	5.0	37.4	9.54	61.2
Clover-grass	6.0	30.4	12.2	40.6
Bermuda grass	8.0	37.6	8.62	33.6
Orchard grass	6.0	29.4	9.08	51.8
Reed canarygrass	6.5	27.0	8.18	29.2
Ryegrass	5.0	33.4	12.2	34.0
Switchgrass	3.0	23.0	4.54	45.6
Tall Fescue	3.5	39.4	9.08	48.0
Timothy	2.5	24.0	9.98	38.0

Table 9-11. Organic nitrogen mineralized factors for different types of effluent.
Amount mineralized (released to crops) during the first year after application of animal manure. Liquids are from pits, lagoons, etc.

Manure	Year 1	Year 2	Year 3	Year 4
Anaerobic liquid				
Swine	0.350	0.175	0.088	0.044
Dairy	0.300	0.150	0.075	0.038
Beef	0.300	0.150	0.075	0.038
Aerobic liquid				
Swine	0.300	0.150	0.075	0.038
Dairy	0.250	0.125	0.063	0.031
Beef	0.250	0.125	0.063	0.031
Deep pit				
Poultry	0.450	0.225	0.113	0.056

Table 9-12. Land required for swine effluent irrigation where 100 pounds of nitrogen is needed per acre (acres per productive sow). Effluent nitrogen levels account for storage and application losses. Based on 16 pigs sold per productive sow per year. Effluent production per productive sow accounts for all animals in the operation such as boars, the sow's pigs in nursery, growing, etc.

Operation	Anaerobic pit	Lagoon
Feeder pig	0.25	0.20
Farrow-to-finish	0.82	0.60
Finishing pigs (50 to 220 lbs)	0.03	0.02

Table 9-13. Land required for cattle effluent irrigation where 100 pounds of nitrogen is needed per acre (acres per 1,000 pounds of live animal). Effluent nitrogen levels account for storage and application losses.

Operation	Anaerobic pit	Lagoon
Dairy cow	0.62	0.27
Dairy heifer	0.31	0.13
Beef feeder	0.31	0.13

management plan is helpful in determining effluent application. Contact the local Cooperative Extension or NRCS office for help in developing a manure management plan.

The fertilizer value of effluent varies according to animal species, handling, and application. Tables 9-12 and 9-13 estimate land requirements for effluent application. Examples 9-1 to 9-4 demonstrate the use of Tables 9-8 to 9-10 and 9-12 to 9-13.

Land Application

There are two common approaches to consider when applying effluent to land. One approach is to **irrigate effluent as a means of disposing of material.** Effluent that has little or no plant value, such as may be the

Example 9-1. Determining N, P_2O_5 and K_2O application rates to meet corn yield goals.

Determine the amount of N, P_2O_5, and K_2O removed by corn than yields 150 bushels per acre.

Solution

From Table 9-8, corn removes 0.902 lbs N per bushel, and an equivalent of 0.356 lbs P_2O_5 and 0.269 lbs K_2O per bushel. The total amount removed is:

$$\text{Total N} = \left(\frac{0.902 \text{ lbs N}}{\text{bu}}\right)\cdot\left(\frac{150 \text{ bu}}{\text{acre}}\right) = 135 \text{ lbs N per acre}$$

$$\text{Total } P_2O_5 = \left(\frac{0.356 \text{ lbs } P_2O_5}{\text{bu}}\right)\cdot\left(\frac{150 \text{ bu}}{\text{acre}}\right) = 53.4 \text{ lbs } P_2O_5 \text{ per acre}$$

$$\text{Total } K_2O = \left(\frac{0.269 \text{ lbs } K_2O}{\text{bu}}\right)\cdot\left(\frac{150 \text{ bu}}{\text{acre}}\right) = 40.4 \text{ lbs } K_2O \text{ per acre}$$

Example 9-2. Determining N, P_2O_5 and K_2O application rates to meet bromegrass yield goals.

Determine the amount of N, P_2O_5, and K_2O utilized by bromegrass yielding 5 tons per acre.

Solution

From Table 9-10, bromegrass yields 37.4 lbs N per tons of harvested material, and an equivalent of 9.54 lbs P_2O_5 and 61.2 lbs K_2O per tons of harvested material. The total amount removed is:

$$\text{Total N} = \left(\frac{37.4 \text{ lbs N}}{\text{ton}}\right)\cdot\left(\frac{5 \text{ tons}}{\text{acre}}\right) = 187 \text{ lbs N per acre}$$

$$\text{Total } P_2O_5 = \left(\frac{9.54 \text{ lbs } P_2O_5}{\text{ton}}\right)\cdot\left(\frac{5 \text{ tons}}{\text{acre}}\right) = 47.7 \text{ lbs } P_2O_5 \text{ per acre}$$

$$\text{Total } K_2O = \left(\frac{61.2 \text{ lbs } K_2O}{\text{ton}}\right)\cdot\left(\frac{5 \text{ tons}}{\text{acre}}\right) = 306 \text{ lbs } K_2O \text{ per acre}$$

Example 9-3. Determining the land requirements to apply effluent from a 50-sow farrow-to-finish operation.

A producer has a 50-sow farrow-to-finish operation. Liquid manure is stored in an anaerobic lagoon and land-applied by irrigation. Determine the number of acres needed to utilize the nitrogen from the manure to grow corn that yields 150 bushels per acre.

Solution

Determining the land required for this operation is a three step process:

1. Determine the number of acres needed to apply 100 pounds of N per acre for a 50-sow farrow-to-finish operation. From Table 9-12, 0.60 acres are needed per sow. The total land required apply 100 pounds of N for the operation is:

$$\text{Acres needed to apply 100 lbs of N} = (50 \text{ sows}) \cdot \left(\frac{0.60 \text{ acres}}{\text{sow}}\right) = 30 \text{ acres}$$

2. Determine the amount of N removed by corn that yields 150 bushels per acre. Example 9-3 shows corn that yields 150 bushels per acre removes 135 pounds of N per acre.

3. Determine the number of acres needed to apply 135 pounds of N per acre for a 50-sow farrow-to-finish operation. From the previous calculations, the number of acres needed to utilize the nitrogen produced from this operation to grow corn that yields 150 bushels per acre is:

$$\text{Total acres needed} = \frac{(30 \text{ acres}) \cdot (100 \text{ lbs N})}{135 \text{ lbs N}} = 22.2 \text{ acres}$$

Apply the lagoon effluent to 22.2 acres to fully utilize the N fertilizer of the manure to grow 150 bushel per acre corn. Applying the manure to meet the N requirements more than adequately meets the crop P_2O_5 and K_2O needs.

case with municipalities, is often irrigated as a means of disposing of material. Because the effluent is very low in nutrient content, the maximum amount of effluent is applied that can be processed by the soil and plants (i.e., soil-plant filter) without causing detrimental effects on surface water or groundwater. The ability of the soil and underlying geological formations to accept, renovate, and transmit water may determine the amount of effluent that can be applied (see Soil Infiltration Rate on page 81). In many cases, however, the application amount is determined by the capability of the crop to remove heavy metals and salts. Near maximum crop production is essential for maximum evapotranspiration and nutrient removal.

The other approach is to **irrigate effluent as a means of increasing crop yields.** Effluent that has value to plants, common with livestock effluent, is applied to provide maximum return for the cropping program. The amount of effluent that can be applied is usually influenced by the amount of nutrients that crop can utilize, but it is possible to irrigate effluent to compensate for a water deficiencies. Apply effluent as close to the planting date as possible so that nutrients will be available to plants, especially in areas of high rainfall and with soils in which nitrogen is lost by leaching or denitrification.

The remaining discussion in this section will give an overview of application sites and rates. A more detailed discussion of land application can be found in the *Livestock*

Example 9-4. Determining land requirements to meet nitrogen yield goals.

A 100 sow farrow-to-finish operation uses deep pits to store effluent. Determine the land required if 200 pounds of nitrogen per acre per year is needed for this operation.

Solution

From Table 9-12, an anaerobic pit for a farrow-to-finish operation requires 0.82 acres per productive sow when 100 pounds of nitrogen is needed per acre. The land required per productive sow to produce when 200 pounds of nitrogen is needed per acre is:

$$(0.82 \text{ acres per sow}) \cdot (100 \text{ lbs N per acre}) = (X) \cdot (200 \text{ lbs N per acre})$$

$$X = (0.82 \text{ acres per sow}) \cdot \left(\frac{100 \text{ lbs N per acre}}{200 \text{ lbs N per acre}}\right)$$

$$X = 0.41 \text{ acres per sow when 200 pounds of nitrogen is needed per acre}$$

Total land required for 100 sows when 200 pounds of nitrogen is needed per acre is:

$$\text{Total Acres} = (0.41 \text{ acres per sow}) \cdot (100 \text{ sows}) = 41 \text{ acres}$$

Waste Handbook, MWPS 18, and the NRCS *Agricultural Waste Management Field Handbook*, Part 651.

Application Site

Unlike water irrigation in which the application site is known and the problem is finding water close enough to make irrigation feasible, the problem with effluent irrigation is finding an application site close enough to make irrigation feasible. Even when a nearby application site is located, there may be a need to acquire additional land for application if the operation expands. There may also be a need to find a new application site due to urban expansion.

Similar to water irrigation, soil infiltration rate, field slope and topography are important considerations when irrigating effluent. Chapter 5 provides a good discussion on soil infiltration rate, field slope and topography.

Because leaching can be a concern, identifying geological formations and knowing the location of groundwater aquifers is important. Do not apply effluent on karst soils. Contact the local NRCS or Cooperative Extension to determine areas that can be susceptible to leaching.

Application Rate and Depth

Three performance characteristics are critical to proper land application of effluent by irrigation:

- Sprinkler application rate.
- Depth of application per irrigation event.
- Total depth of effluent applied annually.

The **sprinkler application rate** is determined by soil infiltration or soil permeability, and by solids content of the effluent. Sprinklers should be selected to be compatible with the soil infiltration rate or soil permeability. If the sprinkler application rate is higher than the soil infiltration rate, the possibility for runoff is increased.

The **depth of application per irrigation event** should be matched to the water-holding capacity of the soil. This is especially important if the effluent is very low in nutrients. The total application amount depends on the water-holding capacity (or moisture deficit) of the soil at the time of application. Effluent high in solids can coat crop leaves, reducing photosynthesis and causing salt burns on leaves. See Chapter 2 for a more details on soil water-holding capacity.

The **total depth of effluent applied annually** should provide the target amount of nutrients to the receiving management plan area. This may be accomplished in a single irrigation event, or may require several separate applications, depending on site conditions and the annual depth of application. The amount of effluent that can be applied to the soil without creating runoff or percolation problems, or salt and heavy metals buildup is also important in determining the total depth of effluent applied annually.

This discussion about **Soil Infiltration Rate** on page 81 of Chapter 5 details the allowable application rate for water. This discussion is applicable to low nutrient effluent with some exceptions. If the total solids content in the effluent is greater than 0.5%, application rates from Table 5-10 should be reduced according to the information in Table 9-14. The reduction coefficients in Table 9-14 are based solely on decreases in hydraulic conductivity because of a layer of effluent (or other solids) that forms on the soil surface during irrigation. Further reductions of application rates are necessary in some situations, such as applications of effluent with salt concentrations sufficient to disperse

Table 9-14. Effluent application rate reduction coefficients.
Based on NRCS Agricultural Waste Management Field Handbook Part 651.

Soil Texture	Solids Content of Effluent (%)						
	0.5	1.0	2.0	3.0	5.0	7.0	10.0
Sand	0.88	0.55	0.31	0.22	0.13	0.10	0.07
Loamy Sand	0.70	0.54	0.37	0.28	0.19	0.14	0.10
Sandy Loam	0.87	0.77	0.63	0.53	0.40	0.32	0.25
Loam	0.97	0.93	0.88	0.83	0.74	0.67	0.59
Silt loam	0.98	0.95	0.91	0.87	0.81	0.75	0.68
Sandy clay loam	0.99	0.97	0.95	0.92	0.87	0.83	0.78
Clay loam	0.99	0.99	0.98	0.97	0.94	0.92	0.89
Silty clay loam	1.00	1.00	0.99	0.99	0.98	0.97	0.96
Sandy clay	1.00	1.00	1.00	1.00	0.99	0.99	0.99
Silty clay	1.00	1.00	1.00	1.00	1.00	1.00	1.00
Clay	1.00	1.00	1.00	1.00	1.00	1.00	1.00

Example 9-5. Determining application rates.

A land user wants to apply effluent with 5% solids content to a silt loam soil that has dense vegetation. The estimated surface storage is 0.2 inches before runoff occurs. The land user would like to apply 1.2 inches per application. Determine the application rate?

Solution:

The land user can apply 0.2 inches before surface runoff starts. The minimum amount that will infiltrate the soil is:

$$\text{Soil Infiltration} = (\text{Application Event}) - (\text{Surface Storage})$$
$$= (1.2 \text{ inches per application}) - (0.2 \text{ inches before runoff})$$
$$= 1.0 \text{ inches}$$

From Table 5-10, the maximum water application rate for a silt loam soil for an application amount of 1.0 inches is 0.8 inches per hour. To determine the application rate for 5% solids content, multiply the soil infiltration by the reduction factor from Table 9-14. The reduction factor from Table 9-14 is 0.81. The application rate for 5% solids content is:

$$\text{Application Rate} = (0.8 \text{ inches per hour}) \cdot (0.81) = 0.65 \text{ inches per hour}$$

clay aggregates. Always determine the salt content of the effluent to assess its effect on the intake rates of the soil.

Designing the Effluent Irrigation System

Various sprinkler systems are described and illustrated in Chapter 4. Although the equipment required for pumping and distributing lagoon effluent may be similar to conventional irrigation equipment, the smaller volume of water handled in most livestock lagoons generally allows the use of smaller and less costly systems. It also is possible to use an application system for both effluent and fresh water irrigation. The type of sprinkler system chosen may depend on the particle size of the solids in the effluent. The system capacity and type selected may depend on available capital and how much time and labor are available for pumping. Table 4-1 gives the labor requirement for irrigating with various systems.

System flow rate usually is determined by the time (and labor) available for pumping. Table 9-15 gives the time required for different flow rates for the average annual pump-down or volumes produced in various livestock operations. The pump-down volumes for lagoons are less than the volumes of water normally handled in conventional irrigation systems.

In some cases, an irrigation reservoir also serves as the lagoon receiving livestock effluent. In such cases, take care that the reservoir does not overflow, or that the overflow does not violate pollution laws and regulations.

If a lagoon and an irrigation reservoir are located in the same vicinity, a properly designed pump intake system may allow simultaneous withdrawal of liquid from both the irrigation reservoir and the lagoon. If the irrigation water source and the lagoon are some distance apart, a pump at the lagoon will probably be needed to inject the lagoon effluent into the pipe downstream from the pump serving the irrigation system. This method also may be used to inject manure pit effluent into an irrigation system. Another alternative is to pump or drain the lagoon effluent into the irrigation reservoir, provided there is sufficient volume available. Check state requirements to determine if draining lagoon effluent into the irrigation reservoir is allowed by law.

Types of Systems

As with water irrigation, there is no one system that is superior over another system. The following systems can be used for effluent irrigation:

- Stationary Volume Gun.
- Traveler.
- Hand-move Sprinkler.
- Center Pivot and Linear.
- Side Roll.
- Solid Set.

Stationary Volume Gun

This system can be used in many small effluent application systems (see Figure 9-1). The system includes a pump and a main line similar to the hand-move system, but with a single large volume gun sprinkler. Advantages of the volume gun system include larger flow rates and a larger wetted area so less labor is required in moving the sprinkler. Some volume guns are wheel-mounted to facilitate moving the unit. Stationary volume guns typically have nozzle sizes that range from 1 to 2 inches, and operate best at pressures of 80 to 120 psi. Coverage areas of 1 to 4 acres can be obtained with proper selection of nozzle size and operating pressure. Although stationary volume guns cost more

Table 9-15. Average pumping time required to irrigate the annual average lagoon pump-down volume (hours).

System flow rate	Volume (acre-inch per year)					
(gpm)	1	5	15	30	50	100
100	4.52	22.6	67.8	136	226	452
300	1.51	7.54	22.6	45.2	75.4	151
500	0.904	4.52	13.6	27.1	45.2	90.4
1000	0.452	2.26	6.78	13.6	22.6	45.2

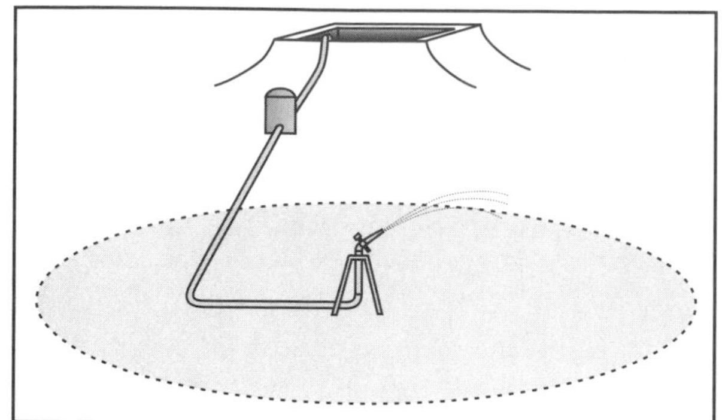
Figure 9-1. Stationary volume gun.

than smaller hand-carry systems, the reduced labor cost and higher flow rates may offset the higher cost.

A typical volume gun that discharges 330 gpm at 90 psi pressure wets a 350-foot diameter circle (2.2 acres) with an application rate of 0.33 inches per hour. The power requirement is about 30 horsepower. This system requires labor for movement from one set or location to another to insure that the soil does not become saturated.

Advantages:

- Few mechanical parts to malfunction.
- Few plugging problems with large nozzle.
- Flexible with respect to land area.
- Pipe requirements are slightly less than with small sprinklers.
- Moderate labor requirement.

Disadvantages:

- Moderate to high initial investment.
- Water application pattern is easily distorted by wind.
- Tendency to over-apply effluents with high nutrient concentrations such as livestock lagoon effluent.

Traveler

The traveler is justifiable on larger acreages irrigated several times a year (Figure 9-2). Refer to Chapter 4 for details on travelers. Generally, a water-driven winch on the machine pulls the traveling gun across the ground by a cable anchored at the end of the field. The gun also can be pulled by a hard hose feeding the water to

Figure 9-2. Traveler with volume gun applying effluent.

the traveling gun. On some models the winch may be driven by a small engine. Such an arrangement prevents the possibility of solids plugging the water turbine. However, plugging problems are minimal when pumping effluent from properly sized and operated lagoons.

Effluent usually is carried from the lagoon to the point of attachment in the field by a portable aluminum pipe or underground plastic pipe.

Large travelers may be costly for pumping livestock lagoons unless they are also used for crop irrigation or custom lagoon pumping. A common size irrigates 10 acres per set compared to about 2 acres for a stationary gun. Most manufacturers of travelers now make small units, which may be competitive in cost with the hand-move or stationary gun sprinklers. These units are adaptable to rolling terrain and are easily transported.

Advantages:

- Moderate labor requirements.
- Few or no plugging problems with the large nozzle.
- Flexible with respect to land area.

Disadvantages:

- Higher initial cost than the previous systems.
- High power requirement.
- More mechanical parts than the other systems, especially with an auxiliary engine.
- Rate of application is high.

Hand-move Sprinkler Systems

The least costly sprinkler systems for effluent irrigation are the hand-move types that require labor to set up and move the system (Figures 4-15 and 4-16, page 60). Although considerable labor input is required, these systems may be desirable for small lagoons. Used hand-move systems may be available, but small nozzles in the sprinklers may not be suited for effluent irrigation. A screened inlet pipe will reduce problems with small nozzles.

An example of such a system might be a 1/4-mile lateral covering 1.8 acres with each 60-foot move. A total of 32 sprinklers would discharge 10 gpm each, for a total 320 gpm pumped through a 5-inch pipe. The application rate would be 0.4 inches per hour.

Nozzle sizes used for moderately to heavily loaded lagoons are generally in the 1/2 to 1-inch range and typically cover 1/2 to 2 acres per sprinkler, depending on nozzle size and system operating pressure.

Advantages:

- Low initial investment especially with a used system.
- Few mechanical parts to malfunction.
- Low power requirement (50 psi at the sprinklers).
- Adaptable to field shape. Different lengths can be set and run almost any direction to get isolated corners.

Disadvantages:

- High labor requirement; individual pipe sections are moved, which can be a very unpleasant task with effluent.
- Small sprinklers can plug.
- Tendency to over-apply effluents with high nutrient concentrations such as livestock lagoon effluent.

Center-Pivot and Linear Irrigation Systems

Center-pivot systems, which move around in a circle like a spoke of a wheel (refer to Chapter 4 for details on center-pivot irrigators), have been used in some large livestock operations, but customarily are used for irrigating effluent from municipal and industrial systems. Center-pivot irrigators are becoming more common for land application of livestock effluent as the size of these operations increase. Center-pivots offer almost complete automation. They have several disadvantages including high cost, small sprinklers, and fixed land area covered. Low-pressure systems in the 20 psi range with nozzles less than 1/4-inch diameter are not recommended for livestock effluent because they could be plugged by solids in the effluent. Linear-move irrigators can apply effluent to rectangular areas without missing the corners but cost considerably more per irrigated acre. Special center-pivot systems are available with volume guns for irrigating undiluted effluent from pits and storage ponds.(Figure 9-3)

Figure 9-3. Center-pivot irrigator designed for high solids contents.

Side-Roll Systems

These systems roll sideways across a rectangular field, but are limited to low-growing crops (Figure 4-17). Crop clearance is slightly less than one-half the diameter of the wheel. These systems use small sprinklers, require rectangular fields, and have several mechanical devices.

Solid-Set Systems

Solid-set systems, (Figures 4-13 and 4-14) permanently installed throughout an entire field are generally too expensive to install. The exceptions would be small systems that utilize small-diameter plastic pipe, or systems that apply effluent year-around. The risers and sprinklers are always in place and must be avoided by machinery and protected from animals. If automation is desired, remote control valves can be programmed to operate separate laterals or blocks in various sequences.

Pumps

The solids content, required pumping pressure, and flow rate are major factors in selecting a pump that can handle effluent. Selecting the correct pump for the job improves performance and reduces repairs. Table 9-17 contains certain characteristics of pumps used for sprinkler irrigation of effluents. Standard centrifugal irrigation pumps are described in Chapter 7.

Solids content of livestock effluent varies with livestock type, housing, bedding, and effluent collection system. Fewer pumping problems result if solids are settled out (as in a lagoon), mechanically separated, or if dilution water is added (such as by rain on a lagoon or feedlot). Table 9-16 shows the typical solids content for various liquid livestock effluent handling systems. Figure 9-4 shows the relative handling characteristics of different types of effluent and the percent total solids for various species. Swine, beef, and dairy manure as excreted contains about 10 to 12% solids, while sheep and poultry manure contains about 25% solids. To be pumped with conventional irrigation pumps, the solids content should be diluted to 4% or less. Due to dilution from rainfall, effluent pumped from lagoons typically contains less than 1 to 2% solids. This should present no special pumping problems if large or fibrous solids are not present. If long, fibrous solids are present in the liquid being pumped, a chopper pump may be required. Larger pumps can pump effluent with larger solids particles.

Pressure requirements vary considerably with the application, however, sprinkler irrigation of effluent requires higher pressures. Generally, pressure must be higher for effluents than for clean water being pumped through the same system. Operating pressures typically are from 40 to 120 psi and higher. Nozzle diameter also affects pressure requirements.

Table 9-16. Solids in liquid effluent handling systems.

System	Solids content (%)
Manure pit	
Swine	4 to 8
Cattle	10 to 15
Holding pond	
Pit overflow	1 to 3
Feedlot runoff	Less than 1
Dairy barn washwater	Less than 1
Lagoon	
Single or first stage	
Swine	1/2 to 1
Cattle	1 to 2
Second stage	Less than 1/2
Flushing effluent	1/2 to 2

Pumping Effluent With Less Than 4% Solids Content

Effluent from lagoons, feedlot runoff holding ponds, and milkhouses (less than 4% solids content) often can be handled by conventional irrigation pumps and equipment. Pipeline transport is most common because it is impractical to transport effluent by slurry tank or truck. Standard, single-stage, closed impeller centrifugal pumps work well. Semi-open impeller pumps are sometimes desirable for their resistance to clogging.

Large solids should be screened out at the pump inlet. The screen area should be as large as possible to minimize velocities into the screen and to eliminate plugging. A trash guard with a 5- to 15-foot radius encircling the screen is often beneficial. The guard can be constructed with small diameter woven or meshed wire. To keep the intake strainer from clogging when large sprinkler nozzles are used, many operators float the intake 18 to 24 inches below the surface and do not use a strainer. This keeps the intake free of floating debris, and above the sludge layer at the bottom of the lagoon. If the lagoon is located in an area where there is a prevailing wind direction, the pump should be located on the upwind side of the lagoon because solids tend to migrate to the downwind side by wind and wave action.

All-steel pumps are less susceptible to corrosion than pumps with brass impellers.

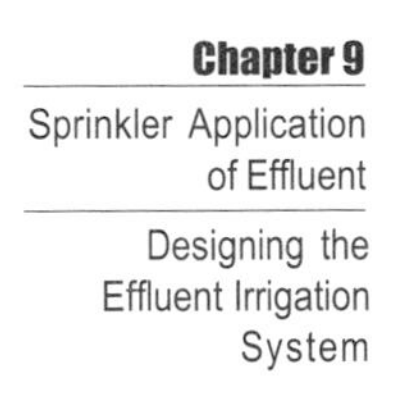

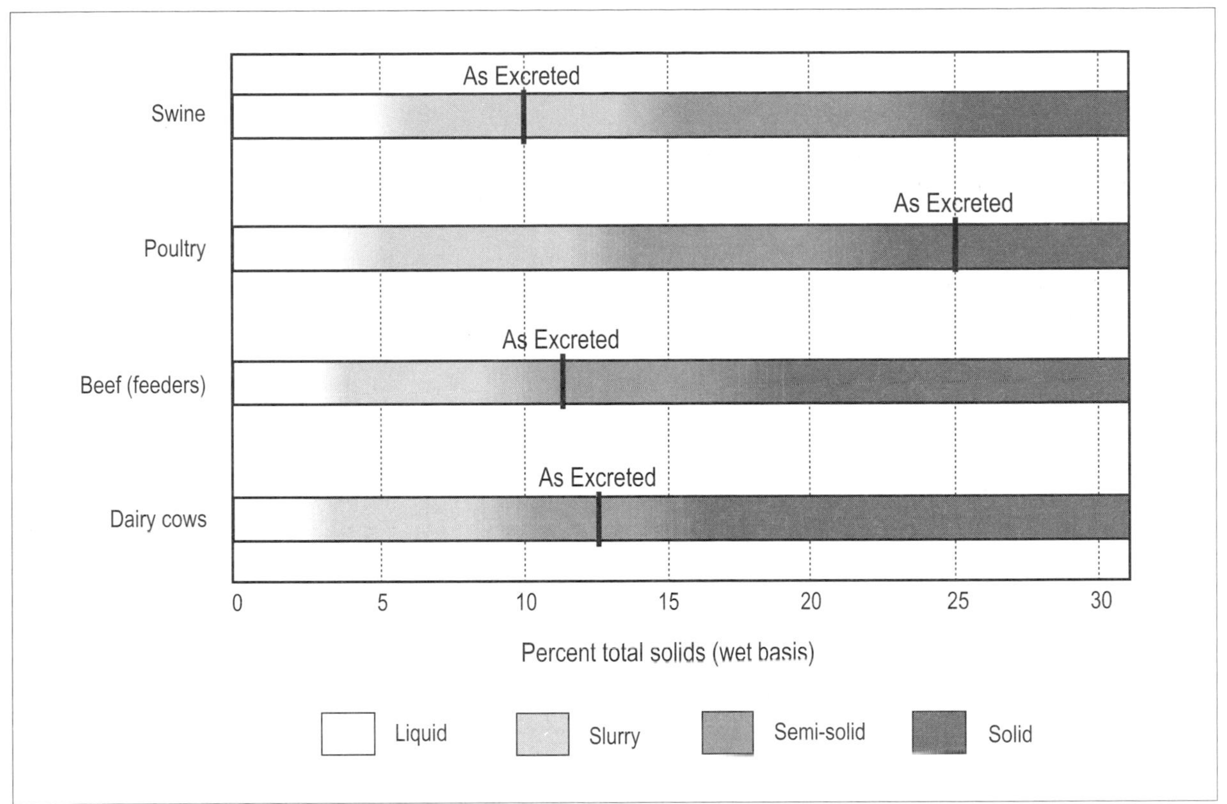

Figure 9-4. Relative handling characteristics of different types of manure for various species.

Table 9-17. Characteristics of semi-solid handling pumps used for sprinkler irrigation of effluent.

Pump Type	Max. solids content (%)	Pumping rate (gpm)	Pumping head (ft of water)
Centrifugal			
Open & semi-open impeller	10 to 12	up to 3,000	25 to 500
Closed impeller	4	500+	200+
Helical screw	4	200 to 300	200+

Flushing the system with clean water after use prolongs equipment life by reducing corrosion.

Pumping Effluent With Up To 15% Solids Content

To minimize clogging problems, an open impeller centrifugal pump may be selected that can handle large solids up to 1 inch in diameter. Open impeller pumps may handle liquids with solids content up to 15%. They often have a sharp rotating blade at the pump inlet to chop large material, such as bedding, hay, or silage. Semi-open impellers can handle water with small solids, e.g., swine lagoon liquid. Large closed impeller pumps are efficient for high-pressure application of effluent without solids, such as municipal lagoons and lagoons where the solids have been separated and removed.

Foot valves of the centrifugal pump can leak, as a result they may need to be reprimed. To avoid priming problems, the pump should be installed below the reservoir level. To prevent possible overland flow in case of a piping failure, the pump may be installed in a wet well. Many custom pumpers of livestock lagoons use a hydraulically-driven, submersible chopper pump to feed a centrifugal pump on the surface. This system eliminates priming and inlet screen problems.

Positive displacement helical-screw pumps can handle liquids with high solids content but they must be free of hard abrasive solids such as nails, stones, and livestock identification tags. Do not operate these pumps dry. A common practice is to add a small stream of fresh water directly into the pump inlet to assure lubrication during operation. When using positive displacement pumps to irrigate directly from liquid manure pits, monitor the line pressure. If irrigation nozzles clog, pressure buildup can burst pipes. A pressure release valve often is used to reduce pressure by allowing excessive liquid to drain back into the lagoon.

Volume gun sprinklers can spread effluents with high solids content with a consistency up to that of thick milk. Slurries that have a high solids content should be agitated before and during pumping to keep the solids suspended. Pumps used for high-pressure pumping of slurries include piston, helical-rotor, and submerged-centrifugal. Centrifugal pumps may have problems with suction starvation if the solids content is higher than 7%. Positive-displacement pumps may have this problem if the solids content of the effluent is over 15%. Large quantities of fibrous solids can clog pumps that are not designed properly.

There are some challenges with pumping effluent high in solids. Liquid may drain from the irrigation pipe when the pump stops, which leaves solids in the pipe that are hard to remove and can corrode aluminum pipe. Flush the pipeline with fresh water after irrigating to remove these solids.

If possible, pump clean water for 10 to 15 minutes before moving the sprinkler. This will clean out the pipe, rinse the pipe near the sprinkler, wash solids off the foliage, and dilute drained liquid effluent at the end.

Avoid applying semi-solid effluent to a growing crop. The crop may need to be rinse-sprayed with clean water about every hour if irrigation is done during the heat of the day. If it is necessary, apply at minimal rates unless fresh water is also applied. Irrigate alfalfa fields after hay-cutting. Do not irrigate corn when the crop is very young or during tasseling or silking.

Pipe

Long-distance pumping of semi-solid effluents is relatively uncommon. The major exception is the land application of liquid dairy effluent through a manure gun or hose-tow ground application. Pipeline design is similar to that for effluents, except that each component is selected to handle more solids.

Table 9-18 shows that for velocities from 1 to 7 feet per second in pipes 6 to 10 inches in diameter, change in friction loss is not significant for solids contents of 8% or less.

Although pipe friction losses might be higher for effluent than for clean water, friction losses generally are a small percentage of the total power requirement in a sprinkler system. When the same pump is used for pumping both slurries and clean water, the pump might operate at two different points on the pump curve for the two liquids. When pumping slurries, look for a marked increase in brake horsepower requirements, a reduction in the head produced, and some reduction in capacity. The system needs more horsepower to overcome the velocity head loss and the pipe friction losses. As a general rule in situations where the friction loss ratio exceeds 1.0 increase the power unit rating by 10% to account for the differences associated with the presence of solids and higher viscosity.

Sprinklers

Important factors in selecting sprinklers to apply effluents are selecting the maximum application rate to prevent runoff and the minimum sized nozzle opening needed to pass the largest solids in the effluent. It is important to prevent clogs because removing them requires labor and time that may not be available when the sprinkler clogs. Sprinklers for effluent with more than 4 percent solids usually are the volume-gun type with 1 inch or larger nozzles, that require 80 psi or more pressure and pumping rates of 100 to 1,000 gpm. Liquid from moderate to heavily-loaded lagoons can be applied through 1/2- to 3/4-inch nozzles. Larger gun-type sprinklers with 3/4- to 2-inch nozzle diameters should be used for lagoons with high concentrations of solids or for liquid sludge irrigation.

Table 9-18. Friction loss ratio for 6- to 10-inch pipe diameters. Ratio is effluent to clean water.

Velocity (ft per sec)	% solids					
	4	5	6	7	8	10
1.0	1.1	1.5	2.1	2.9	4.0	5.3
1.5	1.0	1.2	1.5	2.1	2.5	4.0
2.0	1.0	1.0	1.0	1.6	1.9	3.3
2.5	1.0	1.0	1.0	1.3	1.6	2.9
3.0	1.0	1.0	1.0	1.2	1.5	2.7
3.5	1.0	1.0	1.0	1.1	1.3	2.5
4.0	1.0	1.0	1.0	1.0	1.0	2.4
4.5	1.0	1.0	1.0	1.0	1.0	2.3
5.0	1.0	1.0	1.0	1.0	1.0	2.2
5.5	1.0	1.0	1.0	1.0	1.0	2.1
6.0	1.0	1.0	1.0	1.0	1.0	2.0
6.5	1.0	1.0	1.0	1.0	1.0	2.0
7.0	1.0	1.0	1.0	1.0	1.0	2.0

Example 9-6. Determining friction loss for effluent.

An 8-inch pipeline is to deliver 550 gpm of effluent containing 10% solids. The friction loss for fresh water is 0.19 pounds per square inch per 100 feet of pipe length, and the velocity is 3.42 feet per second. Determine the friction loss for the pipe that will handle effluent.

Solution:

From Table 9-18, the friction ratio loss factor is 2.5 at 3.5 feet per second for 10% solids. The friction loss for the effluent is calculated to be:

$$\left(\frac{0.19 \text{ psi}}{100 \text{ ft}}\right) \cdot (2.5) = 0.48 \text{ psi per 100 feet of pipe length}$$

Smaller sprinklers may be adequate for effluent with less than 3% solids and no large solids. Liquid from lightly-loaded anaerobic lagoons can be applied through sprinkler nozzles 1/4-inch or larger. Single-nozzle, straight-bore sprinklers are recommended. An adequate suction strainer for the pump intake should be used where small sprinkler nozzles are used. Pump intake screens should be sized with openings no larger that the smallest sprinkler orifice. This is especially true for 1/4-inch or smaller nozzles. Single-nozzle sprinklers perform better under windy conditions and clog less than sprinklers with two or more smaller nozzles. One large nozzle is less likely to clog than two smaller nozzles with the same capacity.

Volume guns generally have only one large nozzle, designed for a long throw distance. The wetted diameter varies with nozzle size, pressure, sprinkler angle, and operating conditions such as wind. Check manufacturer's literature for sprinkler and volume guns design. Table 5-2 (page 70) lists estimated output for volume guns.

The application rate varies with distance from the sprinkler or gun. If the sprinkler produces a triangular application pattern (Figure 5-11, page 76) proper spacing should achieve a nearly uniform application depth. Sprinkler application pattern can vary with operating pressure. To attain acceptable application uniformity with multiple sprinkler setups, the sprinkler spacing should be 40 to 80% of the wetted diameter. Overall uniformity can be affected by wind velocity. Try to irrigate when it is sunny and the wind is under 5 mph. Traveler lane spacing variations with wind are given in Table 5-6 on page 77. Higher trajectory sprinklers are used for low wind conditions to obtain maximum distance of throw. Low trajectory sprinklers will give maximum distance of throw and a minimum of pattern distortion in high winds. Use low trajectory sprinklers and low pressures to reduce odor from effluent.

Taper-bore nozzles have the greatest stream integrity, longest throw distance, and minimum wind distortion. Ring nozzles have better stream breakup for lower pressure operation and delicate crops. Ring nozzles catch animal hair on the nozzle lip and plug more often than taper-bore nozzles.

System Application Rate

Effluent system application rate is similar to water system application rate. The discussion and many of the equations in the section on **System Application Rate** on page 79 in Chapter 5 is applicable for effluent irrigation.

The average application rate for a set-type sprinkler system can be determined by Equation 5-2 (page 81). The peak application rate for a center-pivot system can be determined by Equation 5-3 (page 81).

To find the time to operate a system to obtain a given depth of application, use Equation 9-1:

Equation 9-1. Sprinkler Operation Time to Apply a Given Depth of Effluent.

$$TD = \frac{I_d}{R}$$

Where:

TD = Time to operate sprinkler to obtain a given depth of application, hour
I_d = Depth of application, inches
R = Application rate, inches per hour

Use Equation 9-2 to determine the time to operate system to obtain a given amount of nutrient per acre.

Equation 9-2. Sprinkler Operation Time to Apply a Given Amount of Nutrients.

Find the time to operate system to obtain a given amount of nutrient per acre.

$$TN = (0.0368) \cdot \left(\frac{NA}{R \cdot C} \right)$$

Where:

TN = Time to operate sprinkler to obtain a given amount of nutrient per acre
NA = Target nutrient application, pounds per acre
R = Application rate, inches per hour
C = Nutrient content in lagoon effluent, pounds per 1,000 gallons

The sprinklers on the traveling volume gun systems usually are equivalent to the stationary volume guns. However, the sprinkler may be operated part-circle to keep a dry lane ahead of the traveling gun. This increases the average application rate due to the decreased area of application, and also slightly affects the uniformity of application. The application rate for a part-circle gun operation may be found by dividing the rate for full-circle operation, by the fraction of full-circle the gun is operated. The depth of liquid applied by a traveling gun depends on the flow rate, lane spacing, and travel speed (see Table 6-5 in Chapter 6). Table 9-19 shows the number of acres irrigated per set as a function of lane spacing and travel distance. Table 5-6 recommends maximum travel lane spacings for windy conditions.

Use Equation 5-5 (page 81) to determine the average application rate of a traveling gun sprinkler. Use Equation 9-3 to determine the application depth.

Equation 9-3. Effluent Application Depth.

$$I_d = \frac{(0.0368) \cdot NA}{C}$$

Where:

I_d = Depth of water to apply, inches
NA = Target nutrient application rate, pounds per acre
C = Nutrient content in lagoon effluent, pounds per 1,000 gallons

Table 9-19. Acres irrigated per setting by traveling big guns (acres per set). For best watering uniformity, make lane spacing 50% to 70% of the sprinkler wetted diameter.

Lane spacing (ft)	Travel distance 660 ft	Travel distance 1,000 ft
100	1.5	2.3
120	1.8	2.8
140	2.1	3.2
160	2.4	3.7
180	2.7	4.1
200	3.0	4.6
220	3.3	5.1
240	3.6	5.5
260	3.9	6.0
280	4.2	6.4
300	4.5	6.9
320	4.8	7.3
340	5.2	7.8
360	5.5	8.3
380	5.8	8.7
400	6.1	9.2

Use Equation 9-4 to determine the speed to operate a traveling gun sprinkler.

Equation 9-4. Traveler Speed.

$$S = \frac{(1.605) \cdot Q}{S_L \cdot I_d}$$

Where:

S = Travel speed, feet per minute
Q = Flow rate, gpm
SL = Traveling gun lane spacing, feet
I_d = Depth of water applied, inches

Evaluating Application

Discussion on **Application Uniformity** (page 82 and 92) and **Application Efficiency** (page 84) provides a good background to evaluating application effectiveness. The calibration procedures are predicated on the assumed performance of sprinklers operated at certain pressures and sprinkler spacings. To verify that the assumed performance is achieved, catch cans for stationary sprinklers can be spaced as shown in Figure 6-5. Run tests until the average depth in the cans is at least 1 inch; longer tests will reduce errors in measuring small amounts. Catch cans for a traveler can be placed in a line perpendicular to the direction of travel and between two adjacent lanes. Catch cans should not be more than 10 feet apart.

Summary

The design of an effluent irrigation system is similar in many ways to the design of a water irrigation system. The solids in the effluent are the main difference from a system handling standpoint. Knowing the amount of solids in the effluent and selecting the proper equipment to handle the solids is necessary for a successful design. Having effluent and soil tested when matching the test results to crop nutrient uptake is necessary to applying the proper amount of effluent to a site. Also knowing the effects the solids can have on the soil is necessary in determining the proper application rate.

Worksheet Example 10-1

Center Pivot Design with a Well

Situation

A producer near Bismarck, North Dakota, wants to irrigate 160 acres (a square quarter section) of rolling terrain with a center pivot. The field is located over a confined aquifer. Soils are fine sandy loams and loamy fine sands. Three-phase power is available along the county road on the west side of the field. Adjusting prices to include peak monthly kilowatt demand charges, the local Rural Electric Cooperative charges $0.10 per kilowatt-hour (kWh) for regular power and $0.05 per kWh for off-peak power. The producer will raise corn, potatoes, dry beans, and alfalfa on this field.

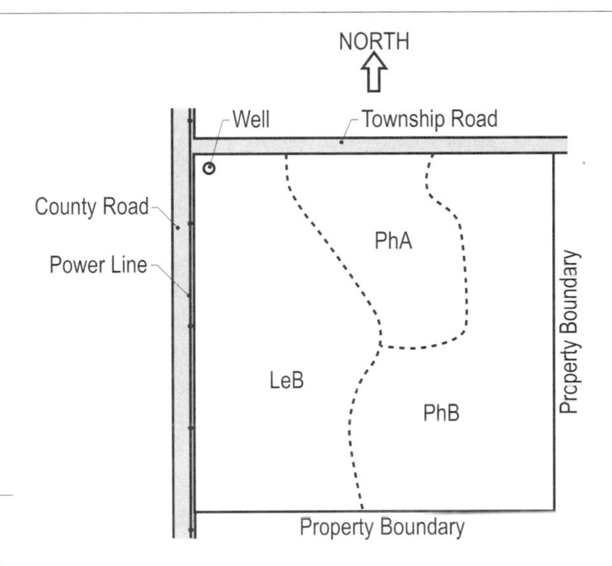

Design Goals and Challenges

The producer's goals are the following:

- Use low-pressure center pivot sprinkler technology.
- Use an end gun to irrigate corners.
- Take advantage of off-peak power savings.
- Apply a weekly fungicide spray by chemigation when potatoes are grown.

Some of the major challenges in developing this system are these:

- Road right-of-ways affecting the west and north edges of the field are 20 feet.
- There is a 5-foot elevation rise from the well to the pivot point.
- A hill with a 15-foot elevation rise is located between the well and the pivot point.
- The mainline pipe will be buried.
- There is a 20-foot elevation rise from the pivot point to the highest point in the field.
- The static water level can drop as much as 20 feet from May to the middle of August.

Design

Field Parameters: Due to the 20-foot road right-of-ways on the west and north, the total pivot length can be only 1,300 feet long. A pivot point was identified. Distances to power poles and other obstructions were measured to make sure the pivot towers and overhang would not be hampered. There are no wetland depression areas in this field. A measurement of elevation changes showed a 20-foot rise from the pivot point to the highest point in the field (near the southeast corner) and a 5-foot rise from the well to the pivot point.

Map ID	Soil type	Slope (%)	Water holding capacity	Field area (%)
LeB	Lihen loamy fine sand	1 to 6	1.6 in/ft	37
PhA	Parshall fine sandy loam	1 to 3	1.9 in/ft (0 to 2 ft)	23
			1.4 in/ft (2 to 5 ft)	
PhB	Parshall fine sandy loam	3 to 6	1.9 in/ft (top 2 ft)	40
			1.4 in/ft (2 to 5 ft)	

The infiltration rate for these soils is between 2 and 6 inches per hour.

Worksheet Example 10-1 (continued)

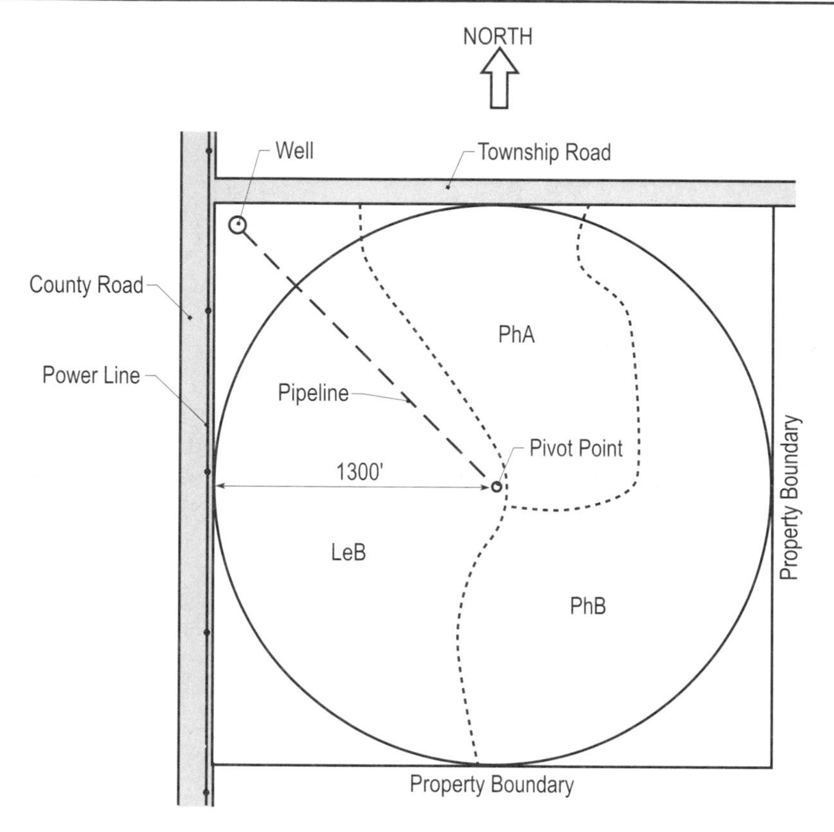

Water Supply: The irrigation well was test pumped at 600 gpm for 72 hours. The static water level at the start was 55 feet below the surface. At the end of the test, the pumping water level was 115 feet or 60 feet of drawdown. The specific capacity of this well is

$$\text{Well Capacity} = \frac{\text{Pumping Rate}}{\text{Pumpdown Level}} = \frac{600 \text{ gpm}}{60 \text{ ft}} = 10 \text{ gpm per foot of drawdown}$$

The top of the screen is at 175 feet. However, consultation with the well driller and other irrigators in the area indicate the static water level can drop as much as 20 feet from May to the middle of August, which means this well could produce a maximum of

$$\text{Maximum Well Yield} = \left[(\text{top of screen}) - (\text{static water level}) - (\text{summer dropdown})\right] \cdot \left[\text{well capacity}\right]$$

$$= \left[(175 \text{ ft}) - (55 \text{ ft}) - (20 \text{ ft})\right] \cdot \left[10 \frac{\text{gpm}}{\text{ft}}\right] = 1{,}000 \text{ gpm}$$

Pivot Area Coverage and Sprinkler Selection: Normal wind speed during the growing season varies from 10 to 15 mph, with periods of a few days where it can be near 30 mph. Because corn will be grown, a rotating spray sprinkler package on 5-foot drop tubes will be installed on the center pivot. The drop tubes will put the sprinkler heads just below the truss rods but above the fully developed corn plants. The rolling terrain requires 15-psi pressure regulators be used. An end gun with a radius of throw of about 100 feet will be used for the corners. Even though the end gun will have a water throw of 100 feet, the end gun will effectively irrigate only an additional 70 feet from the end of the pivot.

Worksheet Example 10-1 (continued)

Due to the low-pressure requirement, a booster pump will be needed for proper end gun operation. The end gun will need to be shut off at those locations where it will throw water onto the pads or neighboring property. The end-gun operation angle is

$$\text{End-gun Operation Angle} = 90 - \left(2 \cdot \cos^{-1}\left(\frac{\text{Lateral Length}}{(\text{Lateral Length}) + (\text{End-gun Throw})}\right)\right)$$

$$= 90 - \left(2 \cdot \cos^{-1}\left(\frac{1{,}300 \text{ ft}}{(1{,}300 \text{ ft}) + (70 \text{ ft})}\right)\right) = 90 - ((2) \cdot (18.5)) = 53^\circ$$

Using the full circle with end-gun Equation B-3 in Appendix B, a 1,300-foot pivot using an end gun that operates for 53 degrees in the corners and effectively irrigates an additional 70 feet will irrigate

$$\text{Irrigated Area} = \left[\left(\frac{3.14 \cdot (R_1)^2}{43{,}560}\right) \cdot \left(\frac{360 - \sum a}{360}\right)\right] + \left[\left(\frac{3.14 \cdot (R_2)^2}{43{,}560}\right) \cdot \left(\frac{\sum a}{360}\right)\right]$$

$$= \left[\left(\frac{(3.14) \cdot (1300 \text{ ft})^2}{43{,}560}\right) \cdot \left(\frac{360 - ((4) \cdot (53^\circ))}{360}\right)\right]$$

$$+ \left[\left(\frac{(3.14) \cdot (1370 \text{ ft})^2}{43{,}560}\right) \cdot \left(\frac{(4) \cdot (53^\circ)}{360}\right)\right] = 130 \text{ acres}$$

The local dealer will order the rotating spray sprinkler package with an end gun from the manufacturer but will need to know the pivot pressure and the flowrate. For proper operation of the end gun's booster pump and the pressure regulators on the last span and overhang, a minimum of 15 psi at the end of the pivot lateral is required. Converting the end-gun pressure into head

$$\text{End-gun Head} = (15 \text{ psi}) \cdot \left(2.31 \frac{\text{ft of head}}{\text{psi}}\right) = 35 \text{ feet of head}$$

The center pivot lateral will be aluminum with an inside diameter of 6-5/8 inches (most common size). For 1,300 feet of aluminum pipe with no outlets and a flow rate of 900 gpm, the friction loss would be

$$\text{Pipe Friction Loss} = \left(\frac{43 \text{ ft head loss}}{1{,}000 \text{ ft pipe}}\right) \cdot (1{,}300 \text{ ft pipe}) = 56 \text{ feet of head,}$$

or converting to pressure loss

$$= (56 \text{ ft of head}) \cdot \left(0.434 \frac{\text{psi}}{\text{ft of head}}\right) = 24 \text{ psi}$$

The lateral friction loss will be less than the mainline friction loss. From the discussion on **Multi-Outlet Submain or Lateral Pressure Loss** in **Chapter 7**, the lateral friction is 54% of the mainline friction loss. The lateral friction loss will be

Worksheet Example 10-1 (continued)

$$\text{Lateral Friction Loss} = (56 \text{ ft of head loss}) \cdot (0.54)$$

$$= 30 \text{ feet of head loss, or converting to pressure loss}$$

$$= (30 \text{ ft of head}) \cdot \left(0.434 \frac{\text{psi}}{\text{ft of head}}\right) = 13 \text{ psi}$$

The elevation rise in land from the pivot point to the highest point is 20 feet. Converting the elevation rise to pressure loss, the result is

$$\text{Elevation Rise Pressure Loss} = (20 \text{ ft of head}) \cdot \left(0.434 \frac{\text{psi}}{\text{ft of head}}\right) = 8.7 \text{ psi}$$

The total design center pivot pressure is

$$\text{Design Center Pivot Pressure} = \text{elevation rise} + \text{pivot end pressure} + \text{lateral friction losses}$$
$$= 8.7 \text{ psi} + 15 \text{ psi} + 13 \text{ psi} = 37 \text{ psi,}$$

or in terms of head

$$\text{Design Center Pivot Head} = 20 \text{ ft} + 35 \text{ ft} + 30 \text{ ft} = 85 \text{ feet of head}$$

Using this information, the local dealer would order a 900-gpm sprinkler package for a pivot 1,300 feet long, using rotating sprays at 10- to 12-foot spacing with 15-psi regulators and a booster-aided end gun with a pivot pressure of 37 psi.

Pumping Rate Needed to Supply Pivot: From Figure 2-9, the average daily ET for the Bismarck, North Dakota, area in July is 0.26 inches per day. Consultation with the Cooperative Extension irrigation specialists indicates this is an accurate estimate. Average yearly precipitation in this area is about 16 inches per year, and growing season rainfall averages about 10 inches. Normal practice in this area is to irrigate 24 hours per day. Because the producer wants to apply a fungicide spray on a weekly basis when potatoes are grown, the irrigation system will be designed to operate 20 hours per day. From Table 2-5, a design water application efficiency of 0.85 will be used. The pumping rate is calculated using Equation 2-1:

$$\text{Pumping Rate} = \frac{453 \cdot (\text{Application Area}) \cdot (\text{Design Evapotranspiration Rate})}{(\text{Pumping Time}) \cdot (\text{Application Efficiency})}$$

$$= \frac{453 \cdot (130 \text{ acres}) \cdot \left(0.26 \frac{\text{in}}{\text{day}}\right)}{\left(20 \frac{\text{hrs}}{\text{day}}\right) \cdot (0.85)} = 900 \text{ gpm}$$

This pumping rate is within the capabilities of the well.

Worksheet Example 10-1 (continued)

Pipeline Design: The elevation difference between the well and the pivot point is 5 feet; however, at two locations between the well and pivot point, the pipeline rises 15 feet. The pipeline will use PVC pipe and be 1,800 feet long. The producer wants to keep the water velocity in the line below 5 feet per second. The nomograph from Figure 7-12 shows a flowrate of 900 gpm will require a 10-inch nominal diameter pipe. Comparing this to the flowrates and sizes listed in Table 7-6 shows that a 10-inch pipe is a good selection. The producer is concerned about water hammer; therefore, a PVC pipe with the highest reasonable pressure rating will be selected. The producer chooses a 10-inch IPS pipe with a pressure rating of 100 psi. Air relief valves will need to be installed at the well and the pivot point and at the highest points of the pipeline. The air relief valves located in the field will be protected by guard posts. Access ports to the pipeline need to be installed near the well and near the pivot point to facilitate pumping out the pipeline for winter.

Pump Requirements: To determine the size of the pump and the electric motor, the flowrate and total pressure head are required. The flowrate has been set at 900 gpm. The total pressure head is the sum of the well lift, friction losses through valves and piping, elevation rise, and pivot operating pressure. The maximum well lift will be 175 feet, which is the distance to the top of the screen and the lowest possible pumping level. However, the pumping water level for 900 gpm will be near 145 feet. For design purposes, the producer will use 160 feet as the design pumping lift. This should compensate for lowering of the static water table during the growing season and partial screen plugging in the coming years. The following converts the design pumping lift to pressure:

$$\text{Design Pumping Lift} = (160 \text{ ft of head}) \cdot \left(0.434 \ \frac{\text{psi}}{\text{ft of head}}\right) = 69 \text{ psi}$$

The sum of the pump column friction loss and the friction losses through the chemigation valve, elbows, flow meter, and other connections will total about 10 feet of head, or 4.3 psi. Using the nomograph in Figure 7-12, a 10-inch PVC pipe with a flowrate of 900 gpm and Hazen-Williams coefficient of 150 will have an approximate friction head loss of 3.4 feet per 1,000 feet of pipe. The total friction loss in the 1,800 feet of pipe is

$$\text{Pipe Friction Loss} = \left(\frac{3.4 \text{ ft head loss}}{1{,}000 \text{ ft pipe}}\right) \cdot (1{,}800 \text{ ft pipe}) = 6 \text{ feet of head,}$$

or converting to pressure loss

$$= (6 \text{ ft of head}) \cdot \left(0.434 \ \frac{\text{psi}}{\text{ft of head}}\right) = 2.7 \text{ psi}$$

The elevation rise from the well to the pivot is 5 feet, resulting in a pressure of 2.2 psi. An additional 12 feet of head, or 5.2 psi, results from the rise of the mainline to the lateral at the pivot. The resulting pressure at the pivot point pressure is 37 psi or 85 feet.

$$\text{Total Head} = \text{design pumping lift} + \text{connection losses} + \text{mainline friction loss}$$
$$+ \text{well to pivot elevation rise} + \text{mainline to lateral elevation rise} + \text{pivot point pressure}$$
$$= 160 \text{ ft} + 10 \text{ ft} + 6 \text{ ft} + 5 \text{ ft} + 12 \text{ ft} + 85 \text{ ft} = 278 \text{ feet of head,}$$

or in terms of pressure loss

$$= 69 \text{ psi} + 4.3 \text{ psi} + 2.7 \text{ psi} + 2.2 \text{ psi} + 5.2 \text{ psi} + 37 \text{ psi} = 120 \text{ psi}$$

A deep-well turbine pump with near maximum efficiency at 278 feet of head and 900 gpm will be ordered from a manufacturer. For added dry season pumping protection, the pump will be installed in the well so the pump intake is 3 to 5 feet above the top of the screen. Locating the pump 3 to 5 feet above the top of the screen means the maximum drawdown will be approximately 175 feet.

Worksheet Example 10-1 (continued)

The water horsepower (WHP) is calculated from Equation 7-13:

$$\text{WHP} = \frac{(\text{Pumping Rate}) \cdot (\text{Total Pumping Head})}{3{,}960} = \frac{(900 \text{ gpm}) \cdot (278 \text{ ft})}{3{,}960} = 63.2 \text{ horsepower}$$

Until the actual pump curve is obtained from the manufacturer, the brake horsepower (BHP) of the electric motor can be estimated with equation 7-14. A drive efficiency of 1.0 and a pump efficiency of 0.75 will be used:

$$\text{BHP} = \frac{(\text{Water Horsepower})}{(\text{Drive Efficiency}) \cdot (\text{Pump Efficiency})} = \frac{63.2 \text{ hp}}{(1.0) \cdot (0.75)} = 84.2 \text{ horsepower}$$

A 90-hp, 3-phase motor will be required; however, this is not a common size, so a100-hp motor will be ordered.

Pumping and Pivot Operation Costs: The estimated pumping costs under regular and off-peak power rates can be calculated using Equation 7-16 and Table 7-13:

$$\text{Hourly Energy Consumption} = \frac{(\text{Brake Horsepower})}{(\text{Electric Value})} = \frac{84.2 \text{ hp}}{1.18 \; \frac{\text{hp-hr}}{\text{kWh}}} = 71.4 \text{ kilowatt-hour per hour}$$

In addition, the pivot control panel, end-gun booster pump, and electric motors on the pivot towers will use an average of 6 kWh per hour. The total irrigation system will use about 77.4 kWh per hour of electrical energy.

For regular power at $0.10 per kWh and off-peak power at $0.05 per kWh, this irrigation system will cost

$$\text{Peak Power Operation Rate} = \left(\frac{\$0.10}{\text{kWh}}\right) \cdot \left(77.4 \; \frac{\text{kWh}}{\text{hr}}\right) = \$7.74 \text{ per hour}$$

$$\text{Off-Peak Power Operation Rate} = \left(\frac{\$0.05}{\text{kWh}}\right) \cdot \left(77.4 \; \frac{\text{kWh}}{\text{hr}}\right) = \$3.87 \text{ per hour}$$

If the pivots operate for 1,000 hours per growing season, the seasonal cost for all 1,000 hours operated at the regular power rate would be

$$\text{Peak Power Seasonal Operation Cost} = \left(\frac{\$7.74}{\text{hr}}\right) \cdot (1{,}000 \text{ hrs}) = \$7{,}740$$

The cost per acre at the peak power rate would be

$$\text{Acreage Cost} = \left(\frac{\$7{,}740}{130 \text{ acres}}\right) = \$59.54 \text{ per acre}$$

If the producer operates all 1,000 hours during the off-peak power time, the seasonal cost would be

$$\text{Off-Peak Power Seasonal Operation Cost} = \left(\frac{\$3.87}{\text{hr}}\right) \cdot (1{,}000 \text{ hrs}) = \$3{,}870$$

Worksheet Example 10-1 (continued)

The cost per acre at the peak power rate would be:

$$\text{Acreage Cost} = \left(\frac{\$3{,}870}{130 \text{ acres}}\right) = \$29.77 \text{ per acre}$$

In the years when potatoes are grown under this pivot, regular power will have to be used. Potatoes have shallow root systems and require relatively light, frequent irrigation. The usual amounts applied each rotation are from 0.5 to 0.75 inches. Using off-peak power on potatoes could jeopardize yield and quality by shorting the potatoes of water at crucial times. Off-peak power can be used with all the other crops (alfalfa, corn, and dry beans).

System Operation: Managing a center pivot requires an understanding of the rotation time to apply a certain amount of water. By rearranging Equation 2-1, it can be used to estimate the rotation time required to achieve the desired depth of application.

$$\text{Rotation Time} = \frac{(453)\cdot(\text{Irrigated Area})\cdot(\text{Application Depth})}{(\text{Pumping Rate})\cdot(\text{Application Efficiency})}$$

For an application depth of 0.5 inches, the rotation time would be

$$\text{Rotation Time} = \frac{(453)\cdot(130 \text{ acres})\cdot(0.5 \text{ in})}{(900 \text{ gpm})\cdot(0.85)} = 38.5 \text{ hours, or } 1.6 \text{ days}$$

For an application depth of 1.0, inch the rotation time would be

$$\text{Rotation Time} = \frac{(453)\cdot(130 \text{ acres})\cdot(1.0 \text{ in})}{(900 \text{ gpm})\cdot(0.85)} = 77.0 \text{ hours, or } 3.2 \text{ days}$$

To determine if there will be runoff when row crops are planted, the peak application rate for a center pivot can be calculated using Equation 5-3 on page 81. The sprinkler type being used has a 30-foot wetted radius:

$$\text{Peak Application Rate} = \frac{(122.6)\cdot(\text{Pumping Rate})}{[(\text{Lateral Length}) + (\text{Endgun Throw})]\cdot[\text{Wetted Radius}]}$$

$$= \frac{(122.6)\cdot(900 \text{ gpm})}{[(1{,}300 \text{ ft}) + (70 \text{ ft})]\cdot[30 \text{ ft}]}$$

$$= 2.7 \text{ inches per hour}$$

Based on the infiltration rates found in the soil survey, this site should have little problem with runoff.

Worksheet Example 10-2

Small Acreage Irrigation

Situation

A producer near Havana, Illinois, plans to irrigate 11 acres of horticulture crops. The field is 600 by 800 feet and slopes gently from north to south. An existing 125-foot deep, 8-inch well with a screen length of 25 feet is located midway on the west side. The static water level is 40 feet, and the level was 60 feet when pumped at 100 gpm. Soils are a combination of sandy loam and loamy sand, 24 inches to sand and gravel.

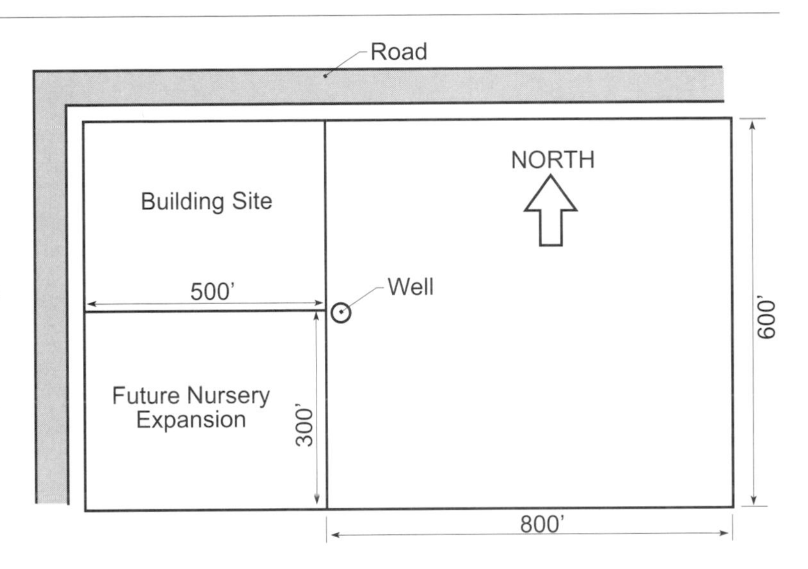

The producer plans to plant half of the field in strawberries with the remaining half in annual crops (melons, corn, and tomatoes, for example). If feasible, frost protection is desired for the strawberries. The producer wants to have a solid-set irrigation system using low-pressure technology and prefers to irrigate only during the eight night hours. The producer wants to use an electric submersible pump and knows that electricity will cost about $0.10 per kilowatt-hour (kWh). A future goal is to be able to irrigate an additional 3.4 acres of nursery stock behind the building site. The task is to design a solid-set irrigation system for the horticulture operation.

Design Goals and Challenges

The goals for this system are the following:

- Use a solid-set system using low pressure technology.
- Irrigate only during the eight night hours.
- Frost-protect the strawberries.

Two of the main challenges in designing this system are these:

- The design needs to allow for the irrigation of an additional 3.4 acres of nursery stock behind the building site in the future.
- Lateral lines will have a 1% downhill slope.

Design

Calculate maximum well yield: First, calculate the water available. Do not pump below the top of the screen. The drawdown available is the depth from the static water level to the top of the screen.

$$\text{Available Drawdown} = (\text{Well Depth} - \text{Screen Length}) - \text{Static Water Level}$$

$$= (125 \text{ ft} - 25 \text{ ft}) - 40 \text{ ft} = 60 \text{ feet}$$

Worksheet Example 10-2 (continued)

Then, dividing the discharge from the pump test by the drawdown gives specific capacity.

$$\text{Specific Capacity} = \frac{(\text{Pumping Rate})}{(\text{Drawdown}) - (\text{Static Water Level})}$$

$$= \frac{(100 \text{ gpm})}{(60 \text{ ft}) - (40 \text{ ft})} = 5 \text{ gpm per foot of drawdown}$$

Finally, multiply the available drawdown by the specific capacity to get maximum well yield.

$$\text{Maximum Well Yield} = (\text{Drawdown}) \cdot (\text{Specific Capacity}) = (60 \text{ ft}) \cdot \left(5 \frac{\text{gpm}}{\text{ft}}\right) = 300 \text{ gpm}$$

Because the producer will be using an existing 8-inch well, the pumping rates listed in Table 3-2 are consulted. The optimal pumping rate for an 8-inch well casing is 75 to 175 gpm. The maximum pumping rate for an 8-inch well casing is 400 gpm. The producer will evaluate water application requirements for frost protection and irrigation then compare these rates to the maximum well yield, optimal pumping rate, and maximum pumping rate.

Water supply for frost protection: The producer wants to protect the strawberries from frost damage. Only half of the field will be in strawberries.

$$\text{Area Planted in Strawberries} = \frac{(600 \text{ ft}) \cdot (800 \text{ ft})}{2} = 240{,}000 \text{ square feet,}$$

converted to acres

$$= \frac{240{,}000 \text{ sq ft}}{43{,}560 \frac{\text{sq ft}}{\text{acre}}} = 5.5 \text{ acres}$$

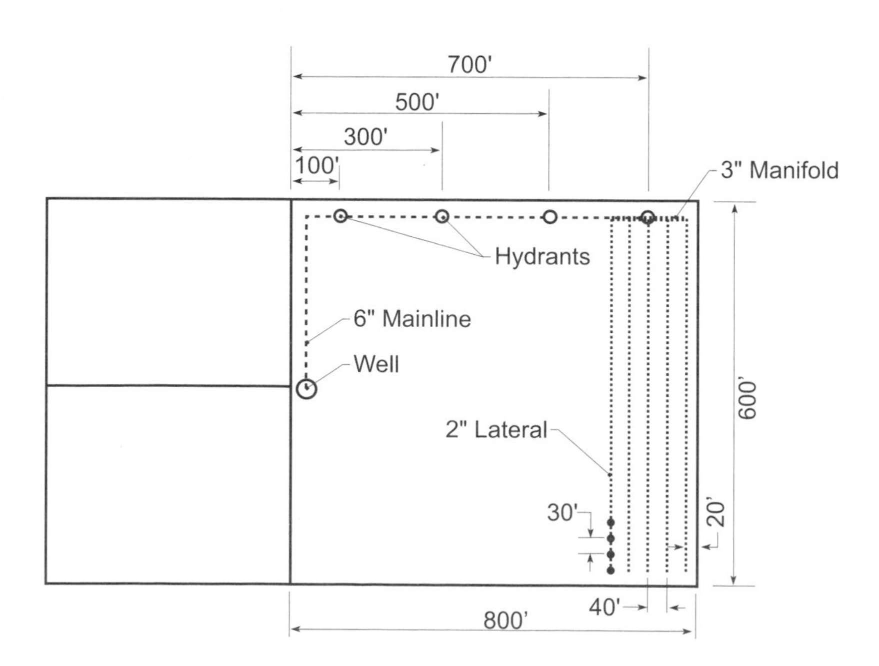

Worksheet Example 10-2 (continued)

After consulting with the area irrigation specialist on frost protection, the producer decides to apply water to the strawberries at a rate of 0.15 inches per hour. The flowrate needed for frost protection is

$$\text{Frost Protection Flowrate} = (\text{Application Rate}) \cdot (\text{Application Area})$$

$$= \left(0.15\ \frac{\text{in}}{\text{hr}}\right) \cdot (5.5\ \text{ac}) \cdot \left(453\ \frac{\text{gpm}}{\text{ac - in per hr}}\right) = 374\ \text{gpm}$$

The frost protection flowrate of 374 gpm is greater than the maximum well yield. One option is to apply water to only a portion of the field. If the producer uses the laterals to irrigate the 600-foot width of the field, then only 400 feet of the field will be planted with strawberries. If the producer applies water at the maximum pump yield and frost protection application rate, the producer can manipulate the frost protection flowrate equation to determine the irrigate areas:

$$\text{Irrigated Area} = \frac{(300\ \text{gpm})}{\left(453\ \frac{\text{gpm}}{\text{ac - in per hr}}\right)\left(0.15\ \frac{\text{in}}{\text{hr}}\right)} = 4.4\ \text{acres}$$

This would result in about 80% of the strawberries being protected from frost based on the rate the irrigation specialist recommended. Another alternative would be to irrigate the entire area at a rate of 0.10 inches per hour. The resulting flowrate would be

$$\text{Frost Protection Flowrate} = \left(0.10\ \frac{\text{in}}{\text{hr}}\right) \cdot \left(453\ \frac{\text{gpm}}{\text{ac - in per hr}}\right) \cdot (5.5\ \text{ac}) = 249\ \text{gpm}$$

This pumping rate is less than the maximum pumping rate. The producer will need to determine irrigation requirements then match a system to meet the irrigation requirements. After selecting a system to meet irrigation requirements, the producer will determine if using the system as a method of frost control is feasible.

Water supply for irrigation: Figure 2-9 shows the average daily ET_d for the Havana area in July to be 0.26 inches per day. The system is being designed to pump only 8 hours per day. The producer has decided that during periods when the daily ET exceeds 0.26 inches per day, the irrigation system will operate longer than 8 hours per day. On the basis of information in Table 2-5, the producer estimates the efficiency of the system to be 0.75. Equation 2-1 provides the irrigation pumping rate for the entire 11 acres.

$$\text{Pumping Rate} = \frac{453 \cdot (\text{Application Area}) \cdot (\text{Design ET Rate})}{(\text{Pumping Time}) \cdot (\text{Application Efficiency})}$$

$$= \frac{453 \cdot (11\ \text{acres}) \cdot \left(0.26\ \frac{\text{in}}{\text{day}}\right)}{\left(8\ \frac{\text{hrs}}{\text{day}}\right) \cdot (0.75)} = 216\ \text{gpm}$$

This pumping rate is less than the maximum well yield and the maximum pumping rate but higher than the optimal pumping rate. The designer will check for submersible pumps that can meet the irrigation pumping requirements.

Worksheet Example 10-2 (continued)

Sprinkler selection and layout: The irrigation dealer identifies three low pressure sprinkler choices:

- Low pressure 23° impact sprinkler operating at 30 psi, producing a wetted diameter of 80 to 90 feet.
- Rotating spray sprinkler operating at 20 psi, producing a wetted diameter of about 50 feet.
- Wobbling type spray sprinkler operating at 20 psi, producing a wetted diameter of about 50 feet.

The producer selects the impact sprinkler because it produces a larger wetted diameter. Based on the discussion of sprinkler spacing and overlap in Chapter 5, the sprinklers are spaced at 40 to 50% of the wetted diameter. The producer decides to space the sprinklers along the lateral at 40% and between laterals at 50% of the wetted diameter:

$$\text{Sprinkler Spacing Along the Lateral } (S_L) = \left(\frac{40\%}{100}\right) \cdot (80 \text{ ft}) = 32 \text{ feet, or approximately 30 feet}$$

$$\text{Sprinkler Spacing Between Lateral } (S_M) = \left(\frac{50\%}{100}\right) \cdot (80 \text{ ft}) = 40 \text{ feet}$$

The producer decides to position the laterals north-south with a mainline running up to the northwest corner of the field then east-west along the north side of the field. This layout provides the desired flexibility for cropping and allows positioning of the mainline such that elevation drop in the laterals can compensate for friction loss. The mainline will be buried, with risers spaced to serve the number of laterals that will be irrigated during one set time. Above-ground aluminum laterals were chosen to reduce investment costs and permit field operations for the annual crops when the irrigation system is not in the field. The figure on page 198 shows only one manifold and set of laterals. An option that would involve additional investment cost is to use buried PVC laterals. This would reduce the labor of placing the aboveground laterals in the field each season but would present some obstacles for preplanting field work.

The number of laterals needed for the field is determined by dividing field length by spacing between laterals.

$$\text{Number of Laterals} = \frac{(\text{Field Length})}{(\text{Spacing Between Laterals})} = \frac{(800 \text{ ft})}{(40 \text{ ft})} = 20 \text{ Laterals}$$

The number of sprinklers needed for each lateral is

$$\text{Number of Sprinklers per Lateral} = \frac{(\text{Field Width})}{(\text{Spacing Between Sprinklers})} = \frac{(600 \text{ ft})}{(30 \text{ ft})}$$

$$= 20 \text{ Sprinklers per Lateral}$$

Table 5-2 on page 70 shows reasonable choices for sprinkler nozzles are 7/64-inch diameter nozzles, discharging 1.86 gpm, and 1/8-inch diameter nozzles, discharging 2.43 gpm. Based on Equation 5-2, the application rate for a 7/64-inch diameter nozzle is

Worksheet Example 10-2 (continued)

$$\text{Average Application Rate} = \frac{(96.3)\cdot(\text{Discharge per Nozzle})}{(\text{Lateral Spacing})\cdot(\text{Sprinkler Spacing})}$$

$$= \frac{(96.3)\cdot(1.86\text{ gpm})}{(40\text{ ft})\cdot(30\text{ ft})} = 0.15\text{ inches per hour}$$

Based on Equation 5-2 (page 81), the application rate for a 1/8-inch diameter nozzle is

$$\text{Average Application Rate} = \frac{(96.3)\cdot(2.43\text{ gpm})}{(40\text{ ft})\cdot(30\text{ ft})} = 0.20\text{ inches per hour}$$

Either size would allow irrigating five lateral lines (100 sprinklers) at one set while remaining below the 300 gpm maximum pump yield pumping rate.

The producer decides to use a total of four sets for the field. The strawberries will be irrigated with two sets. Each set will have a 160-foot manifold. Because the well is located at the extreme west end of the field in the middle of the north south direction, the length of the mainline can be calculated as follows:

$$\text{Mainline Length} = \left((\text{Field Length}) + \frac{(\text{Field Width})}{2}\right) - \left(\frac{(\text{Manifold Length}) + (\text{Lateral Spacing})}{2}\right)$$

$$= \left((800\text{ ft}) + \frac{(600\text{ ft})}{2}\right) - \left(\frac{(160\text{ ft}) + (40\text{ ft})}{2}\right) = 1{,}000\text{ feet}$$

The mainline will have four risers located on the north side of the field and 100, 300, 500, and 700 feet directly east of the well. Each riser will be serving five laterals. A total of nearly 12,000 feet of lateral line in 30-foot lengths and 640 feet of manifold piping and fittings will be needed, along with 400 sprinklers and risers.

The producer selects the 7/64-inch nozzles. This nozzle size more closely matches the application rate desired for frost control and also is better suited to the 8-hour set time. The 100 sprinklers in one set would require 186 gpm. This alternative would allow frost protection on half of the strawberry acreage without any spare parts, inventory, or modification of equipment. The 186 gpm, however, is less than the 216 gpm that was calculated for system capacity. By manipulating Equation 2-1, the producer determines how many inches per day this nozzle will be able to supply

$$\text{ET Rate} = \frac{(\text{Pumping Time})\cdot(\text{Application Efficiency})\cdot(\text{Pumping Rate})}{453\cdot(\text{Application Area})}$$

$$= \frac{\left(8\ \frac{\text{hrs}}{\text{day}}\right)\cdot(0.75)\cdot(186\text{ gpm})}{453\cdot(11\text{ acres})} = 0.22\text{ inches per day}$$

Worksheet Example 10-2 (continued)

The grower considers the alternatives and chooses to pump more hours per day if needed. The following determines the number of hours irrigation will be necessary to meet the design ET:

$$\text{Pumping Time} = \frac{453 \cdot (\text{Application Area}) \cdot (\text{Design Evapotranspiration Rate})}{(\text{Pumping Rate}) \cdot (\text{Application Efficiency})}$$

$$= \frac{453 \cdot (11 \text{ acres}) \cdot \left(0.26 \ \frac{\text{in}}{\text{day}}\right)}{(186 \text{ gpm}) \cdot (0.75)} = 9.3 \text{ hours per day}$$

With the sprinkler nozzles that were designed to meet ET demand selected, the producer is unable to use the system for frost control of the entire strawberry acreage. Using these sprinklers, however, would allow frost control on one-half of the strawberries. The option of having frost control on all of the strawberries could be accomplished only by selecting an application rate lower than the 0.15 inches per hour. For example, 0.1 inches per hour would need 249 gpm. However, the impact sprinkler selected does not have a small enough nozzle available to get 0.1 inches per hour at the 30 foot by 40 foot spacing, and the pump is not big enough to deliver 249 gpm.

Pipeline sizing: The buried mainline and risers should be installed near the field edge to minimize interference with field operations. There should be a shutoff valve at each riser. The mainline will carry only the water serving one manifold at one time; however, the maximum length would be 1,000 feet when serving the most distant manifold. To account for various fittings, 100 feet of equivalent length is added to the 1,000 feet of mainline. Setting a maximum head loss of 11 feet per 1,000 feet of pipe length, Figure 7-12 is used to determine that the minimum pipe diameter for a PVC pipe with a Hazen-Williams coefficient of 150 and a flowrate of 186 gpm is 4.3 inches. Because the inside pipe diameter determined using Figure 7-12 is larger than the inside diameter of a standard 4-inch PVC pipe, a larger size pipe size should be chosen. A 5-inch pipe would be adequate, but this pipe size is not commonly available. A 6-inch pipe is more commonly available, so 6-inch pipe is selected. Another acceptable option would be to use a combination of more than one size of pipe, for example 6 inch and 4 inch, to achieve a desired head loss in line with the lowest investment.

Using Table 7-4, the producer selects Class 100 PIP-PVC pipe with an inside diameter of 5.84 inches. According to information in Figure 7-12, the mainline friction loss for a pipe with an inside diameter of 5.84 inches, a flowrate of 186 gpm, and a Hazen-Williams coefficient of 150 is 2.6 feet of head per 1,000 feet of pipe. The total friction loss is

$$\text{Friction Loss} = (\text{friction loss factor}) \cdot (\text{pipe length} + \text{equivalent fitting length})$$

$$= \left(\frac{2.6 \text{ ft}}{1{,}000 \text{ ft pipe}}\right) \cdot (1{,}000 \text{ ft} + 100 \text{ ft}) = 2.86 \text{ feet of head,}$$

or converting to pressure loss

$$= (2.86 \text{ ft}) \cdot \left(0.434 \ \frac{\text{psi}}{\text{ft of head}}\right) = 1.24 \text{ psi}$$

Worksheet Example 10-2 (continued)

Each manifold location will have a flowrate of 186 gpm. Each manifold pipe will service five lateral lines with one at the center and two laterals on each side; therefore, two-fifths, or 40%, of the flow will be distributed to each side. The remaining one-fifth, or 20%, of the flow will be distributed to the lateral line at the center of the manifold. The flow rate for each lateral will be

$$\text{Lateral Line Flowrate} = \frac{\text{Total Flowrate}}{\text{Number of laterals}} = \frac{186\ \text{gpm}}{5\ \text{laterals}} = 37.2\ \text{gpm per lateral}$$

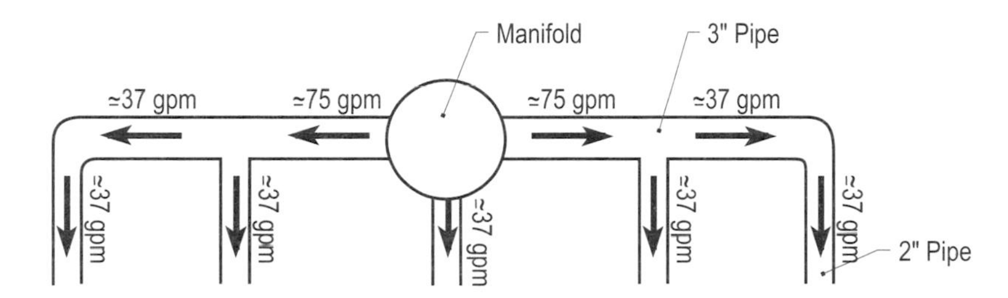

Because each side of the manifold will service two laterals, the flowrate will be about 75 gpm for the first 40 feet and 37 gpm in the last section of the manifold. Using Figure 7-12 for a flowrate of 75 gpm, a 3-inch aluminum pipe with a Hazen-Williams coefficient of 140 will have a friction loss of 11 feet of head per 1,000 feet of pipe. A 3-inch aluminum pipe will have slightly less than 1 foot of head loss for the 80 feet of length.

Each lateral will have 20 sprinklers discharging 1.86 gpm each at a design pressure of 30 psi for a total of 37 gpm. Using Equation 7-12 and Table 7-7, for a flowrate of 37.2 gpm, a Hazen-Williams coefficient of 120, and a pipe diameter of 2 inches, the friction loss in the lateral can be calculated as

$$\text{Friction Coefficient} = \frac{1{,}044 \cdot \left(\dfrac{\text{Flowrate}}{\text{Hazen-Williams Coefficient}}\right)^{1.85}}{(\text{Pipe Diameter})^{4.87}}$$

$$= \frac{1{,}044 \cdot \left(\dfrac{37.2\ \text{gpm}}{120}\right)^{1.85}}{(2\ \text{in})^{4.87}} = \frac{1{,}044 \cdot (0.115)}{(29.24)} = 4.1\ \text{feet per 100 feet}$$

$$\text{Friction Loss in Lateral} = (\text{Friction Coefficient}) \cdot (\text{Pipe Length}) \cdot (\text{Multiple Outlet Coefficient})$$

$$= \left(\frac{4.1\ \text{ft}}{100\ \text{ft}}\right) \cdot (600\ \text{ft}) \cdot (0.38) = 9.3\ \text{feet}$$

To compare with a 3-inch pipeline:

$$\text{Friction Coefficient} = \frac{1{,}044 \cdot \left(\dfrac{37.2\ \text{gpm}}{120}\right)^{1.85}}{(3\ \text{in})^{4.87}} = \frac{1{,}044 \cdot (0.115)}{(210.7)} = 0.57\ \text{feet per 100 feet}$$

Worksheet Example 10-2 (continued)

$$\text{Friction Loss in Lateral} = (\text{Friction Coefficient}) \cdot (\text{Pipe Length}) \cdot (\text{Multiple Outlet Coefficient})$$

$$= \left(\frac{0.57 \text{ ft}}{100 \text{ ft}}\right) \cdot (600 \text{ ft}) \cdot (0.38) = 1.3 \text{ feet}$$

Because the lateral has a 1% downward slope, 6 feet of head will be gained along the lateral. For the 2-inch pipeline, the 6 feet gained will help offset the 9.3 feet of head lost from friction. The difference of 3.3 feet of head is well within the allowable pressure variation for sprinklers along a lateral. The 2-inch pipe would be the most economical size for the system. Without the head gained from slope, however, the 3-inch pipe would have been the optimum size. By laying out the field to take advantage of the slope, substantial cost was saved.

A six-foot sprinkler riser is selected to allow irrigating sweet corn. Shorter risers could be used in the other crops. The producer decided to purchase three-foot risers for the whole field with enough additional extensions to combine where six-foot risers are needed.

Pump calculations: The flowrate and total head loss are required to specify the size of the pump. The flowrate has been selected at 186 gpm. Total head is

$$\text{Total Head} = \text{Well lift} + \text{elevation lift} + \text{sprinkler system head} + \text{total friction loss} + \text{any other losses}$$

Well lift is

$$\text{Well Lift} = (\text{Static Water Level}) + \left(\frac{\text{Pumping rate}}{\text{Specific Capacity}}\right) = (40 \text{ ft}) + \left(\frac{186 \text{ gpm}}{5 \ \frac{\text{gpm}}{\text{ft}}}\right) = 77 \text{ feet}$$

Because the well is 3 feet lower than the highest elevation, the elevation lift is equal to 3 feet plus 6 feet of riser height. Pressure head is 30 psi, which is equal to 69 feet of head loss.

Friction losses from the mainline, manifold, and risers need to be included in determining the total head. Friction losses in the lateral are not included in the calculation because they have been considered in lateral sizing. The total friction losses based on previous calculations and information are

$$\text{Friction Losses} = (\text{Mainline}) + (\text{Manifold}) + (\text{Miscellaneous}) = (2.86 \text{ ft}) + (1 \text{ ft}) + (6 \text{ ft}) = 10 \text{ feet}$$

The total head is

$$\text{Total Head} = \text{Well lift} + \text{elevation lift} + \text{sprinkler system head} + \text{total friction loss} + \text{any other losses}$$

$$= 77 \text{ ft} + 9 \text{ ft} + 69 \text{ ft} + 10 \text{ ft} + 0 \text{ ft} = 165 \text{ ft}$$

The required water horsepower (WHP) and brake horsepower (BHP) for this system, assuming a pump efficiency of 0.75 and drive efficiency of 1.0 are

Worksheet Example 10-2 (continued)

$$\text{WHP} = \frac{(\text{Pumping Rate})\cdot(\text{Total Pumping Head})}{3{,}960} = \frac{(186\ \text{gpm})\cdot(165\ \text{ft})}{3{,}960} = 7.75\ \text{water horsepower}$$

$$\text{BHP} = \frac{(\text{Water Horsepower})}{(\text{Pump Efficiency})\cdot(\text{Drive Efficiency})} = \frac{7.5\ \text{hp}}{(0.75)\cdot(1.0)} = 10.33\ \text{brake horsepower}$$

The size of the motor on the pump is close to the standard 10 horsepower size. The next larger size is 15 horsepower. Most motors over 10 horsepower also require 3-phase power. It would be important to review the pump catalog carefully to find a more efficient pump that would result in a motor size of 10 horsepower or less. For example, an 80% efficient pump would need 9.9 horsepower. Never overload the pump motor. Another consideration could be to use the next size larger motor to provide reserve capacity in the future, if and when needed.

Hourly energy consumption (HEC) can be estimated by using Equation 7-16 and Table 7-13:

$$\text{Hourly Energy Consumption} = \frac{(\text{Brake Horsepower})}{(\text{Electric Value})} = \frac{10.33\ \text{hp}}{1.18\ \frac{\text{hp-hr}}{\text{kWh}}}$$

$$= 8.75\ \text{kilowatt-hours per hour}$$

At $0.10 per kWh, pumping cost would be

$$\text{Operational Cost} = \left(\frac{\$0.10}{\text{kWh}}\right)\cdot\left(8.75\ \frac{\text{kWh-hr}}{\text{hr}}\right) = \$0.88\ \text{per hr}$$

or

$$\text{Operational Cost} = \left(\frac{\$0.88}{\text{hr}}\right)\cdot\left(\frac{453\ \frac{\text{gpm}}{\text{ac-in. per hr}}}{186\ \text{gpm}}\right) = \$2.14\ \text{per ac-in}$$

System operation: Operating each set (five laterals) for 8 hours at an application rate of 0.15 inches per hour would result in 1.2 inches of water being applied per application. At 75% efficiency, a net application of 0.9 inches could be applied every four days, or an average of 0.23 inches per day.

Based on Table 2-1, the soils should have approximately 1.3 inches of available water per foot depth for a total of 2.6 inches available water in the top 2 feet. With a reasonable depletion of 40 to 50% of the available water, sufficient storage capacity is available for the planned irrigation depth (0.9 inches). The average application rate of 0.15 inches per hour is well below the infiltration rate of the sandy soil.

System expansion: Adding the future nursery should pose no problems for the system as designed. The pump can be used during the hours when irrigation is not operating on the main field. No system modifications should be required.

Worksheet Example 10-2 (continued)

List of equipment

Necessary

- Submersible pump, 6-inch; approximately 10 horsepower at 80% efficiency to deliver 186 gpm at 165 feet head.
- 1,000 feet of 6-inch PVC buried mainline, Class 100 PIP, including fittings.
- Four hydrants and valves for connecting to manifolds.
- Four manifolds, total of 640 feet of 3-inch aluminum and fittings.
- 20 lateral lines, approximately 12,000 feet of 2-inch aluminum complete with sprinkler risers and stabilizers.
- 400 low-pressure impact sprinklers with 7/64-inch nozzles, 23° angle.
- Timer on pump control to allow setting time for each set.
- Safety shut down equipment to prevent damage to system in case of malfunction.

Optional

- Permanent system with buried manifolds and laterals.
- Controller and automatic valves on each set to allow automation of irrigation.
- Remote control system to allow monitoring and control of the irrigation from the office or a remote location.

Worksheet Example 10-3

Traveler Irrigation System Design

Situation

The owner of an 80-cow dairy farm near Sleepy Eye, Minnesota, wants to irrigate 60 acres of pasture being used to graze milking cows. The owner is grazing the cows in an intensive rotation management system that moves the cows to paddocks of new grass each morning. Each paddock, which is 180 by 500 feet, or approximately 2 acres, is then rested for 28 to 30 days before the next grazing.

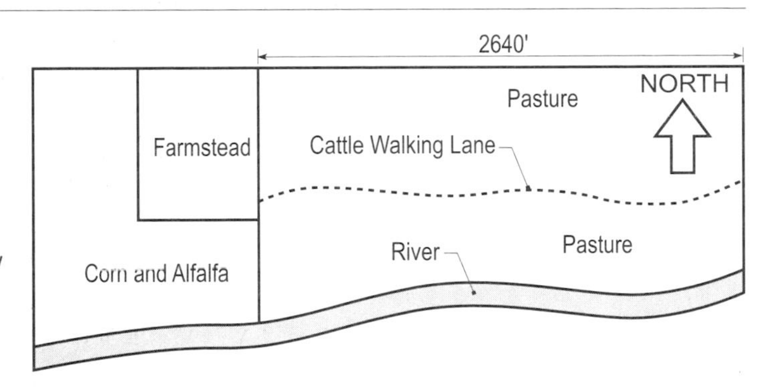

The field has a gently rolling topography with a general slope toward the river of around 2%. Soils in the pasture are mainly loam texture in the top 30 inches. The pasture has a mixture of alfalfa, brome, and orchard grasses.

Design Goals and Challenges

Because of the shape of the field and a desire to best use the available labor, the farmer has the following goals:

- Use a hard hose-tow traveling irrigation system.
- Use a diesel-powered pumping plant that can operate all day when needed.
- Spend no more than 1 hour of labor moving irrigation equipment per day.
- Irrigate each paddock with 1.0 to 1.5 inches of water shortly after it has been grazed.

Some of the design challenges for this system are these:

- The state water appropriation agency has granted the farmer owner an irrigation water withdrawal permit of 480 acre-inches per year and a maximum pumping rate of 0.75 cubic feet per second from the river.
- The river has a minimum flow rate of 10 cubic feet per second during the driest two months of the growing season.
- The land 400 feet north of the potential pump site is about 10 feet above the pump site.
- Land at the far ends of the cattle walking lane is about 15 feet above the potential pump site.
- There is a second rise of more than 20 feet near the north side of the pasture above the pump site.
- The available water supply may be 3 to 5 inches deficient during the driest years.

Design

Field Parameters: The pasture is approximately 2,640 feet in length and approximately 1,000 feet wide. The total land area is

$$\frac{(2{,}640\ \text{ft})\ (1{,}000\ \text{ft})}{43{,}560\ \frac{\text{sq ft}}{\text{ac}}} = 60\ \text{acres (approximately)}$$

Worksheet Example 10-3 (continued)

The owner has a cattle-walking lane running west to east about 500 feet from the north side of the pasture. A measurement of the elevation changes shows that the land surface 400 feet north of the potential pump site is about 10 feet above the pump site. The land surface at each end of the cattle-walking lane is 15 feet above the pump site. The elevation survey also shows a plus 20 feet near the north side of the pasture.

Water supply needs: Table 2-3 suggests that alfalfa can use 20 to 25 inches of water per year, and pastures can use 20 to 28 inches of water per year in Minnesota. Alfalfa in Minnesota will require 11 to 13 inches of supplemental irrigation water per year for the typical irrigated soils. See Table 2-4 (page 12). The 480 acre-inch water permit for 60 acres of irrigated pasture will provide

$$\frac{480 \text{ ac-in}}{60 \text{ ac}} = 8 \text{ acre-inches per acre}$$

Because the water supply available is 3 to 5 inches below what is needed in a typical year, the system may not be able to provide enough water to the pasture during the driest years. Local irrigation experts have told the farmer that because of the higher water-holding capacity, area soils should require only 8 to 10 inches for 8 out of 10 years. The farmer determines that for most years, the water supply looks sufficient. During the driest years, only a portion of the land can be fully irrigated, or the entire area will need to be managed using deficit scheduling techniques.

The farmer obtains a water permit and is allowed a maximum withdrawal rate of 0.75 cubic feet per second, or 340 gpm.

Water supply yield potential and irrigation system design capacity needs: Using Figure 2-9 on page 14, the farmer determines the average crop ET rate to be 0.26 inches per day. From Table 2-5, the farmer determines sprinkler application efficiency to be 0.75. The farmer estimates there will be a little over an hour downtime when moving the traveler between laterals. For design purposes the farmer plans on irrigating for 22 hours.

The farmer would like to determine the pumping rate needed to irrigate 60 acres and compare this rate to the permitted rate of 340 gpm. The farmer can estimate the irrigation system capacity or pumping rate by using Equation 2-1:

$$\text{Pumping Rate for 60 acres} = \frac{453 \cdot (\text{Application Area}) \cdot (\text{Design Evapotranspiration Rate})}{(\text{Pumping Time}) \cdot (\text{Application Efficiency})}$$

$$= \frac{453 \cdot (60 \text{ acres}) \cdot \left(0.26 \frac{\text{in}}{\text{day}}\right)}{\left(22 \frac{\text{hrs}}{\text{day}}\right) \cdot (0.75)} = 428 \text{ gpm}$$

Worksheet Example 10-3 (continued)

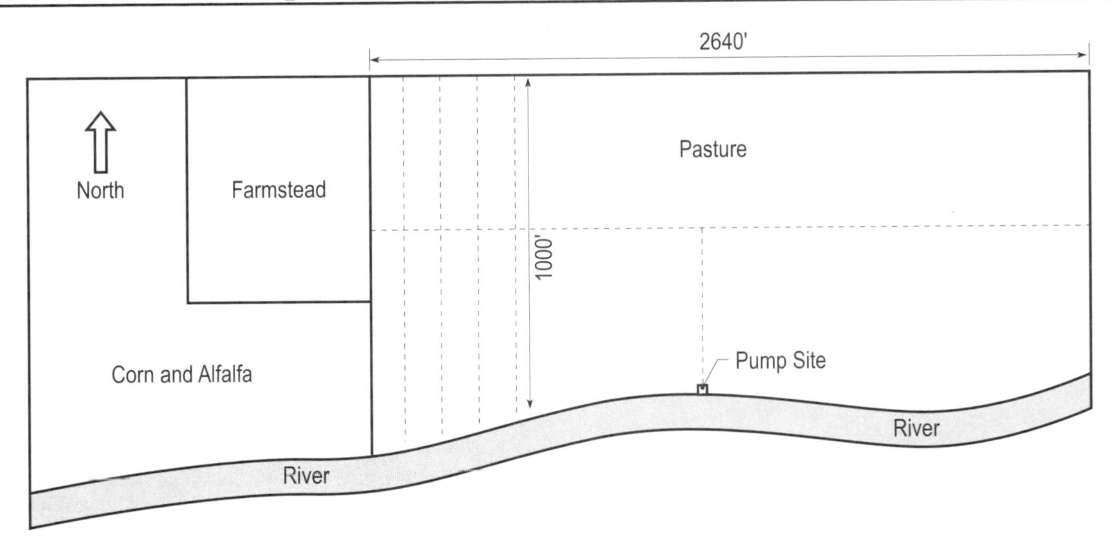

The pumping rate of 428 gpm for 60 acres is the pumping rate needed during the dry years. This pumping rate is about 100 gpm over the maximum permitted pumping rate. The farmer can determine the application area for 340 gpm during the dry years by rearranging Equation 2-1:

$$\text{Application Area} = \frac{(\text{Pumping Time})\cdot(\text{Pumping Rate})\cdot(\text{Application Efficiency})}{453\cdot(\text{Design Evapotranspiration Rate})}$$

$$= \frac{\left(22\ \frac{\text{hrs}}{\text{day}}\right)\cdot(340\ \text{gpm})\cdot(0.75)}{453\cdot\left(0.26\ \frac{\text{in}}{\text{day}}\right)} = 48\ \text{acres}$$

The farmer recognizes that in some years the irrigation system may not keep up with the crop's water needs, and some land may need to be left un-irrigated while the other portion is irrigated for maximum growth.

Sprinkler selection and layout: Because of the arrangement of the 2-acre paddocks, a hard-hose traveler using 500 feet of hose appears to be the best fit. The water supply will be furnished using an above-ground aluminum pipeline along the east-to-west walking lane through the middle of the pasture.

The farmer would like to find a nozzle that can accommodate a flowrate of approximately 340 gpm (water permit maximum) and provide an effective wetted diameter of approximately 180 feet (the field width). After reviewing some manufacturer's literature, the farmer identifies several possible traveler and sprinkler sizes. A traveler with a large-volume gun sprinkler with a 24° trajectory and a 1.1-inch diameter nozzle operating at 80 psi would discharge 335 gpm of water for a wetted diameter of 380 feet. Table 5-6 (page 77) shows the maximum lane spacing for a traveling gun to be 60 to 65% of the wetted diameter during windy conditions (5 to 10 mph). During windy conditions, the sprinkler should be able to produce a good application pattern for a lane spacing of 228 to 250 feet, which is just a little larger than the 180-foot paddock width.

Worksheet Example 10-3 (continued)

If 30-foot pipe sections are used in the mainline, then a lane interval of either 210 or 240 feet would provide for convenient outlet fitting and at the same time provide for sufficient overlap during application. The farmer determines the number of travel lanes for lanes spaced at 240 feet to be

$$\frac{2{,}640 \text{ ft}}{240 \ \frac{\text{ft}}{\text{lane}}} = 11 \text{ lanes}$$

This design provides 335 gpm to an outlet for each hose-tow traveler lane. By rearranging Equation 2-1, the farmer can estimate the approximate travel time for applying a given depth of water per traveler run:

$$\text{Pumping Time} = \frac{453 \cdot (\text{Lane Spacing}) \cdot (\text{Lane Length}) \cdot (\text{Design Application Depth})}{\left(43{,}560 \ \frac{\text{sq ft}}{\text{ac}}\right) \cdot (\text{Pumping Rate}) \cdot (\text{Application Efficiency})}$$

To complete a 500-foot lane length for an application depth of 1.0 inches per day the pumping time would be

$$\text{Pumping Time} = \frac{453 \cdot (240 \text{ ft}) \cdot (500 \text{ ft}) \cdot \left(1.0 \ \frac{\text{in}}{\text{day}}\right)}{\left(43{,}560 \ \frac{\text{sq ft}}{\text{ac}}\right) \cdot (335 \text{ gpm}) \cdot (0.75)} = 5.0 \text{ hours}$$

Other application depths the farmer looked at resulted in the following pumping times:

Application depth (in/day)	Pumping time (hr)
1.0	5.0
1.5	7.5
2.0	10.0

Using the proposed traveler sprinkler size and lane width of 240 feet, the system would be able to irrigate one entire 180-foot paddock width plus about one-third of a paddock width more on each run. This arrangement will allow the farmer to reduce irrigation during low water demand or higher rainfall times by running the system once per day. This sprinkler size also would give the operator the chance to irrigate a previously irrigated paddock in the second part of the day if the soil profile and grass would benefit from an additional irrigation.

Pipeline sizing: Due to the traveler layout and location of the pumping plant, approximately 1,600 feet of aluminum pipe will be needed to supply water to the farthest traveler connection point on the west side. Unless the farmer is willing to move the west-side pipeline after irrigating, an additional 1,200 feet of pipe will be needed to be placed from the first tee to the east side of the pasture.

Table 7-6 (page 121) shows that a 6-inch diameter pipe is required to keep the water velocity at less than 5 feet per second. The nomograph on page 122 shows that a 6-inch aluminum pipe with a Hazen-Williams coefficient of 120 and a flowrate of 335 gpm produces a head loss of 9.5 feet per 1,000 feet of pipe, or an equivalent pressure loss of 4.1 psi per 1,000 feet of pipe. An 8-inch pipe or a combination of 8-inch pipe to the first tee followed by a 6-inch pipe may also be an option. The farmer will need to perform an economic analysis to determine the lowest operating and ownership cost piping arrangement. The Pipe Sizing section of Chapter 7 recommends keeping the pressure losses in a pipeline to under 1 foot of per 100 feet of pipe, or 10 feet per 1,000 feet of pipe. The farmer is

Worksheet Example 10-3 (continued)

not concerned with having pipe on top of the ground, so did not consider buried PVC pipe. Because of availability and price, the farmer selects 6-inch aluminum pipe for the system.

The farmer will assume the equivalent length of the various fittings needed will be 100 feet. Based on Figure 7-12 and Table 7-8 on pages 122 and 124, the estimated total friction loss for a 6-inch pipe and fitting will be

$$\text{Pipe and Fitting Friction Loss} = \left(\frac{9.5 \text{ ft head loss}}{1{,}000 \text{ ft pipe}}\right) \cdot (1{,}600 \text{ ft pipe} + 100 \text{ ft of equivalent pipe})$$

$$= 16 \text{ feet of head,}$$

or converting to pressure loss

$$= (16 \text{ ft of head}) \cdot \left(0.434 \frac{\text{psi}}{\text{ft of head}}\right) = 7.0 \text{ psi}$$

The traveler hard hose also will have some friction losses. The traveler hard hose is 500 feet long. Manufacturer's literature shows a 4-inch inside diameter hose should be used. Table 4-5 shows that this hose will lose about 3 psi per 100 feet, or 30 psi per 1,000 feet. The pressure loss due to the hard hose is

$$\text{Traveler Hose Friction Loss} = \left(\frac{30 \text{ psi}}{1{,}000 \text{ ft pipe}}\right) \cdot (500 \text{ ft hose}) = 15 \text{ psi,}$$

or converting to head loss

$$= (15 \text{ psi}) \cdot \left(2.31 \frac{\text{ft of head}}{\text{psi}}\right) = 35 \text{ feet of head}$$

$$\text{Sprinkler System Pressure Loss} = \text{operating pressure} + \text{hose friction loss} =$$

$$80 \text{ psi} + 15 \text{ psi} = 95 \text{ psi,}$$

or converting to head loss

$$= (95 \text{ psi}) \cdot \left(2.31 \frac{\text{ft of head}}{\text{psi}}\right) = 219 \text{ feet of head}$$

Pump Sizing: To specify the size of the pump, the pumping rate and total head for the system must be known. Based on the Pump Alternatives and Selection section in Chapter 7, a horizontal centrifugal pump should provide the best pumping arrangement. The pumping rate is 335 gpm. The total head is defined by Equation 7-3 as the sum of the suction lift, elevation lift, sprinkler system head, and the pipeline and fitting friction losses.

This pump site will require about 30 feet of 6-inch aluminum suction pipe and a foot valve attached to the pipe inlet to reduce the amount of priming time before each irrigation startup. The suction pipe will have a foot valve. Table 7-8 along with the previous information obtained from Figure 7-12 shows the suction pipe friction loss is

Worksheet Example10-3 (continued)

$$\text{Suction Pipe Pressure Loss} = (\text{friction loss factor}) \cdot (\text{suction pipe length} + \text{equivalent fitting length})$$

$$= \left(\frac{9.5 \text{ ft}}{1{,}000 \text{ ft pipe}} \right) \cdot (30 \text{ ft hose} + 76 \text{ ft equivalent length}) = 1.0 \text{ feet of head,}$$

or converting to pressure loss

$$= (1.0 \text{ ft}) \cdot \left(0.434 \frac{\text{psi}}{\text{ft of head}} \right) = 0.43 \text{ psi}$$

The water permit specifies that the pump and motor must be located approximately 10 feet above the low water level of the river. A survey of the pasture shows that the land surface 400 feet north of the pump site is about 10 feet above the elevation of the pump site. At each end of the proposed mainline, the elevation is 15 feet higher, and near the north side of the pasture, the elevation is about 20 feet above the pump site. Equation 7-3 calculates the total head:

$$\text{Total Head} = \text{suction lift} + \text{elevation lift} + \text{sprinkler system head} + \text{suction pipe friction loss} + \text{pipeline friction loss}$$

$$= 10 \text{ ft} + 20 \text{ ft} + 219 \text{ ft} + 1.0 \text{ ft} + 16 \text{ ft} = 266 \text{ feet of head}$$

Power unit sizing: The brake horsepower (BHP) size of the diesel power unit for the centrifugal pump depends on the pump's discharge rate, total head, and pump and drive efficiency. BHP can be estimated by using Equations 7- 13 and 7-14 (page 134). During early planning, a pump efficiency of 0.75 can be used. This calculation can be refined later when the actual pump is selected. Table 7-10 lists suggested power-unit to pump-drive efficiencies. This design will have a direct shaft connection from the pump to the engine drive shaft and clutch unit.

$$\text{WHP} = \frac{(\text{Pumping Rate}) \cdot (\text{Total Pumping Head})}{3{,}960} = \frac{(335 \text{ gpm}) \cdot (266 \text{ ft})}{3{,}960}$$

$$= 22.5 \text{ water horsepower}$$

$$\text{BHP} = \frac{(\text{Water Horsepower})}{(\text{Drive Efficiency}) \cdot (\text{Pump Efficiency})} = \frac{22.5 \text{ hp}}{(1.0) \cdot (0.75)}$$

$$= 30.0 \text{ brake horsepower}$$

An engine with a continuous shaft output of 50 brake horsepower or greater will most likely be chosen to operate at the required pump shaft speed to enable the selected pump to produce the needed head and pumping rate. The engine should be selected so that the desired pump operating speed will load the engine to at least 50% of its rated brake horsepower at this engine speed.

Pumping Cost: Equations 7-15, 7-16, and 7-17 (pages 142 and 143), and Tables 7-12 and 7-13 (page 141) can be used to estimate the fuel usage and operating cost per acre-inch of water pumped. The engine energy performance for a typical diesel engine operating at full load and peak efficiency is 12.5 water horsepower-hour per gallon. The actual fuel usage for this design project might be slightly greater since the selected engine will not be operating at full load or peak efficiency speed. The proposed diesel pumping plant's hourly energy consumption (HEC) in gallons can be estimated from the following:

Worksheet Example10-3 (continued)

$$\text{Hourly Energy Consumption} = \frac{(\text{Water Horsepower})}{(\text{Fuel Energy Value})} = \frac{22.5\ \text{hp}}{12.5\ \dfrac{\text{hp-hr}}{\text{gal}}} = 1.8\ \text{gallons per hour}$$

At $0.75 per gallon, this pumping plant will cost

$$\text{Operational Cost} = \left(\frac{\$0.75}{\text{gal}}\right)\cdot\left(1.8\ \frac{\text{gal}}{\text{hr}}\right) = \$1.36\ \text{per hour}$$

or

$$\text{Operational Cost} = \left(\frac{\$1.36}{\text{hr}}\right)\cdot\left(\frac{453\ \dfrac{\text{gpm}}{\text{ac-in per hr}}}{335\ \text{gpm}}\right) = \$1.84\ \text{per acre-inch}$$

System Operation: This traveling gun design with 335 gpm does have some limitations in meeting crop water use during the driest years. On the other hand, the traveling gun layout provides the operator with several choices of operation. The traveler could be operated just once per day shortly after each daily grazing with an application of 1 to 2 inches, depending on soil water status of the site. The actual time to operate for a given water depth can be easily determined with the pumping time equation used earlier in this example.

If the operator has the time to run a second irrigation in a day, the traveler could be turned 180, and the large sprinkler gun and cart could be pulled onto the opposite paddock, which was irrigated several days earlier. This approach will require the lowest labor effort from the operator and will help the operator increase the usage of the irrigation system. The traveler also provides for the flexibility of moving it to a paddock even farther away for irrigating land the operator determines needs more soil water.

On the travel runs that have the greatest slopes toward the river, the operator needs to observe if any runoff that might move some animal waste into the river is occurring during an irrigation. If runoff is a possibility, the operator should increase the arc of the rotating gun; run the traveling gun faster thus applying a lighter water application that should reduce the runoff potential. The operator also may need to construct some special vegetative buffer strips between the river and the irrigated land.

The potential for runoff can be analyzed by determining the average application rate of the large-volume gun using Equation 5-5 (page 81) and comparing it to the estimated soil intake rate for the soils in the field. It is typical to operate the rotating gun at 50% of its arc of rotation.

$$\text{Average Application Rate} = \frac{(96.3)\cdot(\text{Discharge per Nozzle})}{(\text{Wetted Diameter})^2\cdot(\text{Fraction of Operation})}$$

$$= \frac{(96.3)\cdot(335\ \text{gpm})}{(380\ \text{ft})^2\cdot(0.50)}$$

$$= 0.57\ \text{inches per hour}$$

The NRCS local soil survey shows the loam texture soils in this field to be part of the 1.0 intake family; therefore, the proposed traveler irrigation system should create none to very little runoff during an irrigation event.

Worksheet Example10-3 (continued)

This farm situation has the potential for using the traveling gun to apply some of its animal wastewater to the pasture or other lands. If this is a real consideration for the operator, the type of traveler and pumping unit need to be reassessed and designed for the usage of wastewater before any purchases are made.

Worksheet Example 10-4

Center Pivot Design for Lagoon Effluent

Situation

A swine producer wants to automate the application of lagoon effluent from a 5,000 head finishing operation by using a center pivot irrigator. The operation needs to irrigate 2,730,000 gallons of effluent per year or 101 acre-inches of effluent per year from the lagoon. The producer owns a square, 160-acre farm (quarter section). The farm is essentially level but has sufficient slope to prevent ponding. The finishing buildings and lagoon are located in the southwest corner of the farm. The farm is bordered on the south and east by county gravel roads. Land adjoining on the west and north is owned by others. The nearest neighbor lives about three-quarters of a mile to the northeast of the location of the center pivot point. Generally, the wind blows from the northwest, a direction away from the nearest neighbor.

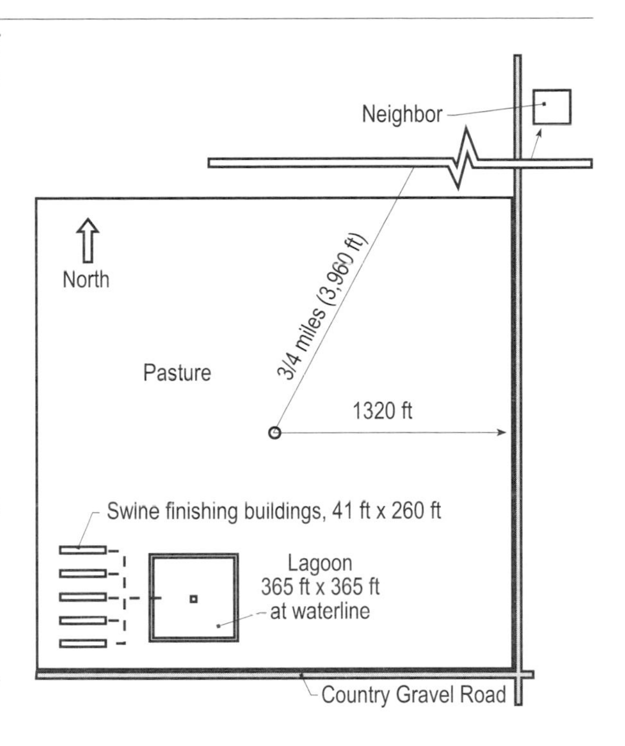

Design Goals and Challenges

The producer's goals are to:

- Apply all the effluent to the quarter section.
- Produce 175-bushel per acre continuous corn on this land.
- Meet as much of the crop nutrient needs as possible with the effluent.
- Not over apply nitrogen or phosphorus.
- Meet as much of the crop water needs as possible.

Two challenges in developing this system are:

- County regulations do not permit effluent irrigation within 100 feet of a public road or neighbor's land.
- The producer wants to minimize odors, drift, and overspray.

System Design and Management

The producer will take the following design approach in designing the system:

1. Analyze yield goals, effluent test, and soil test.
2. Determine application rate.
3. Determine application area.
4. Select equipment.

1. Analyze yield goals, effluent test, and soil test.

The yield goal is for continuous corn of 175 bushels per acre. The results of the soil test showed that to yield 175 bushels per acre, the producer needs to apply 200 pounds of N and 90 pounds of P_2O_5 per acre. The results of the effluent test showed the effluent contained 4.27 pounds of N per 1,000 gallons of effluent. The effluent test

Worksheet Example 10-4 (continued)

showed phosphorus in the fertilizer-equivalent form of P_2O_5 as being 3.06 pounds per 1,000 gallons of effluent. This is equivalent to 116 pounds of N and 83 pounds of P_2O_5 per acre-inch of effluent. The effluent test showed the solids content to be 1%.

2. Determine application rate.

The producer needs to look at two application rates. The first application rate is the rate at which nutrients are to be applied to the land. The second application rate is the soil's rate of liquid (water) intake. After the producer decides on an application rate for nutrients and liquid, a management strategy needs to be developed to combine the two application rates.

Nutrient Application Rate

Because the producer does not want to over apply nutrients to the soil, the producer must consider nitrogen and phosphorus application rates, then choose the lower of the two rates as the design application rate for the system.

a. Application rate based on nitrogen requirements. The nitrogen loss for effluent applied by irrigation is assumed to be 30%. The field has had lagoon effluent applied for the past five years by a custom applicator with a traveling gun. The producer can assume a steady state breakdown of organic matter from this and previous years and 70% of the applied N in a given year. Thus, to have 200 pounds of available N taking into account the effluent test of 116 pounds of N per acre-inch and a 30% volatilization loss, the producer should apply annually

$$\frac{200\ \frac{\text{lbs N}}{\text{acre}}}{\left(116\ \frac{\text{lbs N}}{\text{ac-in}}\right)\cdot(1 - 0.30)} = 2.46 \text{ acre-inches of effluent per acre}$$

b. Application rate based on phosphorus requirements. The producer wants to avoid a buildup of phosphorus. All of the phosphorus will be available for the plant to use in the same year as application. With the effluent test showing 83 pounds of P_2O_5 available per acre-inch, the producer should apply annually

$$\frac{90\ \frac{\text{lbs } P_2O_5}{\text{acre}}}{\left(83\ \frac{\text{lbs } P_2O_5}{\text{ac-in}}\right)\cdot(1 - 0)} = 1.08 \text{ acre-inches of effluent per acre}$$

c. Application rate selection. To avoid over applying nutrients, the producer will apply the effluent at the phosphorus rate of 1.08 acre-inch per acre. At an application rate of 1.08 acre-inch per acre, nitrogen is applied at a rate of

$$\left(1.08\ \frac{\text{ac-in}}{\text{ac}}\right)\cdot\left(116\ \frac{\text{lbs N}}{\text{ac-in}}\right)\cdot(1 - 0.30) = 88 \text{ pounds N per acre}$$

The additional 112 pounds of N per acre will be applied by mechanical means.

Liquid Application Rate

The county soil survey indicates that the limiting soil type in the field is a silt loam with a permeability of 0.6 inches per hour. Table 5-11 shows that a silt loam soil has a steady state water intake rate of 0.6 inches per hour with cover, and a water intake rate of 0.4 inches per hour with bare ground. If the first application is on bare ground, the maximum application rate would be 0.4 inches per hour. From Table 5-7, there is no rate reduction for

Worksheet Example 10-4 (continued)

slopes up to 5%. Table 9-14 shows the effluent application rate reduction coefficient for effluent with 1% solids being applied to a silt loam is 0.95. The reduction coefficient has very little effect on the design water intake rates, so the design will assume a maximum application rate of 0.4 inches per hour on bare ground and 0.6 inches per hour on covered ground. From Table 5-9 (page 81), the producer determines that center pivot peak application rates range from 1 inch per hour to 6 inches per hour for sprinklers and spray nozzles.

Management Strategy

To minimize the risk of runoff or deep percolation, the producer plans on making several applications per year to meet crop needs. The producer plans to apply 50 pounds of available N per acre (an application of 0.62 acre-inches per acre) in the spring. The producer wants to make five additional applications of nitrogen. The producer will apply the remaining effluent with the next application. This will result in 38 pounds of available N being applied (0.46 acre-inches per acre). The remaining 112 pounds of nitrogen will be applied by mechanical means.

The producer will ask the manufacturer to design a center-pivot system that will approach the 0.4 inches per hour and 0.6 inches per hour infiltration rates. The system will be required to have sprinklers and/or spray nozzles set sufficiently high to maintain a wide wetted diameter above the corn canopy and with nozzle diameters large enough to minimize the risk of clogging. The producer will use a commercially available intake screen at the pump. The producer will select the manufacturer's sprinkler package that produces the lowest application rate with no nozzle openings less that 0.25 inch. The center pivot will be equipped with high-speed drives so application depths can be reduced if application depths of 0.62 inch or 0.46 inch produce runoff.

3. Determine application area.

The next step is to determine the area of application. Based on an annual application rate of 1.08 acre-inch per acre, the amount of land required to accommodate the application of 101 acre-inches of effluent per year is

$$\frac{101\ \frac{\text{ac - in}}{\text{yr}}}{1.08\ \frac{\text{ac - in}}{\text{ac}}} = 93 \text{ acres per year}$$

The producer is not permitted to apply effluent within 100 feet of neighboring land or public roads. The producer decides to use a 1,220 foot pivot. Because of the buildings located in the southwest corner of the section, the producer will be able to operate the pivot only 297.5°. The area irrigated by a 1,220 foot pivot rotating 297.5° is

$$\left(\frac{3.14\cdot(1{,}220\text{ ft})^2}{43{,}560\ \frac{\text{sq ft}}{\text{ac}}}\right)\cdot\left(\frac{297.5}{360}\right) = 89 \text{ acres}$$

More land is required to apply the effluent. The producer looks at using an end gun to apply an additional 100 feet in the corners. Based on set back distances to neighboring land or public roads, the end gun can be used for a 42° rotation in each corner. The end gun will be used in three corners for a total of 126°. The standard pivot will be used for 171.5°. Using Equation B-4 in Appendix B, the area irrigated is

Worksheet Example 10-4 (continued)

$$\left(\frac{3.14 \cdot (1320\ \text{ft})^2}{43{,}560\ \frac{\text{sq ft}}{\text{ac}}}\right) \cdot \left(\frac{126}{360}\right) + \left(\frac{3.14 \cdot (1220\ \text{ft})^2}{43{,}560\ \frac{\text{sq ft}}{\text{ac}}}\right) \cdot \left(\frac{171.5}{360}\right) = 96\ \text{acres}$$

For the producer to apply all the effluent and to maintain set-back distances, a 1,220-foot center pivot rotating a total of 297.5° will be used. In the corners, a 100-foot end gun will be used for a 42° rotation. The total amount of land irrigated is 96 acres. Because of variability in knowing the actual volume of effluent in the lagoon in addition to the variability in effluent and soil test results, the producer feels confident that the calculations are close enough to design a system that will not over apply nutrients.

4. Select equipment.

The final step is to select equipment based on pivot requirements, pipeline friction losses, and pump requirements.

Pivot Requirements

The producer has selected a 1,220-foot pivot with end gun to apply 600 gpm with 40 psi (92.4 feet of head) at the pivot point with the pivot on the level. When the pivot is on a 2% upward grade to the end gun, the change in elevation for the 1,220 foot distance is 24.4 feet. The pressure drop at the end gun due to elevation will be approximately 10.6 psi (24.4 feet of head), in addition to the friction loss in the pivot. Pressure regulators will be required for uniform application, resulting in a pressure drop of approximately 5 psi. A booster pump will be required for the end gun. High speed drives will allow the operator to return the pivot dry to the starting point in the least possible time, if operating in a windshield-wiper mode should cause runoff.

Irrigating 96 acres with a system capacity of 600 gpm results in a rate of 6.3 gpm per acre. Table 2-5 shows that with an application efficiency of 0.85, the 6.3 gpm per acre can balance an ET of 0.25 inches per day by pumping slightly more than 20 hours per day, and an ET of about 0.285 inches per day with 24 hours per day operation. This capacity is adequate for 95% of the years at this location.

Pipeline Friction Losses

The producer is using a floating, submerged pump near the center of the lagoon with a direct-connected electric motor. The pump will be positioned approximately 900 feet southwest of the pivot point. The lagoon is 15 feet deep at overflow with a 3-foot average annual pumpdown initiated when the lagoon level is 1 foot below overflow, or lower. The bottom of the lagoon is 14 feet below grade; therefore, pumpdown must occur when the lagoon level is at or below grade and will cease when the level is 3 feet below grade. Surveying determined the pivot point ground elevation to be 4 feet above the ground elevation at the lagoon.

The pump will be connected by 200 feet of 6-inch hose to 860 feet of buried 8-inch PVC pipe to feed the center pivot. From the manufacturer's data, the pressure drop through the 6-inch hose at 600 gpm is 1.1 psi per 100 feet of hose. The total in 200 feet of hose is

$$\left(\frac{1.1\ \text{psi}}{100\ \text{feet of hose}}\right) \cdot (200\ \text{feet of hose}) = 2.2\ \text{psi, or } 5.08\ \text{feet of head for a 6-inch hose}$$

From Table 7-3, the friction loss for PVC pipe at 500 gpm is 0.4 feet of head and at 750 gpm is 0.8 feet of head per 100 feet of pipe. The producer uses the following interpolation formula:

$$\left(\frac{Q_2 - Q_1}{Q_3 - Q_1}\right) = \left(\frac{P_2 - P_1}{P_3 - P_1}\right)$$

Where:

Q_1 = Lowest flowrate with a known pressure loss (500 gpm)
Q_2 = Flowrate for the unknown pressure loss (600 gpm)
Q_3 = Highest flowrate with a known pressure loss (750 gpm)
P_1 = Pressure loss for flowrate Q_1 (0.4 feet of head)
P_2 = Unknown pressure loss for Q_2
P_3 = Pressure loss for flowrate Q_3 (0.8 feet of head)

Interpolating, the producer estimates the friction loss for 600 gpm to be

$$\left(\frac{600 - 500}{750 - 500}\right) = \left(\frac{P_2 - 0.4}{0.8 - 0.4}\right) \Rightarrow \left(\frac{100}{250}\right) = \left(\frac{P_2 - 0.4}{0.4}\right)$$

$$(P_2 - 0.4)\cdot(250) = (0.4)\cdot(100) \Rightarrow P_2 = \left(\frac{40}{250}\right) + 0.4$$

$$P_2 = 0.6 \text{ feet head loss per 100 feet of 8-inch pipe}$$

The total head loss for 860 feet of 8-inch PVC pipe is

$$\left(\frac{0.6 \text{ psi}}{100 \text{ feet of hose}}\right)\cdot(860 \text{ feet PVC pipe}) = 2.2 \text{ psi, or 5.16 feet of head for an 8-inch PVC pipe}$$

Adding the head loss in both the 6- and 8-inch pipe and then adding another 10% for losses through elbows, the total friction loss in the conduit is

$$5.08 \text{ ft} + 5.16 \text{ ft} + (0.1)\cdot(5.08 \text{ ft} + 5.16 \text{ ft}) = 11.26 \text{ ft, or approximately}$$
$$\text{11 feet of head loss through pipes and elbows}$$

Pump Requirements

Head from the pump at 3 feet below grade to the pivot at 4 feet above the grade at the pivot riser, plus 11 feet of friction loss equals 18 feet of head. The total head at the pump is the sum of the pressure loss from the pump to the pivot (18 feet) and the pressure head at the pivot (92.4 feet). The total head at the pump is 110.4 feet of head.

A pump curve for a commercially available pump shows a power requirement of approximately 25 horsepower and an efficiency of about 70% at 600 gpm and 110 feet of head .

Equations 7-13 and 7-14 are used to calculate the water horsepower (WHP) and brake horsepower (BHP) to pump 600 gpm at 110 feet of head and 70% pump efficiency.

Worksheet Example 10-4 (continued)

$$\text{WHP} = \frac{(\text{Flowrate})\cdot(\text{Total Pumping Head})}{3{,}960} = \frac{(600\text{ gpm})\cdot(110\text{ ft})}{3{,}960} = 16.67\text{ horsepower}$$

$$\text{BHP} = \frac{\text{WHP}}{(\text{Pump Efficiency})\cdot(\text{Drive Efficiency})} = \frac{16.67\text{ hp}}{(0.70)\cdot(1.0)} = 23.8\text{ horsepower}$$

Equation 7-15 is used to estimate the pumping time per year as follows:

$$\text{Pumping Time} = \frac{453\cdot(\text{Application Area})\cdot(\text{Application Depth})}{(\text{Pumping Rate})}$$

$$= \frac{453\cdot(96\text{ ac})\cdot(1.08\text{ in})}{(600\text{ gpm})} = 78\text{ hours per year}$$

If the producer assumes effluent will be applied in April, May, and June, the system must be operated 25 hours per month. Or, about 3.5% of the hours during the three-month period.

Adjusting the pump efficiency to 70% from 75%, Equations 7-16 and 7-17 are used to calculate the hourly energy consumption (HEC) and the seasonal energy consumption (SEC).

$$\text{HEC} = \frac{\text{WHP}}{(\text{Pump Efficiency})} = \frac{16.67\text{ hp}}{(0.88)\cdot\left(\frac{70}{75}\right)} = 20.3\text{ kilowatt-hour per hour}$$

$$\text{SEC} = (\text{HEC})\cdot(\text{Pumping Time}) = (20.3\text{ kWh per hr})\cdot(78\text{ hrs per yr})$$

$$= 1{,}583\text{ kilowatt-hours per year}$$

Summary

Depending on the producer's desire to automate the system and budgetary constraints, several options may be available from a given manufacturer. A computerized control system can be set to activate or deactivate the end gun or change the speed of the pivot at any point in the cycle. Settings can be changed at the pivot or from a remote location. Options can include automatic start and restart (requires special equipment at the pump) and shut down if the wind velocity or rainfall during operation exceeds a preset value. The end gun should be controlled by an air-actuated solenoid to prevent solid build-up.

Worksheet Example 10-4 (continued)

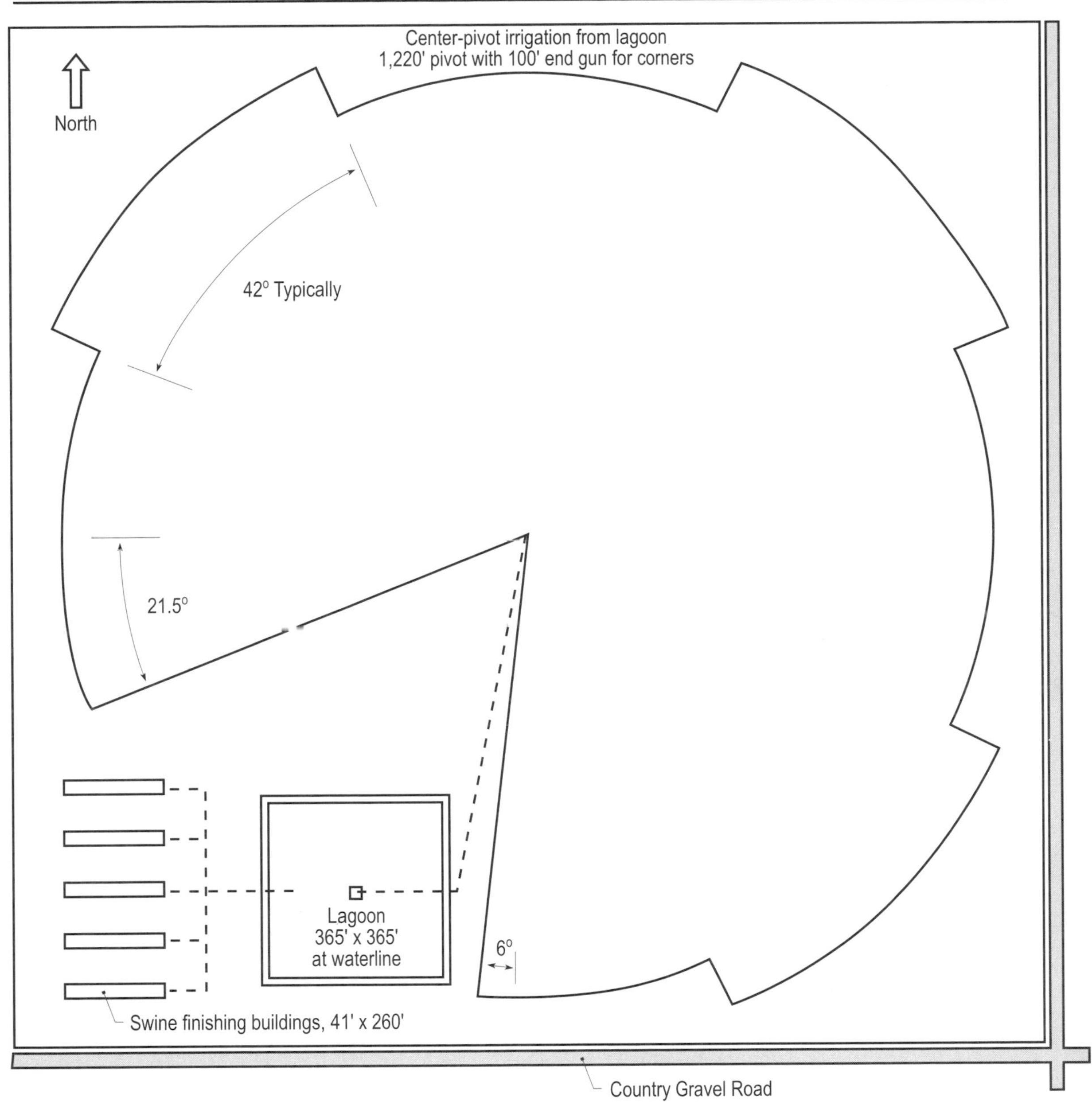

Worksheet Example 10-5

Achieving a Target Nitrogen Application Rate with a Set System

A 200-cow dairy has an annual lagoon pumpdown volume of 400,000 cubic feet. The producer wants to complete pumping out the lagoon in 12 days, pumping 8 hours each day. The producer will use a single-set, stationary gun sprinkler. The effluent will be applied to a medium-textured silty clay soil that is moderately drained. The application area has a vegetative cover, and slopes are in the range of 6 to 8%. Laboratory tests show a nitrogen concentration of 4.0 pounds per 1,000 gallons of effluent. The target nitrogen application is 140 pounds per acre.

Note: This worksheet is to be used to determine a target nutrient application rate only. See Example 10-4 for a total effluent irrigation design.

Worksheet Example 10-5 (continued)

Step 1. Calculating System Flowrate

Based on a pumpdown volume of 400,000 cubic feet and the desire to pump only 8 hours per day for 12 days, the producer determines that a pumping flowrate of 520 gpm is necessary.

Step 2. Selecting a Sprinkler

According to manufacturer's literature, two 27-degree trajectory nozzles listed would give the desired flow rate. A 1.41-inch ring nozzle operating at 110 psi has a flowrate of 525 gpm and produces a wetted diameter of 470 feet. A 1.74-inch ring nozzle operating at 60 psi has a flowrate of 515 gpm and produces a wetted diameter of 430 feet. The producer decides to look at **Step 3** to determine if there could be some problems with the nozzle application rate and the soil infiltration capabilities.

Step 3. Determining Application Rate

In the **Soil Infiltration Rate** section, the producer determines that a medium-textured silty clay soil that is moderately drained and has a vegetative cover has a soil infiltration rate of 0.50 inches per hour. Because the field has a 6 to 8% slope, the soil infiltration rate needs to be multiplied by 0.80. The effluent has less than 1% solids content, and the soils are not sandy or loamy, so the producer does not need to modify the soil infiltration rate based on solids build up. The soil infiltration rate is

$$\text{Soil Infiltration Rate} = (0.50\ \text{in/hr}) \cdot (0.80) = 0.40\ \text{inches per hour}$$

The producer calculates the application rate for the 1.41-inch ring nozzle to be 0.29 inches per hour and the application rate for the 1.74-inch ring nozzle to be 0.34 inches per hour. Application rates for both nozzles are less than 0.40 inches per hour, so application with one stationary gun or a solid-set system using either nozzle would be acceptable. The producer decides to use the 1.41-inch ring nozzle because this nozzle size provides a flowrate closest to the 520 gpm design flowrate, and the nozzle has a slightly lower application rate. The down side of this nozzle is its higher pressure, which may require a larger size pump. If multiple sprinkler locations are used, the producer needs to review **Chapter 5** to avoid over-application of effluent.

The producer returns to **Step 2** and fills out the information before moving on to **Step 4**.

Achieving a Target Nutrient Application Rate

Step 1. Calculating System Flowrate

$$Q = \left(\frac{1}{8}\right) \cdot \frac{\text{(Pumpdown Volume)}}{\text{(Number of pumping days)} \cdot \text{(Average pumping time per day)}}$$

$$= \left(\frac{1}{8}\right) \cdot \frac{(\ 400{,}000 \text{ cu ft})}{(\ 12 \text{ days}) \cdot (\ 8 \text{ hrs})}$$

Desired Flowrate: 520 gpm

Step 2. Selecting a Sprinkler

Use manufacturer's information or **Table 5-2** to select a sprinkler that will give the flowrate calculated in **Step 1**.

Nozzle Diameter: 1.41 inches

Nozzle Pressure: 110 psi

Trajectory: 27 degrees

Nozzle Flowrate: 525 gpm

Wetted Diameter: 470 feet

Step 3. Determining Application Rate

a. Soil Infiltration Rate

Soil Infiltration Rates (inches per hour)

Soil Characteristics	*Covered*	*Bare*
Clay; very poorly drained	0.30	0.15
Silty surface; poorly drained, clay and claypan subsoil	0.40	0.25
Medium textured surface soil; moderate to imperfect drained profile....	0.50	0.30
Silt loam, loam, and very sandy loam, well to moderately well drained.	0.60	0.40
Loamy sand, sandy loam, or peat; well drained	0.90	0.60

Infiltration Reduction Factor

For slopes of:	**6% to 8%**	**9% to 12%**	**13% to 20%**	**over 20%**
Multiply Infiltration Rates by:	0.80	0.60	0.40	0.25

NOTE: See **Chapter 9**, Table 9-14 and Example 9-5, for solids content greater than 5%, or for sandy or loamy soils

Soil Type:

moderately drained, medium texture, silty clay, vegetative cover (0.50 in/hr), 6 to 8% slope (0.80) (0.50 in/hr) · (0.80) = 0.40

Soil Infiltration Rate:

0.40 in/hr

(Be sure to consider the effects of slope and solids content)

b. Calculated Application Rate

$$\text{Application Rate} = (122.6) \cdot \left(\frac{\text{(Nozzle Flowrate)}}{\text{(Wetted Diameter)}^2 \cdot \left(\frac{\text{(Operation Angle)}}{360}\right)} \right)$$

$$= (122.6) \cdot \left(\frac{(\ 525 \text{ , gpm})}{(\ 470 \text{ , ft})^2 \cdot \left(\frac{(\ 360\ )}{360}\right)} \right)$$

Application Rate:

0.29 in/hr

If **Application Rate** is greater than **Soil Infiltration Rate,** return to Step 2 and select a nozzle with a

- Larger application area, and/or
- Lower flowrate

Worksheet Example 10-5 (continued)

Step 4. Determining Water Application Depth

The producer assumes the soil will have a 40% moisture deficit when irrigation takes place, and the applicable root zone depth is 1.5 feet. The water-holding capacity of a silty clay soil is 1.6 inches per foot of depth of soil. The depth of water to apply is calculated to be 0.96 inches.

Step 5. Determining Nutrient Application Depth

The producer wants to apply 140 pounds of available nitrogen per acre, based on an effluent nutrient content of 2.8 pounds of nitrogen per 1,000 gallons. Typically, 15 to 40% of the nitrogen is lost (volitalized) within 4 days of application. The producer estimates about 30% of the nitrogen will be lost from this application. From the formula, the producer needs to apply 1.84 inches.

Step 6. Calculating Total Operation Time per Application

The producer compares the water application depth and nutrient application depth. The smaller of the values from Step 4 and Step 5 will dictate the depth of application per application event. In this case, the nutrient application depth is larger than the water application depth. The producer will need to irrigate effluent on the same land over two different application periods. If water application depth was larger than nutrient application depth, the producer would have multiple application events at different locations to avoid over applying nutrients.

Because the nutrient application depth is larger, the operation time is calculated to be 6.3 hours.

Step 7. Determining Travel Speed (For Moving Systems Only)

This section is not applicable because a set system will be used.

Step 4. Determining Water Application Depth

a. Estimated Water Deficiency:

Moisture Deficiency: 40 %

b. Root Zone Depth:

Root Zone Depth: 1.5 feet

c. Water Holding Capacity (inches per foot of soil depth)

Soil Type	Moisture Capacity
Coarse sands	0.25 to 0.75
Fine sands	0.75 to 1.00
Loamy sands	1.10 to 1.20
Sandy loams	1.25 to 1.40
Fine sandy loam	1.50 to 2.00
Silt loam	2.00 to 2.50
Silty clay loam	1.80 to 2.00
Silty clay	1.50 to 1.70
Clay	1.20 to 1.50

Water Holding Capacity: 1.6 in/ft

d. Application Depth:

$$AD = (\text{Moisture Deficiency}) \cdot (\text{Root Zone Depth}) \cdot (\text{Water Holding Capacity})$$

$$= \left(\frac{40\ \%}{100}\right) \cdot (1.5\ \text{ft}) \cdot (1.6\ \text{in / ft})$$

Water Application Depth: 0.96 in

Step 5. Determining Nutrient Application Depth

$$\text{Application Depth} = (3.68) \cdot \frac{(\text{Nutrients per acre})}{(\text{Nutrients per 1,000 gallons}) \cdot (100 - \%\ \text{loss})}$$

$$= (3.68) \cdot \frac{(140\ \text{lbs nutrient per acre})}{(4.0\ \text{lbs nutrient per 1,000 gallons}) \cdot (100 - 30)}$$

Nutrient Application Depth: 1.84 in

Step 6. Calculating Total Operation Time per Application (Set Systems)

$$\text{Number of Applications} = \frac{(\text{Larger Value Between Step 4 and Step 5})}{(\text{Smaller Value Between Step 4 and Step 5})}$$

$$= \frac{(1.84\ \text{in})}{(0.96\ \text{in})} \quad \text{(round up)}$$

Number of Applications: 2

$$\text{Operation Time} = \frac{(\text{Nutrient Application Depth})}{(\text{Application Rate})}$$

$$= \frac{(1.84\ \text{in})}{(0.29\ \text{in / hr})}$$

Total Operation Time: 6.3 hours

Step 7. Determining Travel Speed (For Moving Systems Only)

$$\text{Travel Speed} = \frac{(160) \cdot (\text{Nozzle Flowrate}) \cdot (\text{Number of Applications})}{(\%\ \text{Spacing}) \cdot (\text{Wetted Diameter}) \cdot (\text{Nutrient Application Depth})}$$

$$= \frac{(160) \cdot (\_\_\_\_\ \text{gpm}) \cdot (\_\_\_\_)}{(\_\_\_\_) \cdot (\_\_\_\_\ \text{ft}) \cdot (\_\_\_\_\ \text{in})}$$

Travel Speed: N/A ft/min

Worksheet Example 10-6

Achieving a Target Application Rate with a Traveler

The producer on the same 200-cow dairy as the one in Example 10-5 wants to look into using a traveler system. The producer assumes a lane spacing of 70% of the wetted diameter. The nozzle will be operated in a 315° part circle to maintain a dry travel lane.

The design approach in this example is very similar to the one used for Example 10-5. The same worksheet is used for both designs. Following are the difference in the two design procedures.

Step 3: Application Rate, Section b

With a 1.41-inch ring nozzle and an operation angle of 315°, the nozzle application rate is

$$\text{Nozzle Application Rate} = (122.6) \cdot \left(\frac{(525 \text{ gpm})}{(470 \text{ ft})^2 \cdot \left(\frac{(315)}{360} \right)} \right) = 0.33 \text{ in/hr}$$

Step 7: Travel Speed

With a lane spacing of 70% of the wetted diameter, the traveler speed is

$$\text{Travel Speed} = \frac{(160) \cdot (525 \text{ gpm}) \cdot (2\ )}{(70) \cdot (470 \text{ ft}) \cdot (184 \text{ in})} = 2.78 \text{ ft/min}$$

Achieving a Target Nutrient Application Rate

Step 1. Calculating System Flowrate

$$Q = \left(\frac{1}{8}\right) \cdot \frac{\text{(Pumpdown Volume)}}{\text{(Number of pumping days)} \cdot \text{(Average pumping time per day)}}$$

$$= \left(\frac{1}{8}\right) \cdot \frac{(\_\_\_\_\_\_\_\_ \text{ cu ft})}{(\_\_\_\_\_\_ \text{ days}) \cdot (\_\_\_\_\_\_ \text{ hrs})}$$

Desired Flowrate:________gpm

Step 2. Selecting a Sprinkler

Use manufacturer's information or **Table 5-2** to select a sprinkler that will give the flowrate calculated in **Step 1**.

Nozzle Diameter: ______ inches
Nozzle Pressure: _________ psi
Trajectory: __________ degrees
Nozzle Flowrate: ________ gpm
Wetted Diameter: ________ feet

Step 3. Determining Application Rate

a. Soil Infiltration Rate

Soil Infiltration Rates (inches per hour)

Soil Characteristics	*Covered*	*Bare*
Clay; very poorly drained	0.30	0.15
Silty surface; poorly drained, clay and claypan subsoil	0.40	0.25
Medium textured surface soil; moderate to imperfect drained profile	0.50	0.30
Silt loam, loam, and very sandy loam, well to moderately well drained.	0.60	0.40
Loamy sand, sandy loam, or peat; well drained	0.90	0.60

Infiltration Reduction Factor

For slopes of:	6% to 8%	9% to 12%	13% to 20%	over 20%
Multiply Infiltration Rates by:	0.80	0.60	0.40	0.25

NOTE: See **Chapter 9**, Table 9-14 and Example 9-5, for solids content greater than 5%, or for sandy or loamy soils

Soil Type:

Soil Infiltration Rate:

______________________in/hr
(Be sure to consider the effects of slope and solids content)

b. Calculated Application Rate

$$\text{Application Rate} = (122.6) \cdot \left(\frac{\text{(Flowrate)}}{\text{(Wetted Diameter)}^2 \cdot \left(\frac{\text{(Operation Angle)}}{360}\right)} \right)$$

$$= (122.6) \cdot \left(\frac{(\_\_\_\_\_\_\_\_\_\_ \text{ gpm})}{(\_\_\_\_\_\_\_\_ \text{ ft})^2 \cdot \left(\frac{(\_\_\_\_\_\_\_\_)}{360}\right)} \right)$$

Application Rate:

______________________ in/hr

If **Application Rate** is greater than **Soil Infiltration Rate,** return to Step 2 and select a nozzle with a
- Larger application area, and/or
- Lower flowrate

Step 4. Determining Water Application Depth

a. Estimated Water Deficiency:

Moisture Deficiency: ___________%

b. Root Zone Depth:

Root Zone Depth: ___________feet

c. Water Holding Capacity (inches per foot of soil depth)

Soil Type	Moisture Capacity
Coarse sands	0.25 to 0.75
Fine sands	0.75 to 1.00
Loamy sands	1.10 to 1.20
Sandy loams	1.25 to 1.40
Fine sandy loam	1.50 to 2.00
Silt loam	2.00 to 2.50
Silty clay loam	1.80 to 2.00
Silty clay	1.50 to 1.70
Clay	1.20 to 1.50

Water Holding Capacity:_______in/ft

d. Application Depth:

$$AD = (\text{Moisture Deficiency}) \cdot (\text{Root Zone Depth}) \cdot (\text{Water Holding Capacity})$$

$$= \left(\frac{\_\_\_\_\_\%}{100}\right) \cdot (\_\_\_\_\_\_\text{ft}) \cdot (\_\_\_\_\_\_\text{in / ft})$$

Water Application Depth:_______in

Step 5. Determining Nutrient Application Depth

$$\text{Application Depth} = (3.68) \cdot \frac{(\text{Nutrients per acre})}{(\text{Nutrients per 1,000 gallons}) \cdot (100 - \%\text{ loss})}$$

$$= (3.68) \cdot \frac{(\_\_\_\_\_\_\text{ lbs nutrient per acre})}{(\_\_\_\_\_\_\text{lbs nutrient per 1,000 gallons}) \cdot (100 - \_\_\_\_)}$$

Nutrient Application Depth:_____in

Step 6. Calculating Total Operation Time per Application (Set Systems)

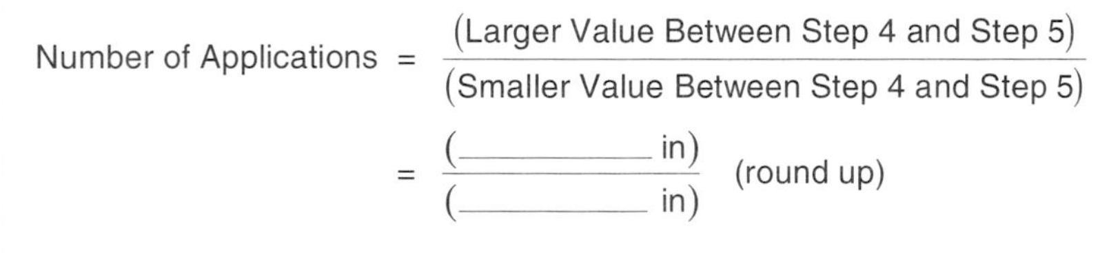

$$\text{Number of Applications} = \frac{(\text{Larger Value Between Step 4 and Step 5})}{(\text{Smaller Value Between Step 4 and Step 5})}$$

$$= \frac{(\_\_\_\_\_\_\_\_\text{ in})}{(\_\_\_\_\_\_\_\_\text{ in})} \quad \text{(round up)}$$

Number of Applications:_________

$$\text{Operation Time} = \frac{(\text{Nutrient Application Depth})}{(\text{Application Rate})}$$

$$= \frac{(\_\_\_\_\_\_\_\_\text{ in})}{(\_\_\_\_\_\_\_\_\text{ in / hr})}$$

Total Operation Time:________hours

Step 7. Determining Travel Speed (For Moving Systems Only)

$$\text{Travel Speed} = \frac{(160) \cdot (\text{Nozzle Flowrate}) \cdot (\text{Number of Applications})}{(\%\text{ Spacing}) \cdot (\text{Wetted Diameter}) \cdot (\text{Nutrient Application Depth})}$$

$$= \frac{(160) \cdot (\_\_\_\_\_\_\text{ gpm}) \cdot (\_\_\_\_\_\_\_\_)}{(\_\_\_\_\_\_) \cdot (\_\_\_\_\_\_\text{ ft}) \cdot (\_\_\_\_\_\_\text{ in})}$$

Travel Speed:_____________ft/min

Appendix A

Conversion Factors

To Convert From	To	Multiply By
acre-feet	gallons (gal)	325,829
acre-inch	gallons (gal)	27,152
acre-inch per hour	gallons per minute (gpm)	453
acres	square feet (sq ft)	43,560
cubic feet (cu ft)	gallons (gal)	7.48
cubic feet (cu ft)	cubic inches (cu in)	1,728
feet per minute (fpm)	miles per hour (mph)	0.0114
feet per second (fps)	miles per hour (mph)	0.682
gallons (gal)	acre-inch	0.00003683
gallons (gal)	acre-feet	0.000003069
gallons (gal)	cubic feet (cu ft)	0.134
gallons per minute (gpm)	acre-inch per hour	0.002212
head (ft)	pressure (psi)	0.433
K	K_2O	1.20
K_2O	K	0.83
miles	feet (ft)	5,280
miles per hour (mph)	feet per minute (fpm)	88
miles per hour (mph)	feet per second (fps)	1.467
P	P_2O_5	2.27
P_2O_5	P	0.44
pressure (psi)	head (ft)	2.31

Appendix B

Irrigated Acreage Formulas for Center-Pivot Systems

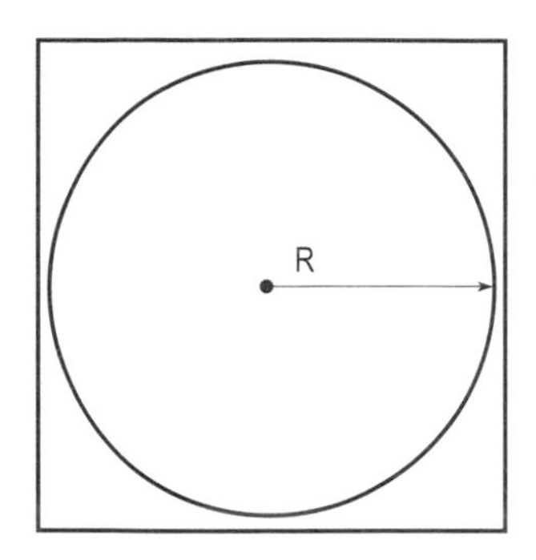

B-1. Full Circle without End Gun

$$\text{Area (acres)} = \frac{3.14 \cdot (R)^2}{43{,}560}$$

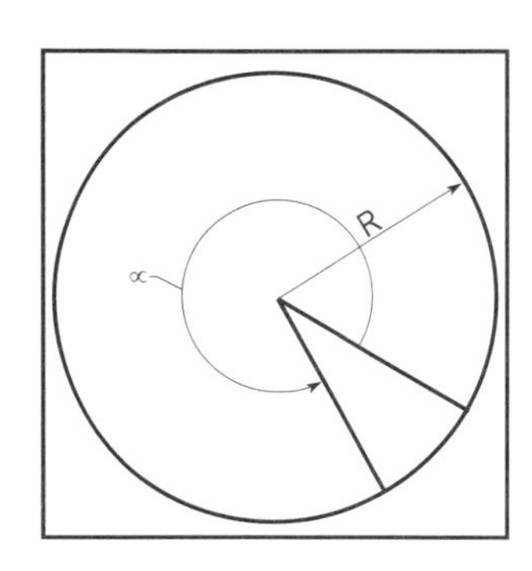

B-2. Part Circle without End Gun

$$\text{Area (acres)} = \left(\frac{3.14 \cdot (R)^2}{43{,}560}\right) \cdot \left(\frac{\alpha}{360}\right)$$

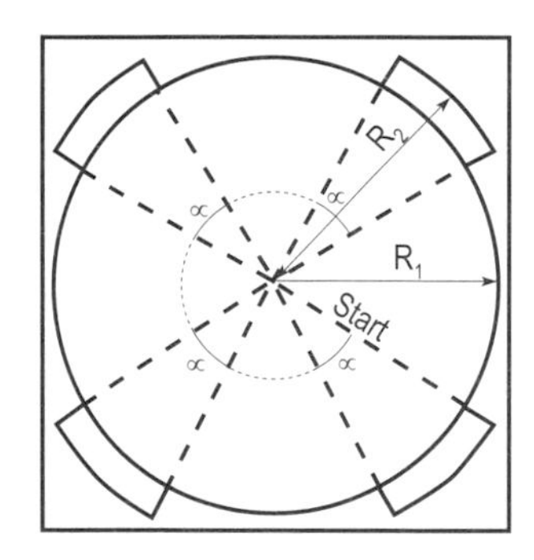

B-3. Full Circle with End Gun

$$\text{Area (acres)} = \left[\left(\frac{3.14 \cdot (R_1)^2}{43{,}560}\right) \cdot \left(\frac{360 - \sum\alpha}{360}\right)\right] + \left[\left(\frac{3.14 \cdot (R_2)^2}{43{,}560}\right) \cdot \left(\frac{\sum\alpha}{360}\right)\right]$$

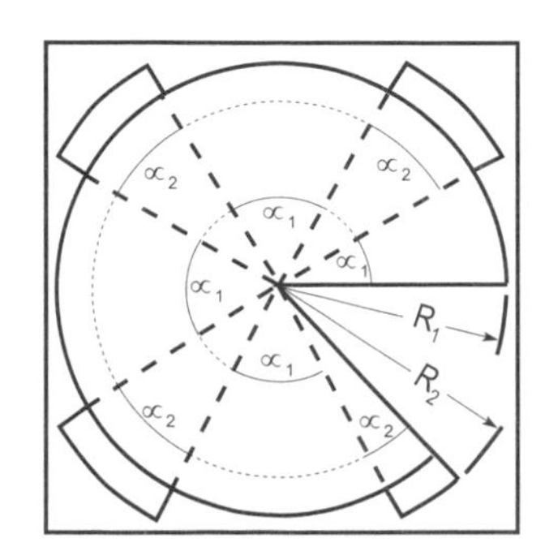

B-4. Part Circle with End Gun

$$\text{Area (acres)} = \left[\left(\frac{3.14 \cdot (R_1)^2}{43{,}560}\right) \cdot \left(\frac{\sum\alpha_1}{360}\right)\right] + \left[\left(\frac{3.14 \cdot (R_2)^2}{43{,}560}\right) \cdot \left(\frac{\sum\alpha_2}{360}\right)\right]$$

$$\text{End-gun Operation Angle} = 90 - \left(2 \cdot \cos^{-1}\left(\frac{\text{Lateral Length}}{(\text{Lateral Length}) + (\text{End-gun Throw})}\right)\right)$$

Appendix B

Irrigated Acreage for Center Pivots with Corner Attachments

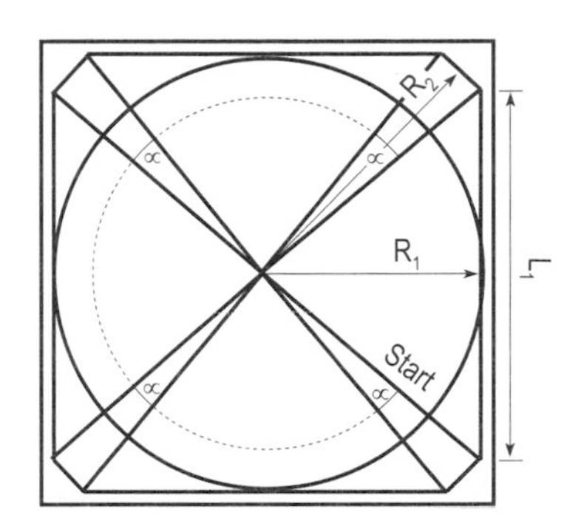

B-5. Full Circle, Four Corners

$$\text{Area (acres)} = \left[\frac{4 \cdot \left(\frac{R_1 \cdot L}{2}\right)}{43{,}560}\right] + \left[\left(\frac{3.14 \cdot (R_2)^2}{43{,}560}\right) \cdot \left(\frac{\sum \alpha}{360}\right)\right]$$

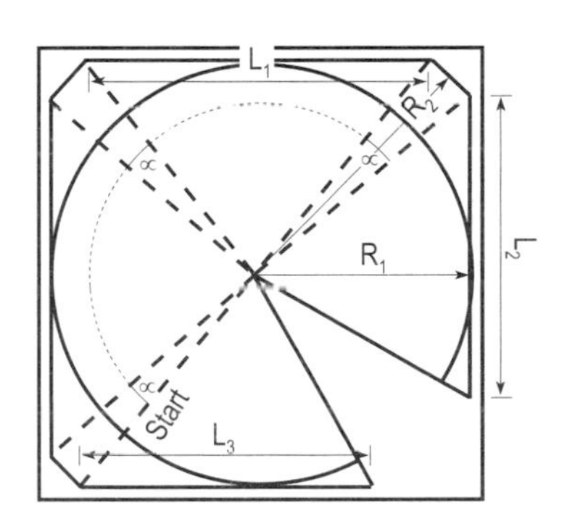

B-6. Part Circle

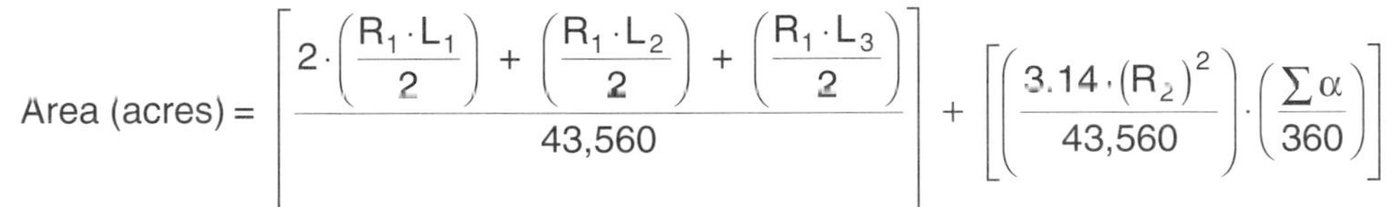

$$\text{Area (acres)} = \left[\frac{2 \cdot \left(\frac{R_1 \cdot L_1}{2}\right) + \left(\frac{R_1 \cdot L_2}{2}\right) + \left(\frac{R_1 \cdot L_3}{2}\right)}{43{,}560}\right] + \left[\left(\frac{3.14 \cdot (R_2)^2}{43{,}560}\right) \cdot \left(\frac{\sum \alpha}{360}\right)\right]$$

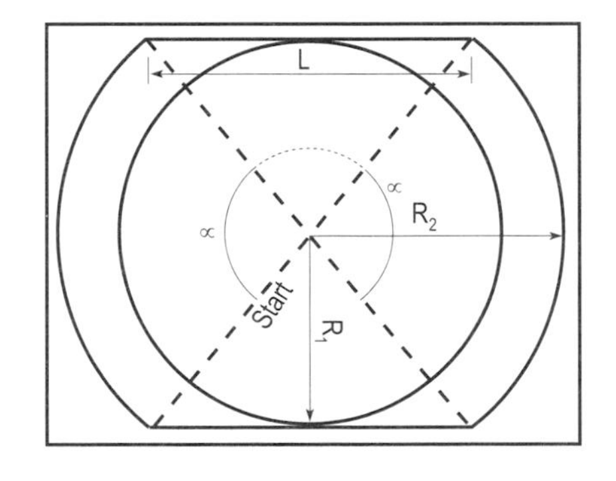

B-7. Rectangular Field

$$\text{Area (acres)} = \left[\frac{2 \cdot \left(\frac{R_1 \cdot L}{2}\right)}{43{,}560}\right] + \left[\left(\frac{3.14 \cdot (R_2)^2}{43{,}560}\right) \cdot \left(\frac{\sum \alpha}{360}\right)\right]$$

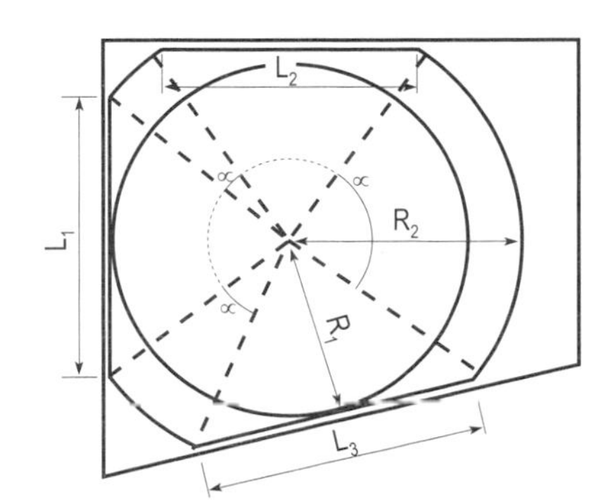

B-8. Odd-shaped Field

$$\text{Area (acres)} = \left[\frac{\left(\frac{R_1 \cdot L_1}{2}\right) + \overbrace{\left(\frac{R_1 \cdot L_2}{2}\right) + \left(\frac{R_1 \cdot L_3}{2}\right)}^{\text{As Required}}}{43{,}560}\right] + \left[\left(\frac{3.14 \cdot (R_2)^2}{43{,}560}\right) \cdot \left(\frac{\sum \alpha}{360}\right)\right]$$

Glossary

Acre-inch (ac-in): The volume of water covering an area of one acre by one inch deep.

Allowable pumping time (T_p): The average number of hours the irrigation system can be expected to apply water each day.

Area irrigated: The surface area to be irrigated.

Application depth: The depth of irrigation water applied during one irrigation event or cycle. The net application depth is the depth of water applied after losses due to application efficiency are subtracted from the water that is pumped.

Aquifer: A water-saturated geologic unit or system that yields water to wells or springs at a sufficient rate that the wells or springs can serve as practical sources of water.

Artesian aquifer (flowing and non-flowing): A well where the water rises above the surface of the water in the aquifer. It is a flowing artesian well if the water rises above the surface of the earth.

Biochemical Oxygen Demand (BOD): The indirect measure of the concentration of biodegradable substances present in an aqueous solution. Determined by the amount of dissolved oxygen required for the aerobic degradation of the organic matter at 68 F. BOD_5 refers to that oxygen demand for the initial five days of the degradation process.

Break horsepower (BHP): The power output of the engine crankshaft required to drive a pump.

Cable-tow traveler: A gun sprinkler mounted on a wheel cart to which a soft hose is connected. A cable anchored at the end of the traveler run winds on a cart-mounted winch.

Cavitation: The formation and collapse of low-pressure vapor cavities in a flowing liquid, often resulting in serious damage to pumps, propellers, etc.

Center pivot: A self-propelled system consisting of a lateral mounted on A-frame towers that rotates around a center pivot point. Sprinklers are mounted along the lateral.

Centrifugal pump: A pump consisting of a rotating impeller enclosed in a housing and used to impart energy to a fluid through centrifugal force.

Chemical Oxygen Demand (COD): An indirect measure of the biochemical load exerted on the oxygen content of a body of water when organic wastes are introduced into the water. It is determined by the amount of potassium dichromate consumed in a boiling mixture of chromic and sulfuric acids. The amount of oxidizable organic matter is proportional to the potassium dichromate consumed. If the wastes contain only readily available organic bacterial food and not toxic matter, the COD values can be correlated with BOD values obtained from the same wastes.

Chemigation: The use of an irrigation system to apply chemicals to the irrigated area.

Deficit Irrigation: Irrigation management with a supply of water less than the seasonal ET requirements of the crop. Relys on stored soil moisture in the root zone to provide the difference and requires filling the root zone soil profile to field capacity before or early in the growing season.

Deep-well turbine pump: A pump having one or more stages, each consisting of an impeller on a vertical shaft, surrounded by stationary and usually symmetrical guide vanes in a pump bowl assembly. Power is delivered to the impeller by a shaft, and a column carries the water upward.

Design evapotranspiration rate (ET_d): The average daily peak evapotranspiration rate for the proposed crop for seven days.

Discharge quantity, Q: The rate at which an irrigation system can apply water, gpm or gpm/ac.

Drawdown: The dropping of the water level when water is pumped from a well.

Encrustation: Carbonate deposits near the well and in the screen openings due to the velocity of the water entering the well.

Evaporation: The process where liquid water is converted to a gas. In irrigation, evaporation is considered to be when water is evaporated from the soil or plant surface.

Evapotranspiration (ET): The combination of water transpired from the plant and evaporated from the soil and plant surfaces.

Extension Service: The outreach division of the land grant universities and USDA.

Field capacity: The amount of water remaining in a soil when the downward water flow due to gravity becomes negligible.

Fixed solids: The part of the total solids remaining after volatile gases are driven off at 1,112 F; ash. Fixed solids are measured by determining the weight (mass) of residue after volatile solids have been removed as combustible gases when heated at 1,112 F for at least 1 hour.

Flow rate: The quantity of water available and/or needed per minute, per hour, or per day.

Friction head: The energy required to overcome friction caused by fluid movement through pipes, fittings, and valves, around corners, or by pipe size changes.

Gallons per minute (gpm): The volumetric flow rate.

Gravitational water: Soil water that moves into, through, or out of the soil under the influence of gravity.

Hand-move system: An irrigation system designed to be moved manually.

Head: A measure of water pressure, in feet of water or pounds per square inch (psi). 1 psi = 2.31 ft of water.

Hose-tow traveler: A gun sprinkler mounted on a wheel cart to a hard hose connected to a large stationary hose reel. The hose pulls the sprinkler through the field.

Impeller: The rotating component of the pump that moves water through the pump.

Irrigation system capacity: The rate at which an irrigation system can apply water to a given land area, generally expressed as gallons per minute per acre.

Linear-move traveling lateral: A self-propelled system consisting of a lateral mounted on A-frame towers that moves in a straight line across the field. Water is fed into the lateral at any point and can come from a ditch, risers, or drag hose.

Maximum practical suction lift (MPSL): An estimate of the lift required to move water on the suction side of the pump.

Net positive suction head (NPSH): The pump manufacturer's specified suction head at sea level, ft.

Permanent wilting point: Soil water content below which plants cannot readily obtain water and permanently wilt.

Piezometric head: Combined elevation and pressure head as measured from a reference plane.

Pressure head: The pressure energy necessary at the distribution system entrance to overcome losses and operate sprinklers at the design pressure.

Propeller pump: A pump having one or more stages, each consisting of an impeller on a vertical shaft. The impeller develops head by the propelling or lifting action on the water. Power is delivered to the impeller by a shaft.

Pumping lift: The sum of the static lift and the drawdown.

Side-roll system: A lateral line mounted on wheels where the lateral serves as the axle. The system rolls across the field which is square or rectangular.

Solid-set system: A complete, above-ground aluminum or buried pipe system that is placed in the field to be irrigated prior to the start of the growing season and left in place throughout the growing season.

Spoil: Material removed in excavating.

Submersible pump: A turbine pump with a submersible electric motor that is located below the water level in the well.

Total (or operating) head (TH): The energy required to pump water from its source to the point of discharge; equal to the total static head, plus pressure head, plus friction head, plus velocity head. Also commonly called Total Dynamic Head.

Total lift: The sum of all head components for irrigation system design (also total head).

Total solids: The total amount of solids in a waste, both in solution or suspension. Measured by evaporating free water on steam table and drying in an oven at 217 F for 24 hours or until a constant weight.

Total static head: The potential energy due to the vertical distance between the pumping water level and the entrance to the distribution system.

Transpiration: The biological process of plants by which water is converted from the liquid state at the leaf surface to a gas, generally considered to govern the energy balance of the leaf surface.

Velocity head: The energy required to put water in motion.

Volatile Solids (VS): Readily vaporized solids. Those solids that are combustible at 1,112 °F.

Water application efficiency (E_A): The ratio of the average depth of irrigation water stored in soil during irrigation to the average depth of irrigation water pumped.

Water hammer: The sharp rise in pressure that occurs when a valve is closed and a column of moving liquid is suddenly halted.

Water horsepower (WHP): Useful work done by a pump, which is a function of pump discharge and Total Head.

Index

References

Additional Reading

1. Water quality for agriculture, R. S. Ayers, D. W. Westcot, Food and Agriculture Organization of the United Nations, 1976.
2. Groundwater and wells, F. G. Driscoll., St. Paul, MN: Johnson Division, 1986.

Extension Departments with Irrigation Publications

1. Agricultural Engineering, Box 2120, South Dakota State University, Brookings, SD 57007.
2. Extension Agricultural Engineering, Box 5626, North Dakota State University, Fargo, ND 58105.
3. Biological Systems Engineering Department, 219A LW Chase Hall, University of Nebraska, Lincoln, NE 68583-0727.

NRCS, USDA/NRCS, P.O. Box 2890, Washington, D.C. 20013

1. Irrigation Chapter 11 Sprinkler Irrigation, National Engineering Handbook.
2. Agricultural Waste Management Field Handbook, Part 651, National Engineering Handbook.

ASAE, 2950 Niles Road, St. Joseph, MI, 49085-9659

1. Design, Installation and Performance of Underground, Thermoplastic Irrigation Pipelines, ANSI/ASAE S376.
2. Wiring and Equipment for Electrically Driven or Controlled Irrigation Machines, ANSI/ASAE S362.
3. Electrical Service and Equipment for Irrigation, ANSI/ASAE S397.
4. Designing and Constructing Irrigation Wells EP400.
5. Test Procedures for Determining the Uniformity of Water Distribution of Center-Pivot, Corner-Pivot, and Moving Lateral Irrigation Machines Equipped with Spray or Sprinkler Nozzles, ASAE S436.

MWPS, 122 Davidson Hall, Iowa State University, Ames IA 50011-3080

1. Designing Facilities for Pesticide and Fertilizer Containment, MWPS-37.
2. Livestock Waste Handbook, MWPS-18.
3. Structures and Environment Handbook, MWPS-1.

Reviewers

Ted Funk
University of Illinois

Philip Goodrich
University of Minnesota

Rick Koelsch
University of Nebraska

Jake LaRue
Valmont Irrigation

Jim Lindley
North Dakota State University

Leonard Massie
University of Wisconsin

Danny H. Rodgers
Kansas State University

Tom Spofford
NRCS, Oregon

Dean Steele
North Dakota State University

Allen Thompson
University of Missouri

MWPS thanks Leonard Massie, Leroy Cluever, Carroll Drablos, Don Jones, Ted Louden, Darnell Lundstrom, Harry Manges, Stu Melvin, Ed Monke, Rolland Wheaton, Darrell Watts, Melvin Palmer, Leland Hardy, Delynn Hay and Glenn Church for contributions that provided the framework for this publication and for reviews of previously written material.